Diagnosability, Security and Safety of Hybrid Dynamic and Cyber-Physical Systems

Moamar Sayed-Mouchaweh
Editor

Diagnosability, Security and Safety of Hybrid Dynamic and Cyber-Physical Systems

Springer

Editor
Moamar Sayed-Mouchaweh
Institute Mines-Telecom Lille Douai
Douai, France

ISBN 978-3-030-09114-9 ISBN 978-3-319-74962-4 (eBook)
https://doi.org/10.1007/978-3-319-74962-4

Softcover re-print of the Hardcover 1st edition 2018

Printed on acid-free paper

This Springer imprint is published by the registered company Springer International Publishing AG part of Springer Nature.
The registered company address is: Gewerbestrasse 11, 6330 Cham, Switzerland

Preface

Cyber-physical systems (CPS) are characterized as a combination of physical (physical plant, process, network) and cyber (software, algorithm, computation) components whose operations are monitored, controlled, coordinated, and integrated by a computing and communicating core. The interaction between physical and computational components in CPS is intensive. They cover an increasing number of real life applications such as autonomous vehicles, aircrafts, smart manufacturing processes, surgical robots and human robot collaboration, smart electric grids, home appliances, air traffic control, automated farming, and implanted medical devices. The interaction between both physical and cyber components requires tools allowing analyzing and modeling both the discrete (discrete event control, communication protocols, discrete sensors/actuators, scheduling algorithms, etc.) and continuous (continuous dynamics, physics, continuous sensors/actuators, etc.) dynamics. Therefore, many CPS can be modeled as hybrid dynamic systems in order to take into account both discrete and continuous behaviors as well as the interactions between them.

Many critical infrastructures, such as power generation and distribution networks, water networks and mass transportation systems, autonomous vehicles and traffic monitoring, are CPS. Such systems, including critical infrastructures, are becoming widely used and covering many aspects of our daily life. Therefore, the security, safety, and reliability of CPS is essential for the success of their implementation and operation. However, these systems are prone to major incidents resulting from cyberattacks and system failures. These incidents affect significantly their security and safety. Attacks can be represented as an interference in the communication channel between the supervisor and the system intentionally generated by intruders in order to damage the system or as the enablement, respectively disablement, of actuators events that are disabled, respectively enabled, by the supervisor. In general, intruders hide, create, or even change intentionally events that transit from one device (actuator, sensor) to another in a control communication channel. These attacks in a supervisory control system can lead the plant to execute event sequences entailing the system to reach unsafe or dangerous states that can damage the system.

Therefore, reliable, scalable, and timely fault diagnosis is crucial in order to improve the robustness of CPS to failures. In addition, it is primordial to detect intrusions that exploit the vulnerabilities of industrial control systems in order to alter intentionally the integrity, confidentiality, and availability of CPS. These cyberattacks affect the control commands of the controller [Programmable Logic Controller (PLC)], the reports (sensors readings) coming from the plant as well as the communication between them. Moreover, a thorough understanding of the vulnerability of CPS' components against such incidents can be incorporated in future design processes in order to better design such systems. Finally, the timely fault diagnosis can help operators to have better situation awareness and give them ample time to implement correction (maintenance) actions.

However, guaranteeing the security and safety of CPS requires verifying their behavioral or safety properties either at design stage such as state reachability, diagnosability, and predictability or online such as fault detection and isolation. This is a challenging task because of the inherent interconnected and heterogeneous combination of behaviors (cyber/physical, discrete/continuous) in these systems. Indeed, fault propagation in CPS is governed not only by the behaviors of components in the physical and cyber subsystems but also by their interactions. This makes the identification of the root cause of observed anomalies and predicting the failure events a hard problem. Moreover, the increasing complexity of CPS and the security and safety requirements of their operation as well as their decentralized resource management entail a significant increase in the likelihood of failures in these systems. Finally, it is worth mentioning that computing the reachable set of states of HDS is an undecidable matter due to the infinite state space of continuous systems.

This edited Springer book presents recent and advanced approaches and techniques that address the complex problem of analyzing the diagnosability property of CPS and ensuring their security and safety against faults and attacks. The CPS are modeled as hybrid dynamic systems using different model-based and data-driven approaches in different application domains (electric transmission networks, wireless communication networks, intrusions in industrial control systems, intrusions in production systems, wind farms, etc.). These approaches handle the problem of ensuring the security of CPS in presence of attacks and verifying their diagnosability in presence of different kinds of uncertainty (uncertainty related to the event occurrences, to their order of occurrence, to their value etc.).

Finally, the editor is very grateful to all authors and reviewers for their very valuable contribution allowing setting another cornerstone in the research and publication history of studding the diagnosability, security, and safety of CPS modeled as hybrid dynamic systems. I would like also to acknowledge Mrs. Mary E. James for establishing the contract with Springer and supporting the editor in any organizational aspects. I hope that this volume will be a useful basis for further fruitful investigations and fresh ideas for researcher and engineers as well as a motivation and inspiration for newcomers to address the problems related to this very important and promising field of research.

Douai, France Moamar Sayed-Mouchaweh

Contents

Chapter 1
Prologue

Moamar Sayed-Mouchaweh

1.1 Cyber-Physical Systems as Hybrid Dynamic Systems

Cyber-physical systems (CPS) [1] are characterized as a combination of physical (physical plant, process, network) and cyber (software, algorithm, computation) components whose operations are monitored, controlled, coordinated, and integrated by a computing and communicating core. The interaction between physical and computational components in CPS is intensive. They cover increasing number of real life applications such as autonomous vehicles, aircrafts, smart manufacturing processes, surgical robots and human robot collaboration, smart electric grids, home appliances, air traffic control, automated farming and implanted medical devices.

Hybrid dynamic systems (HDS) [2] are systems in which the discrete and continuous dynamics cohabit. The discrete dynamics is described by discrete state variables while the continuous dynamics is described by continuous state variables. HDS exhibit different continuous dynamic behavior depending on the current operation mode q as follows:

$$\dot{X} = A_{(q)}\,X + B_{(q)}\,u$$

where X is the state vector and u is the input vector. In the case of linear systems, $A_{(q)}$ and $B_{(q)}$ are constant matrices of appropriate dimensions.

The interaction between both physical and cyber components requires tools allowing analyzing and modeling both the discrete (discrete event control, communication protocols, discrete sensors/actuators, scheduling algorithms, etc.) and continuous (continuous dynamics, physics, continuous sensors/actuators, etc.)

M. Sayed-Mouchaweh (✉)
Institute Mines-Telecom Lille Douai, Douai, France
e-mail: moamar.sayed-mouchaweh@imt-lille-douai.fr

M. Sayed-Mouchaweh (ed.), *Diagnosability, Security and Safety of Hybrid Dynamic and Cyber-Physical Systems*, https://doi.org/10.1007/978-3-319-74962-4_1

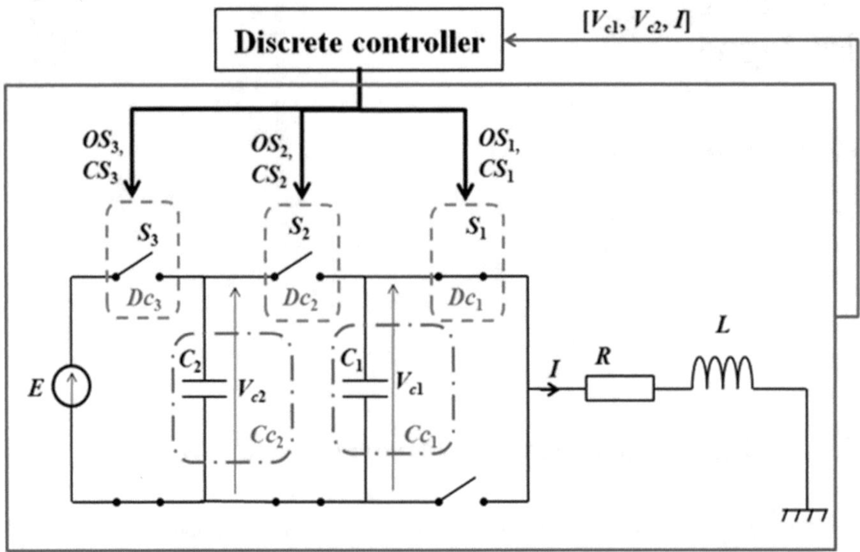

Fig. 1.1 Three-cell power converter as discretely controlled continuous system (DCCS) where capacitors C_1 and C_2 represent the continuous components (Cc) and switches S_1, S_2 and S_3 the discrete components (Dc)

dynamics. Therefore, many CPS can be modeled as hybrid dynamic systems [3, 4] in order to take into account both discrete and continuous behaviors as well as the interactions between them.

There are different classes of HDS, e.g., autonomous switching systems [5], discretely controlled switching systems [2], pricewise affine systems [6], discretely controlled jumping systems [7]. Many complex systems are embedded in the sense that they consist of a physical plant with a discrete controller. Therefore, the system has several discrete changes between different configuration modes through the actions of the controller exercised on the system plant (e.g., actuators). This kind of HDS is called discretely controlled continuous or switching systems (DCCS) [7]. Piecewise affine systems [6] are another important class of HDS where complex nonlinearities are substituted by a sequence of simpler piecewise linear behaviors.

The three-cellular power converter [8], depicted in Fig. 1.1, presents an example of DCCS. The continuous dynamics of the system is described by state vector $X = [Vc_1 \quad Vc_2 \quad I]^T$, where Vc_1 and Vc_2 represent, respectively, the floating voltage of capacitors C_1 and C_2 and I represents the load current flowing from source E towards the load (R, L) through three elementary switching cells S_j, $j \in \{1, 2, 3\}$. The latter represent the system discrete dynamics. Each discrete switch S_j has two discrete states: S_j opened or S_j closed. The control of this system has two main tasks: (1) balancing the voltages between the switches and (2) regulating the load current to a desired value. To accomplish that, the controller changes the switches' states from opened to closed or from closed to opened by applying discrete commands

"CS_j" or "OS_j" to each discrete switch $S_j, j \in \{1, 2, 3\}$ (see Fig. 1.1) where CS_j refers to "close switch S_j" and OS_j to "open switch S_j." Thus, the considered example is a DCCS.

There are three major modeling tools widely used in the literature to model HDS. These tools are hybrid Petri nets [9], hybrid bond graphs [10], and hybrid automata [11].

Hybrid Petri nets (HPN) model HDS by combining discrete and continuous parts. HPN is formally defined by the tuple:

$$\text{HPN} = \{P, T, h, \text{Pre}, \text{Post}\}$$

where $P = P_d \cup P_c$ is a finite, not empty, set of places partitioned into a set of discrete places P_d, represented as circles, and a set of continuous places P_c, represented as double circles. $T = T_d \cup T_c$ is a finite, not empty, set of discrete transitions T_d and a set of continuous transitions T_c represented as double boxes. $h : P \cap T \rightarrow \{D, C\}$, called "hybrid function," indicates for every node whether it is a discrete node (D) or a continuous node (C). $\text{Pre} : P_c x T \rightarrow \mathbb{R}^+$ or $\text{Pre} : P_d x T \rightarrow \mathbb{N}$ is a function that defines an arc from a place to a transition. $\text{Post} : P_i x T_j \rightarrow \mathbb{R}^+$ or $\text{Post} : P_d x T \rightarrow \mathbb{N}$is a function that defines an arc from a transition to a place.

Hybrid bond graph is a graphical description of a physical dynamic system with discontinuities. The latter represent the transitions between discrete modes. Similar to a regular bond graph, it is an energy-based technique. It is directed graphs defined by a set of summits and a set of edges. Summits represent components. The latter are: (1) passive components which transform energy into potential energy (C-components), inertia energy (L-components), and dissipated energy (T-components), (2) active components that can be source of effort or source pf flow. The edges, called bonds (drawn as half arrows), represent ideal energy connections between the components. The components interconnected by the edges construct the model of the global system. This model is represented by 1 junction for components having a common flow, 0 junction for common effort and transformers and gyrators to connect different kinds of energy. In order to take into account the information during the transitions between discrete modes, hybrid bond graph is extended by adding controlled junctions (CJs). The latter allow considering the local changes in individual component modes due to discrete transitions. The CJs may be switched ON (activated) or OFF (deactivated). An activated CJ behaves like a conventional bond graph junction. Deactivated CJs turn inactive the entire incident junction and hence do not influence any part of the system.

Hybrid automata are a mathematical model for HDS, which combines, in a single formalism, transitions for capturing discrete change with differential equations for capturing continuous dynamics. A hybrid automaton is a finite state machine with a finite set of continuous variables whose values are described by a set of ordinary differential equations. A hybrid automaton is defined by the tuple:

$$G = (Q, \Sigma, X, \text{flux}, \text{Init}, \delta)$$

where Q is the set of states, Σ is the set of discrete events, X is a finite set of continuous variables describing the continuous dynamics of the system, flux : $Q \times X \to \mathbb{R}^n$ is a function characterizing the continuous dynamics of X in each state q of Q, Init $= (q \in Q, X(q), \text{flux}(q))$ is the set of initial conditions and $\delta : Q \times \Sigma \to Q$ is the state transition function. A transition $\delta(q, e) = q^+$ corresponds to a change from state q to state q^+ after the occurrence of discrete event $e \in \Sigma$.

1.2 Diagnosability, Security, and Safety in Cyber-Physical Systems: Problem Formulation, Methods, and Challenges

A fault can be defined as a non-permitted deviation of at least one characteristic property of a system or one of its components from its normal or intended behavior. Fault diagnosis is the operation of detecting faults and determining possible candidates that explain their occurrence. Online fault diagnosis is crucial to ensure safe operation of complex dynamic systems in spite of faults affecting the system behaviors. Consequences of the occurrence of faults can be severe and result in human casualties, environmentally harmful emissions, high repair costs, and economical losses caused by unexpected stops in production lines. Therefore, early detection and isolation of faults is the key to maintaining system performance, ensuring system safety, and increasing system life.

Faults may manifest in different parts of the system, namely, the actuators (loss of engine power, leakage in a cylinder, etc.), the system (e.g., leakage in the tank), the sensors (e.g., reduction of the displayed value relative to the true value, or the presence of a skew or increased noise preventing proper reading), and the controller (i.e., the controller does not respond properly to its inputs sensor reading). Faults can be abrupt (e.g., the failed-on or failed-off of the pump and the stuck opened or stuck closed of the valve), intermittent or gradual (degradation of a component). Faults also may occur in a single or a multiple scenario. In the former, one fault candidate explains the observations (is responsible for the fault behavior). In the latter, several fault candidates are responsible for the fault behavior.

In HDS, faults can occur as a change in the nominal values of parameters characterizing the continuous dynamics, and are called parametric faults. Faults can also occur in the form of abnormal or unpredicted mode-changing behavior and are called discrete faults. Therefore, two types of faults should be considered for HDS depending on the dynamics that is affected by faults (parametric or discrete). Discrete faults are related to faults in actuators and usually exhibit great discontinuities in system behavior, whilst parametric faults are related to tear and wear and introduce faults with much slower dynamics. For parametric faults, after the fault detection and isolation (determining the fault candidate), a fault identification phase is required in order to estimate the amplitude (e.g., the section of leakage of a tank) of the fault, its time of occurrence, its importance, etc.

Fault types	*Fault labels*	*Fault event - Fault description*
Discrete faults	F_1	f_{s1so} - S_1 stuck opened
		f_{s1sc} - S_1 stuck closed
	F_2	f_{s2so} - S_2 stuck opened
		f_{s2sc} - S_2 stuck closed
	F_3	f_{s3so} - S_3 stuck opened
		f_{s3sc} - S_3 stuck closed
Parametric faults	F_4	f_{C1} – Abnormal change in the nominal values of C_1 due to C_1 ageing
	F_5	f_{C2} - Abnormal change in the nominal values of C_2 due to C_2 ageing

Fig. 1.2 Faults for the diagnosis of three-cell converters

For the example of three-cellular converters, eight faults can be considered for the diagnosis [7] as it is depicted in Fig. 1.2. Parametric faults (abnormal deviation of the nominal value of capacitors) are principally due to the effect of aging or pollution. The discrete faults (switch stuck-on or stuck-off) are more frequent and their consequences are more destructive. For instance, in open-circuit (stuck-off) failure, the system operates in degraded performance. However, unstable load may lead to further damage on the system. Therefore, the fault diagnosis of these faults is necessary to ensure the system safety and quality.

The fault diagnosis task [12, 13] is generally performed by reasoning over differences between desired or expected behavior, defined by a model, and observed behavior provided by sensors. This task can be performed offline or online. Offline diagnosis assumes that the system is not operating in normal conditions but it is in a test bed, i.e., ready to be tested for possible prior failures. The test is based on inputs, e.g. commands, and outputs, e.g. sensors readings, in order to observe a difference between the resulting signals with the ones obtained in normal conditions. In online diagnosis, the system is assumed to be operational and the diagnostic module is designed in order to continuously monitor the system behavior, isolate and identify failures. Within these methods, we can distinguish between active diagnosis that uses both inputs and outputs, and passive diagnosis that uses only system outputs. The diagnosis can also be non-incremental (i.e., the diagnosis inference engine is built offline) or incremental (the diagnosis inference engine is built online in response to the observation).

Diagnosability notion [13] aims at verifying if the system model is rich enough in information in order to allow the diagnosis inference engine, generally called diagnoser, to infer the occurrence of parametric and discrete faults within a bounded delay after their occurrence. The diagnosability notion was initially defined for discrete event systems (discrete faults) but it can be extended for HDS (parametric and discrete faults). In general, there are two categories of methods to build fault diagnosis inference engine allowing taking into account both the discrete and continuous dynamics in HDS as well as the interactions between them. In the first category, [14], the system's model is an extension of the continuous model by adding the system discrete modes. The fault-free continuous behavior is defined

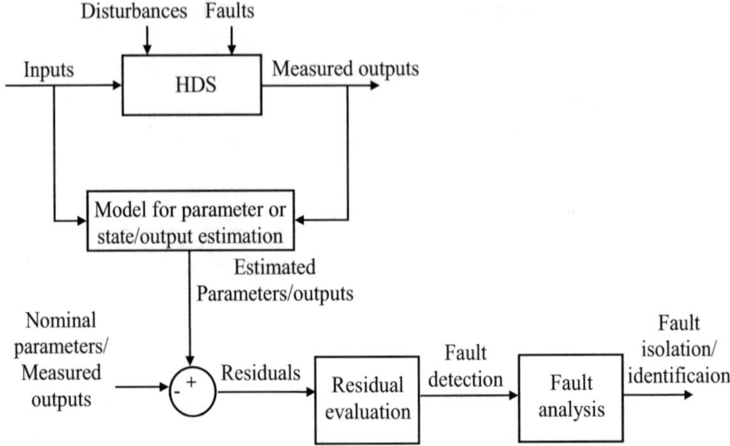

Fig. 1.3 Internal methods for fault diagnosis

in each discrete mode by relations over observable variables. These relations are used in order to generate residuals sensitive to a certain subset of faults. A fault is diagnosed when the value of the sensitive residuals to this fault is different from zero. In the second category, the discrete model is extended or enriched by adding events generated by the abstraction of system's continuous dynamics.

There are numerous methods in the literature that are used to build the fault diagnosis inference engine in HDS. They can be divided into internal, or model-based, and external, or data-driven, methods. The internal methods (see Fig. 1.3) use a mathematical or/and structural model to represent the relationships between measurable variables by exploiting the physical knowledge or/and experimental data about the system dynamics. They can be categorized into residual-based and set-membership [15] approaches. In residual-based approaches, the response of the mathematical model is compared to the observed values of variables in order to generate indicators used as a basis for the fault diagnosis. Generally, the model is used to estimate the system state, its output or its parameters. The difference between the system and the model responses is monitored on the basis of residual generation. Then, the trend analysis of this difference can be used to detect changing characteristics of the system resulting from a fault occurrence. Set-membership based fault diagnosis techniques are used for the detection of some specific faults. Generally, they discard models that are not compatible with observed data, in contrast to the residual-based approaches which identify the most likely model.

The external methods [6, 16–19] (see Fig. 1.4) consider the system as a black box, in other words, they do not need any mathematical model to describe the system dynamical behaviors. They use exclusively a set of measurements or/and heuristic knowledge about system dynamics to build a mapping from the measurement space into a decision space. They include expert systems and machine learning and data mining techniques.

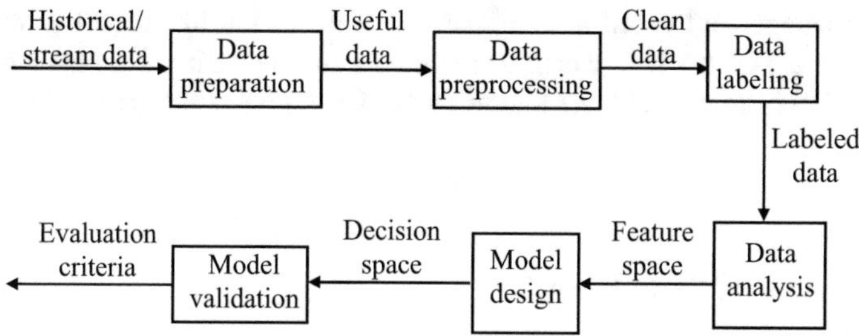

Fig. 1.4 External methods for fault diagnosis

Many critical infrastructures, such as power generation and distribution networks, water networks and mass transportation systems, autonomous vehicles and traffic monitoring, etc., are CPS. Such systems, including critical infrastructures, are becoming widely used and covering many aspects of our daily life. Therefore, the security, safety, and reliability of CPS is essential for the success of their implementation and operation. However, these systems are prone to major incidents resulting of cyberattacks and system failures. These incidents affect significantly their security and safety. Attacks can be represented as an interference [20, 21] in the communication channel between the supervisor and the system intentionally generated by intruders in order to damage the system or as the enablement, respectively disablement, of actuators events that are disabled, respectively enabled, by the supervisor. In general, intruders hide, create, or even change intentionally events that transit from one device (actuator, sensor) to another in a control communication channel. These attacks in a supervisory control system can lead the plant to execute event sequences entailing the system to reach unsafe or dangerous states that can damage the system. Therefore, reliable, scalable, and timely fault and attack diagnosis is crucial in order to improve the robustness of CPS to failures and adversarial attacks. Moreover, a thorough understanding of the vulnerability of CPS' components against such incidents can be incorporated in future design processes in order to better design such systems. Finally, the timely fault diagnosis can help operators to have better situation awareness and give them ample time to implement correction (maintenance) actions.

However, guaranteeing the security and safety of CPS requires verifying their behavioral or safety properties either at design stage such as state reachability, diagnosability, and predictability or online such as fault detection and isolation. This is a challenging task because of the inherent interconnected and heterogeneous combination of behaviors (cyber/physical, discrete/continuous) in these systems. Indeed, fault propagation in CPS is not only governed by the behaviors of components in the physical and cyber sub-systems but also by their interactions. This makes the identification of the root cause of observed anomalies and predicting the failure events a hard problem. Moreover, the increasing complexity of CPS and the security and safety requirements of their operation as well as their decentralized resource management entail a significant increase in the likelihood of failures in these systems. Finally, it is worth to mention that computing the reachable set of states of HDS is an undecidable matter due to the infinite state space of continuous systems.

1.3 Contents of the Book

This edited Springer book presents recent and advanced approaches and techniques that address the complex problem of analyzing the diagnosability property of cyber-physical systems and ensuring their security and safety against faults and attacks. The CPS are modeled as hybrid dynamic systems using different model-based and data-driven approaches in different application domains (electric transmission networks, wireless communication networks, intrusions in industrial control systems, intrusions in production systems, wind farms, etc.). These approaches handle the problem of ensuring the security of CPS in presence of attacks and verifying their diagnosability in presence of different kinds of uncertainty (uncertainty related to the event occurrences, to their order of occurrence, to their value, etc.).

1.3.1 Chapter 2

This chapter treats the problem of fault diagnosis of complex dynamic systems and its application to the aid of conditional maintenance of wind turbines. The proposed approach is an abductive model-based diagnosis where the system behavior is described logically by a set of propositions as premises (hypotheses) that entail conclusions (diagnoses). The set of propositions, describing how failures affect the system variables (i.e., components), represents the knowledge base that the diagnosis engine uses to provide an explanation for an observation. The proposed approach is based on three main phases: offline/online model development, online fault detection, and fault identification. The offline portion of the model is automatically built by exploiting failure assessments (e.g., Failure Mode Effect Analysis (FMEA) allowing failure modes characterizations and their manifestations or symptoms). The online portion of the model is built based on the fault detection phase. When the latter detects a new, unrecorded, abnormal behavior, the knowledge base of the model is updated in order to integrate this new abnormal behavior. The fault identification engine is triggered when an incorrect behavior is detected. It uses both the observed symptoms and the offline constructed model to compute the abductive diagnoses or explanations. The latter are continuously refined over time thanks to the discovered new symptoms and to the interactions with the maintenance technicians. The interactions with the latter are achieved through an interface allowing supporting the service technicians in preparing all spare parts and tools necessary before traveling to a wind turbine to ensure minimal downtime. In addition, on site, this interface provides contextual information as well as interactive questions in order to facilitate the diagnosis refinement or/and the discovering of new failures/abnormal behaviors. The advantage of the proposed approach is mainly related to its capacity to explain the occurrence of a failure using taxonomy adapted to the mode reasoning of technicians, operators, and supervisors. It is also related to its capacity to interact with technicians in order to facilitate the refinement of diagnosis candidates and the discovering of new failures/abnormal behaviors. The latter lead to enrich continuously the model/knowledge base over time. However, the computational complexity of the model/fault identification computation remains an issue to be solved in particular for large scale systems.

1.3.2 Chapter 3

This paper presents a solution in order to perform abductive fault diagnosis using the popular modeling language Modelica. The latter is an object-oriented, open, and multi-domain language used for representing hybrid cyber-physical systems. It allows generating intuitively models that are used to simulate the system normal behavior. The presented solution enriches Modelica models by creating a knowledge base (i.e., rules) used for abductive diagnosis. The knowledge base is built by extracting cause-effect rules from Modelica models. These rules are intuitive to designers familiar with failure mode and effect analysis (FMEA). When the difference is significant between a simulation of the normal (fault-free) model and the one where individual fault's effects are integrated, a rule is extracted automatically. This rule is used for the identification of this individual fault. Therefore, the set of these extracted rules represents the behavior deviations of a system in response to a set of pre-defined individual faults. To compute the difference (deviation) between the signals representing normal (fault-free) and faulty behaviors, three different approaches are used: the average values (reference signal) and pre-defined tolerances, temporal band sequences, and the Pearson correlation coefficient. The time when a significant deviation is detected represents the fault detection time. The chapter uses two examples (voltage divider circuit and a switch circuit with a bulb and capacitor) in order to illustrate the proposed solution. The switch circuit is an example of a hybrid dynamic system where the continuous dynamics is represented by the current and voltage of the capacitor and resistor and the discrete dynamics is represented by the switch state (on/off). The represented solution is flexible in the sense that the augmented Modelica models can be reused for other components. The adaptation of these Modelica models to perform the abductive diagnosis allows them to be used in decision support tools to aid human operators to make decisions as the conditional/predictive maintenance. However, the robustness of the proposed solution against outliers, noises, environment, and load variations as well as other types of uncertainties need to be improved.

1.3.3 Chapter 4

This chapter presents a data-driven modeling approach in order to perform the fault diagnosis of a production system represented as a multi-sensor network. The sensors are spatially distributed and measuring one or more channels. The different data streams, generated by the sensors, are gathered in a central data sink together with the discrete control events coming from the process control system. The fault diagnosis is performed by investigating the causal relations and dependencies between the system variables (channels). These causal relations and dependencies characterize the system states (normal/faulty) and are represented as causal relation network where nodes and vertices denote the channels and input/output variables, respectively. The use of such causal relation network may help operators and experts to gain insights into the interpretation and understanding of a failure mode development and causes. The reference models, representing the causal relations and dependencies, in fault-free conditions are used to compare the measurements issued from current conditions with the normal ones. This comparison generates residuals that are used to form a feature space. In the latter, the data measurements corresponding to normal operation conditions occupy restricted zones. When a fault occurs, the measurements occupy spaces far from these "normal" zones. The reference models are regularly updated in order to include the novelties of the system (new normal operation conditions, fault operation conditions) and thus to omit high false alarms and missed fault detection due to wrong predictions and quantifications on new samples. This update leads to more reliable fault detection and more stable model updates. A drift detection mechanism is used in order to detect a fault in early stage before becoming more severe or downtrends in the quality of components or products. The main advantages of the proposed approach are its capacity to update the generated models in response to the novelties and changes in the system environment and internal state and the use of drift detection mechanism to detect a fault in its early stage. However, updating the causal relation network (vertices and nodes) is a time-consuming task in particular for a hybrid dynamic system with multiple discrete modes.

1.3.4 Chapter 5

This chapter presents an approach for detecting intrusions in industrial control systems (ICS). These intrusions exploit the vulnerabilities of ICS in order to alter intentionally the integrity, confidentiality, and availability of the production system. These cyberattacks affect the control commands of the controller (Programmable Logic Controller PLC), the reports (sensors readings) coming from the plant as well as the communication between them. The proposed approach is based on the use of two filters: control and report filters. The control filter aims at verifying the consistency of control commands according to the current state and the sensor outputs while the report filter checks the integrity of the measurements (reports) according to the current state and the predicted (sent) control commands. Both filters are integrated out the production system and communicate through an independent and secured network in order to limit the cyberattack surface. They are based on three steps. The first and second steps aim at identifying offline the critical and prohibited states as well as the sequences (actions/sensor outputs) required to reach the prohibited states from any other normal (initial) state. The third state aims at detecting the cyberattacks online by observing a deviation from the normal behavior. The latter is described by the sequences of the pairs (state/action). This deviation is computed as a distance between the current state and the prohibited one. This distance is characterized by the minimal number of actions that can be applied on the process in order to reach a prohibited state. The main advantages of this approach are: (1) it is a non-intrusive approach since it does not need to install new probes in the production system and (2) it limits the cyberattack surface since the control and report filters are installed out the PLC and the plant and they communicate through an independent and secure network. However, the proposed approach cannot adapt to new cyberattacks (new prohibited and dangerous states) and its computation complexity grows exponentially for large scale systems with multiple discrete states.

1.3.5 Chapter 6

This chapter presents an approach to fault diagnosis of switched systems modeled by a Mealy machine (an automaton with inputs and outputs). Some transitions of this automaton, including those corresponding to faults, may occur in the absence of a control input and therefore are unobservable. Consequently, several states of the diagnoser (the model performing the fault diagnosis) are uncertain in the sense that they contain several fault labels (indicating several faulty components) and therefore they cannot isolate the responsible component of the fault occurrence. The diagnosis, performed by the proposed approach, is active in the sense that it ensures simultaneously the control and the diagnosability of the system. In order to remove the diagnosis uncertainty, the proposed approach computes the fault isolating sequence that leads to reach a certain diagnosis state with one fault label indicating the fault component. The proposed approach considers the initial conditions of the system are known and the initial mode is without fault. It considers also that only control input that drives the evolution of the system is represented by the switching function. This function specifies the active mode. Furthermore, discrete outputs are also available, as a result of each mode transition, in order to detect and isolate the fault. When a fault is detected, the nominal control is suspended for safety reason. The proposed approach requires building the system model with all its nominal and fault discrete states. This model, called diagnose, is then associated with the minimal fault isolating sequences for all uncertain states in the diagnoser. The proposed approach is illustrated and tested using a multicellular power converter in order to diagnose the discrete faults (switch blocked-on/blocked-off). The discrete dynamics of the power converter is represented by the on/off switches, while the continuous dynamics are related to the charging/discharging of the capacitors. The advantage of the proposed approach is related to its improvement of the system diagnosability by applying the fault isolating sequences. However, the proposed approach does not scale well according to the system size in particular with multiple discrete modes.

1.3.6 Chapter 7

This chapter studies and investigates the security issues for hybrid dynamic systems in presence of attacks. The latter are represented by compromised sensor measurements exchanged by the means of a wireless communication network. The challenge to ensure the system security and safety is to correctly estimate or reconstruct instantaneously or within a finite time interval the system internal state despite the presence of corrupted sensor measurements by external hackers. This chapter formalizes the required conditions in order to perform a secure state estimation despite the malicious attacks. To this end, the continuous dynamics is abstracted in order to generate discrete events that are used to enrich the system model. Only a subset of sensors of fixed size is considered to be corrupted. However, the sensors of this subset are unknown. Then, the link between diagnosability notion, developed for discrete event systems, and the secure state estimation is explored. The goal is to determine the conditions that allow distinguishing normal states from corrupted ones. The main advantage of the proposed approach is its capacity to estimate the secure state in the context of hybrid dynamic states where problems of decidability and computational complexity arise because of the cohabitation of both discrete and continuous dynamics. However, the proposed approach is restricted to linear non-Zeno systems with fixed size of corrupted sensors.

1.3.7 Chapter 8

This chapter proposes a hierarchical component-based approach for fault diagnosis and prognosis as well as failure mitigation in critical cyber-physical systems (CPS). The latter is composed by the physical system (plant), the actuators, the protection devices, and the discrete controllers. The latter try to arrest the failure effect if detected. The proposed approach uses temporal causal diagrams in order to describe the consequences of fault occurrence and propagation in physical and cyber components. The built model accounts for normal and fault behaviors. Each behavior is represented by a sequence of states and events. A state is characterized as safe or harmful. When a primary fault occurs, its propagation enforces the protection devices to be activated. This leads the system to a blackout state. The fault prognosis is based on the determination whether the system at its current state satisfies the constraints to reach a blackout state. If the answer is yes, then the trajectory (set of states and transitions) reachable from the current state is computed. The proposed approach is used for fault diagnosis and prognosis as well as failure mitigation of power systems, in particular electric transmission networks. The physical system consists of generators, buses, transmission lines and loads, while actuators are the circuit breakers and the protection devices are the relays. The latter cause system reconfiguration by instructing actuators to change their state. Two types of faults are considered: phase to phase faults and phase to ground fault. The faults of the breakers (breaker stuck-closed and breaker stuck-opened) and distance relays (missed detection faults and false alarms) are considered. The observable events in the case of power transmission system are commands sent by relays to breakers, messages sent by relays to each other, state change of breakers, physical fault detection alarms, etc. The faults are the unobservable events. The proposed approach is modular in the sense that each component has its proper model and its proper local diagnoser and a reasoner is used to compute the global diagnosis decision. However, a global model is needed to build the diagnosers and the reasoner.

1.3.8 Chapter 9

This chapter presents a model-based and data-driven approach to perform the fault diagnosis of cyber-physical systems that are safety critical, yet prone to system failures. Fault refers to any fault, attack, or anomaly. The nominal behaviors and the fault modes are represented by hidden-mode switched affine models with time-varying parametric uncertainty subject to process and measurement noise. The proposed approach is based on three steps: model invalidation, fault detection, and fault isolation. Model invalidation aims at determining whether an input–output sequence over a horizon T is compatible with a switched affine model. If the data is not compatible with the nominal model, then the model is invalidated and a fault is detected. When a fault is detected, the fault isolation step aims at uniquely determining which specific fault model is validated or rather, not invalidated, based on the measured input–output data. The proposed approach is evaluated for the simple and multiple fault diagnosis of the Heating, Ventilating, and Air Conditioning (HVAC) system. The considered faults are: faulty fan (the fan rotates at half of its nominal speed), faulty chiller water pump (the pump is stuck and spins at half of its nominal speed), and faulty humidity sensor (the humidity measurements are biased by an amount of +0.005). The advantage of this approach is its ability to diagnose simple and multiple faults either discrete or continuous (parametric) by taking into account the noises and parameter uncertainties. However, it requires an important effort and depth knowledge to build offline the nominal and fault models. Moreover, it needs to determine the horizon T (time delay) required to distinguish between normal and fault behaviors (T-detectability) and between two different fault models (I-isolability).

1.3.9 Chapter 10

This chapter proposes an extension of the diagnosability notion proposed for discrete event systems (DES) for hybrid dynamic systems (HDS) with uncertain observation. This chapter provides the answer to the question: can diagnosability be achieved even if the observation is uncertain? that is, when the order of the observed events and/or their (discrete) values are partially unknown. Indeed, in many applications, the temporal order of the observable events that have occurred within the DES is not always known, in particular when they occur in a short time span. In addition, the occurrence of some events is not certain in the sense that they may have occurred or not. Therefore, the developed diagnosability notion in this chapter considers the combination of these two types of uncertainties. However, the time delay required to verify the diagnosability for HDS is not proved to be bounded.

1.3.10 Chapter 11

This chapter proposes a diagnosability notion adapted to hybrid dynamic systems (HDS). It proposed an algorithm able to verify at design state if a fault that would occur at runtime could be unambiguously detected within a given finite time using only the allowed observations. The proposed algorithm is based on the abstractions that discretize the infinite state space of the continuous variables into finite sets. It starts by generating the most abstract discrete event system (DES) model of the HDS and checking diagnosability of this DES model. A counterexample that negates diagnosability is provided based on the twin plant. This counterexample is obtained when there is a path in the twin model with at least one ambiguous state (state with two different diagnosis labels) cycle. The model is then refined in order to try to invalidate the counterexample and the procedure repeats as far as diagnosability is not proved. If the counterexample is validated, then the system is not diagnosable. If there is no validated counterexample, then the system is diagnosable. The refinement is based on the understanding of the causes that entailed the refusal of the counterexample. Then, this spurious counterexample or any close spurious counterexamples will be eliminated next time. This will make the best out of computation. The proposed algorithm was illustrated using two examples. The first example is a server with 4 buffers. Each buffer is assigned a workflow and the switching between buffers is controlled by a user input. The second example is a classical thermostat with two different faults. The first fault is discrete fault due to a bad calibration of the temperature sensor, while the second fault is parametric due to a problem in the heater. The proposed algorithm has the advantage to account explicitly for the hybrid dynamic nature of the system and to verify the diagnosability in cost-effectiveness analysis. However, the time delay required to verify the diagnosability is not proved to be bounded.

References

1. R. Rajkumar, I. Lee, L. Sha, J. Stankovic, Cyber-physical systems: the next computing revolution, in *Proceedings of the 47th Design Automation Conference* (ACM, 2010), pp. 731–736
2. A.J. Van Der Schaft, J.M. Schumacher, *An Introduction to Hybrid Dynamical Systems*, vol 251 (Springer, London, 2000)
3. A. Benveniste, T. Bourke, B. Caillaud, M. Pouzet, in *Hybrid systems modeling challenges caused by cyber-physical systems, cyber-physical systems (CPS) foundations and challenges*, ed. by J. Baras, V. Srinivasan. Lecture Notes in Control and Information Sciences (2013)
4. Y. Yalei, X. Zhou, Cyber-physical systems modeling based on extended hybrid automata, in *5th IEEE International Conference on Computational and Information Sciences (ICCIS)* (2013)
5. M.S. Branicky, V.S. Borkar, S.K. Mitter, A unified framework for hybrid control: model and optimal control theory. IEEE Trans. Autom. Control **43**(1), 31–45 (1998)
6. L. Rodrigues, S. Boyd, Piecewise-affine state feedback for piecewise-affine slab systems using convex optimization. Syst. Control Lett. **54**(9), 835–853 (2005)
7. H. Louajri, M. Sayed-Mouchaweh, Decentralized diagnosis and diagnosability of a class of hybrid dynamic systems, in *Informatics in Control, Automation and Robotics (ICINCO), 2014 11th International Conference on*, vol. 2 (2014)
8. M. Shahbazi, E. Jamshidpour, P. Poure, S. Saadate, M.R. Zolghadri, Open-and short-circuit switch fault diagnosis for nonisolated dc dc converters using eld programmable gate array. IEEE Trans. Ind. Electron. **60**(9), 4136–4146 (2013)
9. R. David, H. Alla, *Discrete, Continuous, and Hybrid Petri Nets* (Springer, Berlin, 2010)
10. D. Wang, S. Arogeti, J.B. Zhang, C.B. Low, Monitoring ability analysis and qualitative fault diagnosis using hybrid bond graph. IFAC Proc. **41**(2), 10516–10521 (2008)
11. T.A. Henzinger, The theory of hybrid automata, in *Verification of Digital and Hybrid Systems*, (Springer, Berlin, 2000), pp. 265–292
12. M. Sayed-Mouchaweh, E. Lughofer, Decentralized fault diagnosis approach without a global model for fault diagnosis of discrete event systems. Int. J. Control. **88**(11), 2228–2241 (2015)
13. M. Sayed-Mouchaweh, *Discrete Event Systems: Diagnosis and Diagnosability* (Springer, New York, 2014)
14. T. Kamel, C. Diduch, Y. Bilestkiy, L. Chang, Fault diagnoses for the Dc filters of power electronic converters, in *Energy Conversion Congress and Exposition (ECCE)* (IEEE, 2012), pp. 2135-2141
15. M. Tabatabaeipour, P.F. Odgaard, T. Bak, J. Stoustrup, Fault detection of wind turbines with uncertain parameters: a set-membership approach. Energies **5**(7), 2424–2448 (2012)
16. L. Hartert, M. Sayed-Mouchaweh, Dynamic supervised classification method for online monitoring in non-stationary environments. Neurocomputing **126**, 118–131 (2014)
17. M. Sayed-Mouchaweh, N. Messai, A clustering-based approach for the identification of a class of temporally switched linear systems. Pattern Recogn. Lett. **33**(2), 144–151 (2012)
18. H. Toubakh, M. Sayed-Mouchaweh, Hybrid dynamic data-driven approach for drift-like fault detection in wind turbines. Evol. Syst. **6**(2), 115–129 (2015)
19. M. Sayed-Mouchaweh, Diagnosis in real time for evolutionary processes in using pattern recognition and possibility theory. Int. J. Comput. Cognit. **2**(1), 79–112 (2004)
20. L.K. Carvalho, Y.C. Wu, R. Kwong, S. Lafortune, Detection and prevention of actuator enablement attacks in supervisory control systems, in *13th International Workshop on Discrete Event Systems* (2016), pp. 298–305
21. D. Thorsley, D. Teneketzis, Intrusion detection in controlled discrete event systems, in *45th IEEE Conference on Decision and Control* (2006) pp. 6047–6054

Chapter 2
Wind Turbine Fault Localization: A Practical Application of Model-Based Diagnosis

Roxane Koitz, Franz Wotawa, Johannes Lüftenegger, Christopher S. Gray, and Franz Langmayr

2.1 Introduction

The increasing complexity and magnitude of technical systems is leading to a demand for effective and efficient automatic diagnosis procedures to identify failure-inducing components in practice. This is especially true in application areas experiencing excessive service costs and idle time revenue loss. In the industrial wind turbine domain operation and maintenance constitute significant factors in terms of turbine life expenditure. Given the remote locations onshore and offshore of wind turbine installations, accurate fault identification is essential for reducing costs and risks of component failures as well as turbine downtime [14]. Wind turbine diagnosis is complicated, however, since their overall reliability is affected by a multitude of failure modes concerning all major sub-systems and environments and furthermore the load conditions change regularly [36]. While electrical and control systems account for most wind turbine failures, other sub-systems, such as gearboxes, cause extensive downtimes due to the complexity of maintenance and thus pose a higher cost risk [35]. Unfortunately, currently implemented standard alarm systems deliver a large number of false alarms and thus are not suitable for standalone fault detection and identification [14].

Wind turbine operators often rely on time-based maintenance, where turbines are inspected periodically to assess their condition. This practice may lead to unnecessary turbine downtime for healthy systems, while failure-inducing conditions between services remain unnoticed. Due to these disadvantages predictive

R. Koitz (✉) · F. Wotawa · J. Lüftenegger
Graz University of Technology, Graz, Austria
e-mail: rkoitz@ist.tugraz.at; wotawa@ist.tugraz.at; jlueften@ist.tugraz.at

C. S. Gray · F. Langmayr
Uptime Engineering GmbH, Graz, Austria
e-mail: c.gray@uptime-engineering.com; f.langmayr@uptime-engineering.com

M. Sayed-Mouchaweh (ed.), *Diagnosability, Security and Safety of Hybrid Dynamic and Cyber-Physical Systems*, https://doi.org/10.1007/978-3-319-74962-4_2

and condition-based maintenance have become increasingly popular. Both rely on condition monitoring software and diagnosis methods [23]. Condition monitoring software utilizes the signals transmitted from the sensors integrated within the turbine and further processes the information to derive health information of critical components, e.g., gearboxes and main bearings. Some specification indicators of the subsequent failure are observable prior to around 99% of equipment fault occurrences [35]. Unnecessary maintenance activities can be avoided by scheduling repair or replacement of components based on their present or impending failure risks [1].

Numerous approaches exist for wind turbine diagnosis. Signal processing techniques analyze the multidimensional turbine data without considering an a priori developed mathematical model to extract faults based on spectral analysis or trend checking techniques, while machine learning methods, such as neural networks, can rely on historic data for failure identifications [16]. Zaher and McArthur [41] introduce a fault and degradation detection system for entire wind turbine installations based on supervised learning of the nominal behavior. By collecting data from various downtimes their system computes an overall turbine operational behavior model. Schlechtingen et al. [29] also adopt machine learning by creating neural networks based on the normal wind turbine behavior in combination with fuzzy rules representing expert knowledge on faults. Their approach requires the availability of Supervisory Control And Data Acquisition (SCADA) operation logs, which provide 10-min values of various measurements, such as power output, rotor speed, or gearbox oil temperature [40]. Anomalies can be detected by comparing the normal-behavior model with the actual performance. The fuzzy inference system then automatically identifies the faulty components. Gray et al. [14, 15] describe a combination of diagnostic and prognostic techniques exploiting the relations between operational and environmental loads as also damage accumulation rates. Based on an analysis of potential component-based failure modes and their damage driving physics, a mathematical model is computed that can be used to calculate the rate at which damage accumulates in response to the operating environment. This method offers a means for determining current and projected failure probabilities based on the derived damage model and statistical failure model. Statements about absolute remaining useful life cannot be made, since the load capacity would have to be known in advance to a high degree of accuracy. This is very rarely the case, and considerable variation occurs due to, e.g., variations in material quality, tolerances in component manufacture, influence of transport, installation and configuration. Therefore the prognostic method focuses on quantification of the applied loads instead, and uses probabilistic methods to relate the said loads to damage accumulation.

Model-based diagnosis (MBD) techniques have been developed in the Fault Detection and Isolation (FDI) and the Artificial Intelligence (AI) community. Approaches stemming from the FDI field usually depend on quantitative models, while the AI methods utilize qualitative/symbolic representations of the underlying system to draw conclusions regarding the state of the system and its components

[1, 6, 9, 28]. For instance, Echavarria et al. [10] utilize qualitative physics to formalize the behavior of wind turbines. Combined with a solver, their model-based system is capable of detecting and identifying faults.

In this chapter, we focus on the AI variation of MBD. MBD relies on a description of the system, which together with abnormal observations of the current system state is exploited to derive root causes [9, 28]. Multiple applications have been developed for diverse fields, e.g., the automotive industry [32] or environmental decision support systems [39]. Even though MBD can look back on decades of research, a widespread dissemination in practice is still lacking. The reasons for this include the effort associated with the development of diagnosis models [4] and the difficulties of adequately integrating MBD into current industrial work processes [12, 24, 31].

In cooperation between Graz University of Technology and Uptime Engineering[1] a project was initiated with the aim of providing a methodology and framework for MBD in the industrial domain. Due to Uptime Engineering's many years of experience and expertise in the field of wind power plant maintenance, industrial wind turbines constitute an ideal test bed and application area for MBD. In this chapter, we present some of the results of the collaboration as well as the ongoing realization of MBD in practice. With this work we aim to bridge the gap between the theory of MBD and its practical application. We first discuss the foundations of MBD and present a general process for integration of this approach in real-world fault identification. Subsequently, we introduce an application designed to facilitate diagnosis in the industrial wind turbine domain. In particular, the application's graphical user interface (GUI) is presented, which has been created taking the needs, work processes, and environments of the maintenance personnel into consideration. Subsequently, we discuss the current status of the integration of an MBD engine in the industrial wind turbine domain and provide some concluding remarks.

2.2 Model-Based Diagnosis

Model-based reasoning fosters the idea of reusing knowledge by relying on a formalization of the system under consideration. The model together with a set of observed symptoms can be exploited to obtain diagnostic hypotheses for the observations. Two variations have emerged in the literature: consistency-based and abductive MBD. Consistency-based diagnosis utilizes a description of the correct system behavior and identifies root causes through inconsistencies arising from the model in combination with the given symptoms. A diagnosis is then a set of abnormality assumptions about the system such that the observations and assumptions are consistent [9, 28]. In contrast, abductive diagnosis is based on the notion of logical entailment [26]. A set of premises ψ logically entails a

[1]Uptime Engineering GmbH provides consulting services as well as software tools in the field of technical reliability.

conclusion ϕ if and only if for any interpretation in which ψ holds ϕ is also true. We write this relation as $\psi \models \phi$ and call ϕ a logical consequence of ψ. A set of abnormality assumptions entailing the observations constitutes an abductive diagnosis or explanation. To derive causes for observed anomalies by utilizing this type of inference, the abductive MBD approach depends on a model representing the links between faults and their manifestations. Even though both variations are based on different reasoning techniques, Console et al. [5] showed the close relation between consistency-based and abductive diagnosis. In the upcoming portion of the chapter, we focus on abductive MBD. First, we describe an abductive diagnosis problem and its solution based on a subset of propositional logic, namely Horn clauses. Subsequently, we discuss a process facilitating the incorporation of abductive MBD in real-world applications.

2.2.1 Propositional Horn Clause Abduction

A Horn clause is defined as a disjunction of literals featuring at most one positive literal and can be described by a rule, e.g., $\{\neg a_1, \ldots, \neg a_n, a_{n+1}\}$ can be written as $a_1 \wedge \ldots \wedge a_n \rightarrow a_{n+1}$. Similar to Friedrich et al. [13], we define a knowledge base (KB) representing the abductive diagnosis model in the context of propositional Horn clause abduction.

Definition 1 A knowledge base (KB) is a tuple $(A, \mathrm{Hyp}, \mathrm{Th})$ where A denotes the set of propositional variables, $\mathrm{Hyp} \subseteq A$ the set of hypotheses, and Th the set of Horn clause sentences over A.

A hypothesis, also referred to as an assumption, is a propositional variable for which we can presume a certain truth value. Hypotheses are the propositions which can be part of a diagnosis, while the Horn theory depicts the relationships between the variables.

Example 1 Gearbox lubrication is an essential aspect of industrial wind turbine reliability as it protects the contact surfaces of gears and bearings from excessive wear and prevents overheating. Considering a simplified scenario, insufficient lubrication can be caused by a damaged oil pump, which leads to loss of oil pressure and therefore a reduction in the flow rate of oil through the system. Furthermore, a blockage of the filter in the oil cooling system may cause overheating of the oil, which also negatively affects the lubrication due to a reduction in the film thickness at the bearing and gear contacts.

(continued)

Example 1 (continued)

Starting from this description, we can identify two root causes of insufficient lubrication: a blocked filter or a damaged oil pump. These causes constitute the faults we want to identify during diagnosis and thus their corresponding variables form the set of hypotheses:

$$\text{Hyp} = \{\textit{damaged\_pump}, \textit{blocked\_filter}\}$$

The set of propositional variables A comprises all hypotheses as also propositions representing effects:

$$A = \left\{ \begin{array}{c} \textit{damaged\_pump}, \textit{blocked\_filter}, \textit{reduced\_pressure}, \textit{overheating}, \\ \textit{reduced\_film\_thickness\_bearing\_contacts}, \\ \textit{reduced\_film\_thickness\_gear\_contacts}, \textit{poor\_lubrication} \end{array} \right\}$$

Given the set of propositional variables, the circumstances leading to an insufficient greasing of the gearbox can be represented by a Horn theory:

$$\text{Th} = \left\{ \begin{array}{c} \textit{damaged\_pump} \rightarrow \textit{reduced\_pressure}, \\ \textit{reduced\_pressure} \rightarrow \textit{poor\_lubrication}, \\ \textit{blocked\_filter} \rightarrow \textit{overheating}, \\ \textit{overheating} \rightarrow \textit{reduced\_film\_thickness\_bearing\_contacts}, \\ \textit{overheating} \rightarrow \textit{reduced\_film\_thickness\_gear\_contacts}, \\ \textit{reduced\_film\_thickness\_bearing\_contacts} \wedge \\ \textit{reduced\_film\_thickness\_gear\_contacts} \rightarrow \textit{poor\_lubrication} \end{array} \right\}$$

Definition 2 Given a knowledge base (A,Hyp,Th) and a set of observations Obs $\subseteq A$ then the tuple (A,Hyp,Th,Obs) forms a propositional Horn clause abduction problem (PHCAP).

A diagnosis problem involves a KB plus a set of observations for which the explanations are to be computed. In our context, these observables may only be a conjunction of propositions and not an arbitrary logical sentence. The solution to a PHCAP or diagnosis Δ is a set of hypotheses explaining the propositions in Obs, i.e., entailing them together with the theory Th. In other words, the observations are a logical consequence of the failure relations described in the theory and the determined explanation. An additional requirement is that only consistent diagnoses are permitted, thus solutions leading to a contradiction are disregarded. Imposing a parsimonious criterion on the solutions is a principle commonly used in diagnosis. From our practical point of view only subset minimal explanations are of interest.

Definition 3 Given a PHCAP (A,Hyp,Th,Obs). A set $\Delta \subseteq$ Hyp is a solution if and only if $\Delta \cup \text{Th} \models \text{Obs}$ and $\Delta \cup \text{Th} \not\models \bot$. A solution Δ is parsimonious or minimal if and only if no set $\Delta' \subset \Delta$ is a solution. Δ-Set contains all solutions obtained from a PHCAP.

Example 1 (continued) Considering the PHCAP and assuming we detect an insufficient lubrication of the gearbox, i.e., Obs = $\{poor\_lubrication\}$, we can derive two minimal explanations; either the oil pump is faulty inducing a reduction of the oil flow ($\Delta_1 = \{damaged\_pump\}$) or a blocked filter causes poor cooling which leads to insufficient greasing of the bearing and gear contacts ($\Delta_2 = \{blocked\_filter\}$), i.e., $\Delta\text{-Set} = \{\{damaged\_pump\}, \{blocked\_filter\}\}$.

While abductive reasoning provides an intuitive approach for fault localization, its computational complexity for general propositional theories is located within the second level of the polynomial hierarchy [11]. Focusing on a less expressive modeling languages, such as in our case Horn clause models, reduces the complexity. Yet, Friedrich et al. [13] showed that computing the solution to a PHCAP is still NP-complete. Thus, for practical applications efficient solvers are essential to compute diagnoses in a reasonable time frame.

2.2.2 *Incorporating Diagnosis into Practice*

While MBD offers several attractive features, such as allowing the reuse of already created system models and a clean separation between the problem description and its solving mechanism, the dissemination of implementations in practice is limited [31]. The integration of MBD in industrial applications is impeded by two main drawbacks associated with this type of reasoning. First, the computational complexity as mentioned in the previous section discourages the practical use in cases where diagnoses are to be computed within short periods of time. Second, model-based reasoning techniques always demand the existence of a system description, whether it is of the correct behavior in consistency-based diagnosis or how failures affect system variables in the context of the abductive variation. Developing a model is associated with an initial effort and acquiring a technical description of a system suitable for diagnostic purposes can be challenging and is often hindered by organizational issues. Further, a lack of tools facilitating the model generation and integrating it into existing work processes complicates the modeling phase [2].

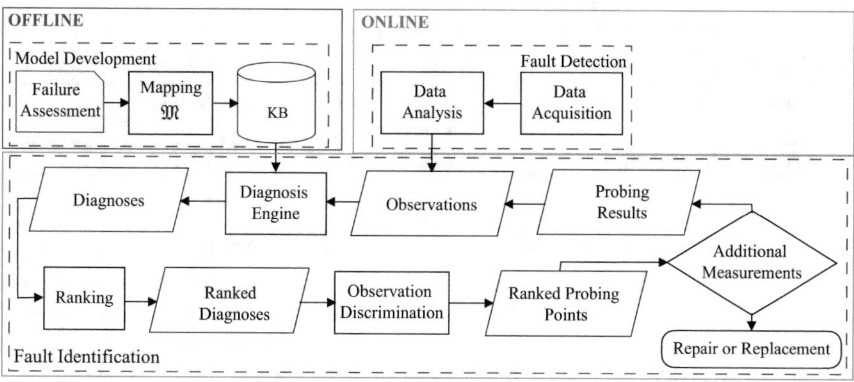

Fig. 2.1 Abductive MBD process (adapted from [18])

To counteract these factors Koitz and Wotawa [18] have defined a process based on discussion with Uptime Engineering for applying abductive MBD to real-world applications. The method relies on an automated model creation on the one hand and on the other hand ensures an efficient diagnosis computation by limiting the models to Horn logic. A graphical representation of the activities is depicted in Fig. 2.1. The process is divided into three phases: model development, fault detection, and fault identification. In order to lower the entrance barrier for implementing the MBD approach, models are automatically built by exploiting failure assessments frequently used in practice. Such analyses must characterize failures and their manifestations to be mapped to a KB required for abductive reasoning. A wide-spread tool which can be used for this purpose is Failure Mode Effect Analysis (FMEA) incorporating expert knowledge on component failures [38]. Constructing the model from a fault assessment only needs to be performed once—given that there are no updates to the analysis—and can be accomplished offline. The online portion of the MBD process is prompted once the presence of a fault has been detected by a mechanism discovering the existence of an incorrect system behavior such as a condition monitoring system. Based on the observed symptoms and the offline constructed model, the fault can be identified by deriving the abductive diagnoses. Various approaches are capable of computing abductive explanations [17] and further additional refinements to the initial diagnoses can be made via additional observations and prioritization of the results in regard to certain objectives, e.g., diagnosis likelihood or maintenance cost.

2.2.2.1 Model Development

Automatic generation of a suitable knowledge base using information available a priori is an essential feature of the proposed diagnosis process as it reduces additional modeling efforts. While different assessments can be utilized, we show

Table 2.1 *Example 2*: FMEA excerpt (adapted from [27])

Component	Failure mode	Failure effect	Likelihood	Severity
Yaw drive	Fails to rotate	No yaw, safety system failure, decrease of efficiency	2.2E−5	V
Yaw drive	Drive shaft blocked	No yaw, decrease of efficiency	1.3E−5	IV

here an example of a conversion based on FMEA [38]. FMEA is an established standardized reliability approach, in which an expert group analyzes a system and determines potential component-based single faults. Each failure mode is examined in regard to its causes, consequences, and various other characteristics [3]. Table 2.1 depicts an excerpt of an FMEA for the yaw drive of a wind turbine presenting two failure modes.

While in general an FMEA can feature more columns than the ones shown in Table 2.1 depending on the standard followed, the modeling methodology as proposed by Wotawa [38] focuses on three aspects which have to be considered within the analysis: the components (COMP), the failure modes (MODES), and moreover, the existence of propositions PROPS corresponding to observable failure effects.

Definition 4 An FMEA is a set of tuples (C, M, E) where $C \in$ COMP is a component, $M \in$ MODES is a failure mode, and $E \subseteq$ PROPS is a set of effects.

As the FMEA typically contains a description of how each fault affects a set of system variables, we can convert this information in a straightforward manner to a logical KB, where the hypotheses comprise the component-based failures and the theory consists of propositional Horn clause sentences describing the cause-effect relation depicted in the FMEA. A variable *mode(C,M)* is constructed for each component-failure mode pair in the analysis, where C is the component and M is the failure mode. These propositions compose the set of hypotheses:

$$\text{Hyp} =_{def} \bigcup_{(C,M,E)\in\text{FMEA}} \{\text{mode}(C,M)\} \tag{2.1}$$

To form the set of all variables A, the union over all hypotheses as well as propositions representing effects is constructed:

$$A =_{def} \bigcup_{(C,M,E)\in\text{FMEA}} E \cup \{\text{mode}(C,M)\} \tag{2.2}$$

Each record in the FMEA describes the effects of a single fault, thus, the relations between defects and their manifestations can be transformed into a Horn model in a straightforward way. Let HC be the set of Horn clause sentences, then the mapping function $\mathfrak{M} : 2^{\text{FMEA}} \mapsto \text{HC}$ generates a set of Horn clauses which are a subset of HC for each record in the FMEA.

Definition 5 Given an FMEA, the function $\mathfrak{M}$ is defined as follows:

$$\mathfrak{M}(\text{FMEA}) =_{def} \bigcup_{t \in \text{FMEA}} \mathfrak{M}(t) \tag{2.3}$$

where

$$\mathfrak{M}(C, M, E) =_{def} \{\text{mode}(C, M) \rightarrow e \,|e \in E\} \tag{2.4}$$

Example 2 (continued) The FMEA in Table 2.1 features the two component-fault mode pairs (Yaw Drive, Fails to rotate) and (Yaw Drive, Drive shaft blocked). Their corresponding propositional variables are added to Hyp.

$$\text{Hyp} = \left\{ \begin{array}{c} mode(Yaw\_Drive, Fails\_to\_rotate), \\ mode(Yaw\_Drive, Drive\_shaft\_blocked) \end{array} \right\}$$

The set of all propositions then contains the hypotheses as well as all variables corresponding to effects.

$$A = \left\{ \begin{array}{c} mode(Yaw\_Drive, Fails\_to\_rotate), \\ mode(Yaw\_Drive, Drive\_shaft\_blocked), \\ no\_yaw, safety\_system\_failure, decrease\_of\_efficiency \end{array} \right\}$$

For each manifestation contained in a record, a rule is built such that the single hypothesis representing the component-fault mode pair implies the effect. The theory is then simply a union over all these Horn clauses.

$$\text{Th} = \left\{ \begin{array}{c} mode(Yaw\_Drive, Fails\_to\_rotate) \rightarrow no\_yaw, \\ mode(Yaw\_Drive, Fails\_to\_rotate) \rightarrow safety\_system\_failure, \\ mode(Yaw\_Drive, Fails\_to\_rotate) \rightarrow decrease\_of\_efficiency, \\ mode(Yaw\_Drive, Driveshaft\_blocked) \rightarrow no\_yaw, \\ mode(Yaw\_Drive, Drive\_shaft\_blocked) \rightarrow decrease\_of\_efficiency \end{array} \right\}$$

Due to the structure of an FMEA, the resulting logical system description is acyclic[2] and consists of bijunctive Horn clauses, i.e., implications always lead from one hypothesis to a single effect variable. This results in an efficient diagnosis computation and a system considering the single fault assumption [19].

[2] A logical theory is acyclic in cases when it can be represented by a directed acyclic graph where each proposition is represented as a node and an edge is drawn from a proposition to another in case the former directly implies the latter.

A shortcoming of FMEA is that it does not take into account the potential interdependencies between various manifestations, which might be essential from a practical point of view to describe how a fault affects the system. Thus, other failure assessments, such as fault trees, can be used as the basis of the automatic modeling [20]. Depending on the underlying failure analysis type, the resulting diagnosis system description may feature different characteristics, such as being a non-bijunctive Horn model.

Independent of the assessment type, the accuracy and composition of the failure analysis largely impact the quality of the automatically generated diagnosis model. It is apparent that failures and manifestations disregarded in the failure review are missing from the system description and thus cannot be considered during fault identification. Hence, to achieve precise diagnoses, model completeness is an essential premise [24]. Furthermore, manifestations must be detectable in order to be useful in a diagnostic context and effects as well as failures have to be coherently reported throughout the assessment to allow automatic processing.

2.2.2.2 Fault Identification

Abductive diagnosis, even in the case where the system description is restricted in expressiveness to Horn sentences, is at least NP-complete [13]. Hence, efficient methods for deriving diagnoses are required in practice. There are various techniques capable of computing abductive diagnoses such as SAT-based approaches [17], consequence finding procedures [30], or the well-known Assumption-based Truth Maintenance System (ATMS) [8]. Internally the ATMS operates on a directed graph representing the logical relations contained in the theory, where propositions and the contradiction are nodes and implications determine the edges. Each node is equipped with a label recording for the corresponding variable the sets of assumptions, i.e., hypotheses, it can be derived from. Thus, the ATMS documents the entailment relations characterized within the theory [22] and ensures that labels are consistent and minimal with respect to subsumption. To compute the abductive diagnoses an additional implication is added to the ATMS such that $o_1 \wedge \ldots \wedge o_n \rightarrow ex$, where $\{o1, \ldots, o_n\} = Obs$ and ex represents a new propositional variable not yet contained within A. The label of ex then comprises all solutions to the PHCAP.

MBD may yield an exponential number of explanations in the worst case. Thus, techniques assisting in distinguishing diagnoses are required to allow for an effective decision making in regard to repair and replacement activities. Subsequently, we present two methods aiming at supporting fault identification: observation discrimination and diagnosis ranking.

Observation Discrimination Probing has been proposed as a means to decrease the solution space by supplying additional facts to the diagnostic reasoner. While Friedrich et al. [13] propose an interleaved process between diagnosis, probing and repair, Wotawa [37] suggests computing all explanations and subsequently adding new symptoms, which allows either the removing or confirming of diagnoses.

Definition 6 Given a PHCAP (A,Hyp,Th,Obs) and two diagnoses Δ_1 and Δ_2. A new observation $o \in A \setminus$ Obs discriminates two diagnoses if and only if Δ is a diagnosis for (A,Hyp,Th,Obs $\cup$ $\{o\}$) but Δ_2 is not.

Once discriminating observations have been selected, probes are taken and the fault identification process is restarted with the additional measurement information. Determining an ideal probing point is essential in order to converge to a plausible solution efficiently. The best new observation o is the probing point with the highest entropy $H(o)$. Entropy represents the information gain; thus, a higher entropy value indicates a measurement with a greater discrimination capability [9].

$$H(o) = -p(o) \times \log_2 p(o) - (1 - p(o)) \times \log_2(1 - p(o)) \tag{2.5}$$

Equation (2.5) defines the entropy value for an observation o, where $p(o)$ is the probability of o defined as the ratio between the diagnoses entailing the symptom together with the theory and the total number of explanations:

$$p(o) = \frac{|\{\Delta \mid \Delta \in \Delta\text{–Set}, \Delta \cup \mathrm{Th} \models \{o\}\}|}{|\Delta\text{–Set}|} \tag{2.6}$$

Diagnosis Ranking Depending on the underlying logical theory and set of observations, there might not be a single solution available. Thus, in these cases a prioritization of the diagnosis results can be useful to initiate appropriate maintenance activities. A common strategy is to exploit probabilities. Considering Bayes rule for conditional probability, we can define the probability of an explanation Δ given an observation o as

$$p(\Delta \mid o) = \frac{p(o \mid \Delta)p(\Delta)}{p(o)} \,. \tag{2.7}$$

Under the presumption that there is no uncertainty in the measurement, i.e., the data has not been subjected to errors or noise, we can state for any $o \in Obs$ that $p(o) = 1$. As we known from the entailment relation required by abductive diagnosis that the explanation logically implies the observation, we can assign $p(o \mid \Delta) = 1$. Consequently from these two assignments and Eq. (2.7) it follows that $p(\Delta \mid o) = p(\Delta)$. Assuming independence amongst faults, the probability of each diagnosis Δ can be computed based on the a priori probabilities $p(h)$ of the hypotheses:

$$p(\Delta) = \prod_{h \in \Delta} p(h) \prod_{h \notin \Delta} (1 - p(h)) \tag{2.8}$$

Given a PHCAP's solutions we compute $p(\Delta)$ for all diagnoses in Δ-Set and subsequently assign ranks accordingly. FMEA, for instance, holds additional information such as failure likelihoods, which can be utilized for prioritization. Other criteria, such as repair and replacement costs or fault and diagnosis severity, i.e., seriousness of consequences in regard to safety or monetary considerations, could also be considered instead [34].

2.3 Industrial Wind Turbine Diagnosis

Wind turbine reliability presents a very interesting use-case for the application of diagnostic methods. The cost of electrical energy produced depends strongly on the operational efficiency of the machines as also on the availability. Component faults leading to unplanned downtime have been shown to impact the overall energy production significantly, the financial motivation for optimization in this respect is thus high. The use of remote detection and diagnostic technology is an area which is receiving an increasing level of focus, in particular in the offshore wind energy industry, where turbine failures are even more critical due to the difficulties related to accessing and repairing the machines in potentially harsh environmental conditions.

All modern wind turbines use sensors, data acquisition, and on-board processing as part of the closed-loop control system. Furthermore, a range of diagnostic functions is typically included within the system controller, so that at least basic status information can be provided in case of faulty operation. However, such on-board diagnostics are limited by the computing resources of the turbine controller and the absence of instant access to a long-term historical database.

The turbine continuously stores operational data logs (SCADA logs), which can be retrieved and transferred to a central data store. The use of such data stores for detailed performance analysis and diagnostic work is becoming standard in the wind industry, since the data is readily available and provides information about a number of systems and components within the turbine.

Today most medium to large scale operators of wind turbine fleets have installed centralized data management systems to collect and store such SCADA logs. Uptime Engineering has developed a software application that is capable of performing automated and continuous analysis of such data, typically with the aim of detecting anomalies in the behavior of individual turbines. Continuous advances have been made in the capabilities of the analytic models, and it is now possible to detect outlying behavior with a high degree of sensitivity. The results of such analysis are combined with the above-mentioned on-board diagnostic results together with general information concerning the turbine age, type and build status, in order to support the turbine operator in efficiently reacting to detected anomalies.

However, such analysis activities often produce a high volume of information (multiple turbines monitored, multiple alarms originating from many systems and accompanied by a range of heterogeneous supporting information). The main challenge facing the user of such a system is efficient interpretation of the results and the derivation of an effective response strategy. Therefore a strong need has been identified to provide the software user with "decision support"; i.e., an additional layer of intelligence built in to the software, which combines all generated observations and produces clear recommendations for action. MBD is a highly relevant solution, due to the strong capability of the approach in combining state information from a multitude of sources and identifying the root cause with the highest likelihood.

The project between Uptime Engineering and Graz University of Technology aims at integrating an abductive MBD engine, created by the university, taking into account the process shown in Sect. 2.2.2 into Uptime Engineering's wind turbine condition monitoring software. Uptime Engineering continuously extends and updates a comprehensive failure assessment of industrial wind turbines, providing a structured evaluation of faults and their manifestations. This analysis can be exploited by the MBD engine as the basis of the model development phase to construct a suitable diagnostic system description. Once an anomaly has been detected by Uptime Engineering's condition monitoring software, an MBD computation is triggered taking into account of the abductive KB created and the symptom discovered. Given the results of the diagnosis, the MBD engine then provides additional information on the next best measurement based on entropy values. To ensure a suitable integration into the actual wind turbine maintenance workflow, we collaborate with an energy provider employing Uptime Engineering's condition monitoring. In this section, we first describe the interface and interaction design of the diagnosis engine and how it will be incorporated into the work processes of the maintenance personnel of the energy provider. We then describe the phases of the integration and its current status.

2.3.1 Abductive Model-Based Diagnosis Prototype

To enable MBD in industrial practice as proposed in Sect. 2.2.2, the necessary failure information must be available to automatically extract a suitable diagnostic model and an anomaly detection method is needed to initiate the fault identification phase. Furthermore, in order to yield benefits from deploying such a system, solutions need to be computed efficiently[3] and effectively reflecting defects present in the system. We argue, however, that these technical features are not the only deciding factors determining the success of a newly integrated diagnosis software. While current research frequently focuses on developing and improving reasoning techniques, the suitable integration of MBD in operational processes is rarely addressed [24, 33]. In addition, it is well known that the acceptance of new technology is tightly linked to the perceived usefulness of the product as well as its perceived ease of use [7]. The former refers to the benefits for the users and other stakeholders in regard to the performance of work tasks, whereas the latter is on par with the usability of a product.

Hence, in developing an MBD application for use in the field within our project, we focus not only on the technical aspects of feasibility but further account for the human factor. An interface and interaction design was incrementally developed

[3]Here, efficiency is subjective to the application domain, e.g., in the context of wind turbines deriving explanations in minutes is sufficient, while for automotive on-board diagnosis this computation time is unacceptable.

for an abductive MBD engine, which should function as a template for the actual implementation of the tools which will be integrated into Uptime Engineering's software. Various prototypes were created iteratively, starting from a low-fidelity paper mock-up to a clickable prototype depicting a usual fault identification scenario. These prototypes reflect the above-described general process of abductive MBD in the context of wind power plants. Particular attention was paid to respecting current work processes and accounting for a usable design.

The design process started with eliciting the requirements of the diagnosis application in consideration of the stakeholders involved in the project, who were:

- the service technicians, who are the users, will operate the diagnosis software for troubleshooting from the service center as also in the field, and are responsible for performing the turbines' planned maintenance, repair as well as replacement activities
- the management of a wind energy provider, planning on extending their self-maintenance activities for their wind turbine plants in the future
- Uptime Engineering, who currently develops condition monitoring software for wind turbines and will extend their portfolio with usable and extendable diagnosis software

2.3.1.1 Requirements

A list of requirements in regard to the final diagnosis application was established during the course of the various design iterations. The three distinct stakeholder groups have differing requests, which were analyzed in order to resolve conflicts and prioritize the resulting requirements. Since the success of the application depends to a great extent on being used by the service technicians, special attention was given to their suggestions and needs.

An important observation is that current fault detection activities performed by the service personnel typically rely on visual inspection. Hence, in order to support diagnosis, images should be used for easier recognition. Once a fault has been identified, the repair or replacement task is executed according to the wind turbine manufacturer's instruction manuals. Therefore, such documents need to be easily accessible via the software. After the maintenance activities have been completed, the service technicians are required to create a report of the task and the actions performed. The software should thus support automation of the reporting step to reduce the overall effort. The working environment inside a wind turbine is often uncomfortable and limited in space, and work is performed under time pressure in potentially difficult weather conditions. The user interface therefore needs to be intuitive in use and must guide the user through a strictly defined sequence with minimal user interactions. Considering the overall work process, the software should feature a desktop software part operated in the service center as well as a mobile application, which should be used within the turbine itself.

The management of the energy provider is interested in promoting digitalization as well as increasing the productivity and safety of their wind operations in use. On the one hand, the software should support the service technicians in preparing all spare parts and tools necessary before traveling to a wind turbine to ensure minimal downtime, while on the other hand given the hazardous environment in the field it should support the safety processes already in place, e.g., the service technicians personal safety equipment. In addition to the user and management requirements, the specifications of Uptime Engineering needed to be satisfied. To extend and update the knowledge base, i.e., abductive model, the users should be able to report new fault modes, which have not been previously contemplated. Further, the user interface should be extendable and adaptable to satisfy other customers as well as other domains for future projects.

2.3.1.2 Design Process

To ensure a user friendly end product, the diagnosis engine GUI was developed using an iterative process [25]. Each iteration starts with a definition or adaptation of the requirements, a design is then created and subsequently a prototype is implemented. This prototype is evaluated by users from the target group to determine usability issues, which must be fixed in the design of the proceeding iteration. According to Nielsen [25] due to the various repetitions of this cycle, this type of design process allows gaining sufficient insight into usability issues even given a limited number of test users.

In our case, the first iteration was kicked-off with a meeting between Graz University of Technology and Uptime Engineering to elicit the first set of requirements. One of the main goals identified was that the software should be designed in a way that supports the service technicians' current work processes without causing additional effort. Facilitating the service personnel's work tasks is essential as this assures usefulness, which is a key aspect in technology acceptance [7]. Furthermore, we defined the overall workflow for the application, the general structure for an initial paper mock-up, and a small set of features, which should be realized. The initial paper prototype and all following designs were evaluated at meetings with the management of the energy provider and service technicians. At these meetings the current prototype was presented and a more detailed knowledge about the users and their work process was gained, usability issues could be uncovered and useful features, which would aid the maintenance personnel throughout the fault correction process, were identified. During the first iterations predominant usability problems were detected and some more drastic changes to the design were introduced, while in the later iterations only minor issues were found and as a result only slight GUI alternations were necessary. The product of the design phase is a clickable prototype which has undergone a small-scale qualitative usability test involving five service technicians. In the test scenario the users performed a mock-up fault identification process from start to finish. The design of the final prototype is presented below together with the overall application workflow.

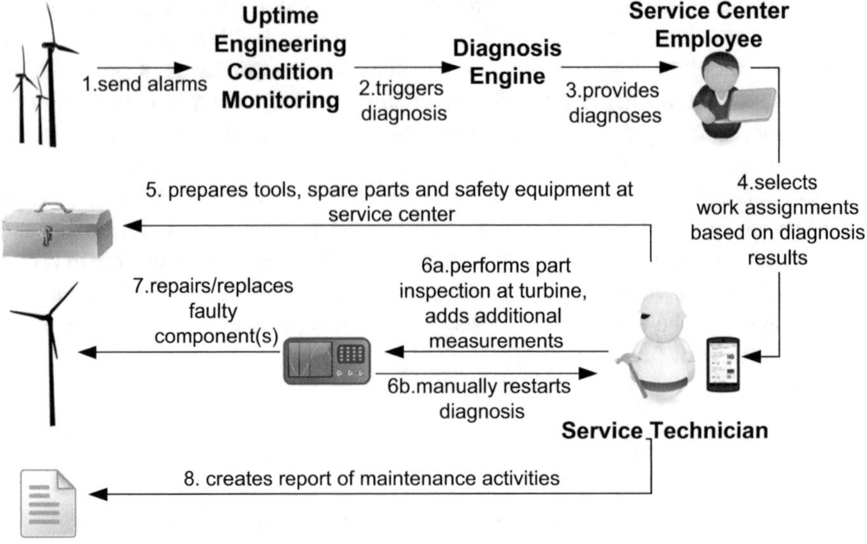

Fig. 2.2 Workflow of the diagnosis application (adapted from [21])

2.3.1.3 Workflow and GUI Design

The workflow of the diagnosis application was created in consideration of the current functionality of Uptime Engineering's condition monitoring software, the maintenance process of the energy provider, and the general abductive diagnosis procedure. Figure 2.2 depicts the identified activity sequence of the diagnostic process. The interface and interaction design decisions of the application were taken based on the workflow and requirement analysis. As mentioned in the previous section, a diagnosis computation is invoked once an anomaly has been encountered. Each wind turbine includes a set of sensors and a basic on-board system that triggers alarms whenever measurements fall outside certain limits (Step **1** in Fig. 2.2). Uptime Engineering's condition monitoring software extends and refines the fault detection by further processing the available sensor information.

Once a symptom of a faulty turbine has been identified, the Uptime Engineering's software triggers the root cause identification by supplying the previously created system description as well as the observations to an MBD engine (Step **2** in Fig. 2.2). After the computation, the results are accessible to the employees at the service center (Step **3** in Fig. 2.2). The diagnosis results are displayed as part of Uptime Engineering's web interface, i.e., at the *Operations Center*, which is depicted in Fig. 2.3 and designed for desktop or laptop computers. The results are available per turbine instance and displayed as collapsible panels. For each triggering symptom,

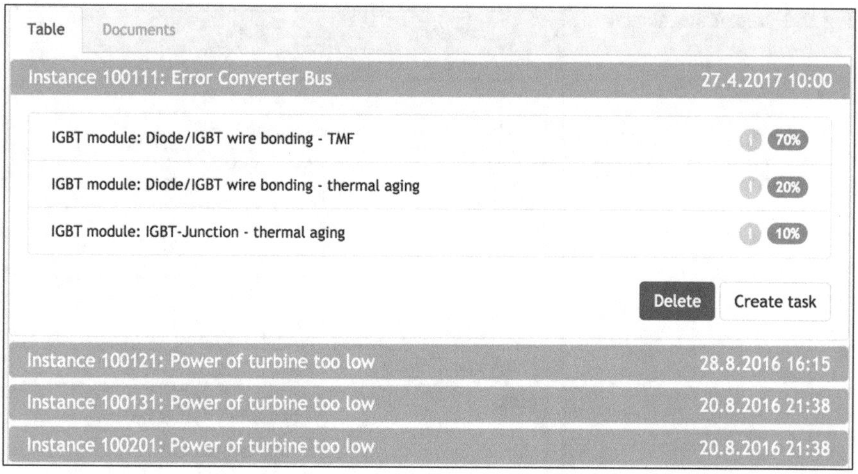

Fig. 2.3 *Operations Center* [21]

e.g., *Error Converter Bus*, the panel contains the possible root causes[4] as well as diagnosis likelihood expressed as percentages.

Based on the outcome, the service center employee can create and assign repair tasks for the service technicians preselecting some of the possible faults for consideration during the field work (Step **4** in Fig. 2.2). Each repair task is either preformed in conjunction with the next planned maintenance, scheduled, or immediately executed. Figure 2.4 depicts an example for a scheduled repair tasks, consisting of the anomaly and the corresponding error codes. The service center employee can then define a trouble shooting task, schedule the activity under consideration of the time table depicting the availability of service technicians, assigning both a supervisor and a team for the task, add the corresponding parts and tools to the work assignment depending on the failures proposed by the engine, and provide additional auxiliary information to the work assignment such as previous issues with the targeted wind turbine. Several repair tasks can be scheduled for the same day and the same maintenance team.

In the context of diagnosis within the field, we concluded that the software would be most usable on a mobile device since the technicians prefer not to carry a laptop. Thus, once work assignments have been created, the rest of the diagnosis process is conducted by the service technician teams over a mobile application. An essential aspect of the software design, was to follow guidelines and best practices for mobile user interfaces to ensure an easy-to-use application. A flat navigation was thus chosen for the prototype featuring little nesting of sub-levels, and thus warranting minimal user interaction and proving a defined role in the work process of the

[4]In the case of wind turbines, there is generally a strong single fault assumption. Thus, each depicted root cause in this example only consists of a single failure, e.g., *IGBT module: Diode/IGBT wire bonding—TMF*. Yet, the diagnosis engine is of course capable of determining multiple fault diagnoses.

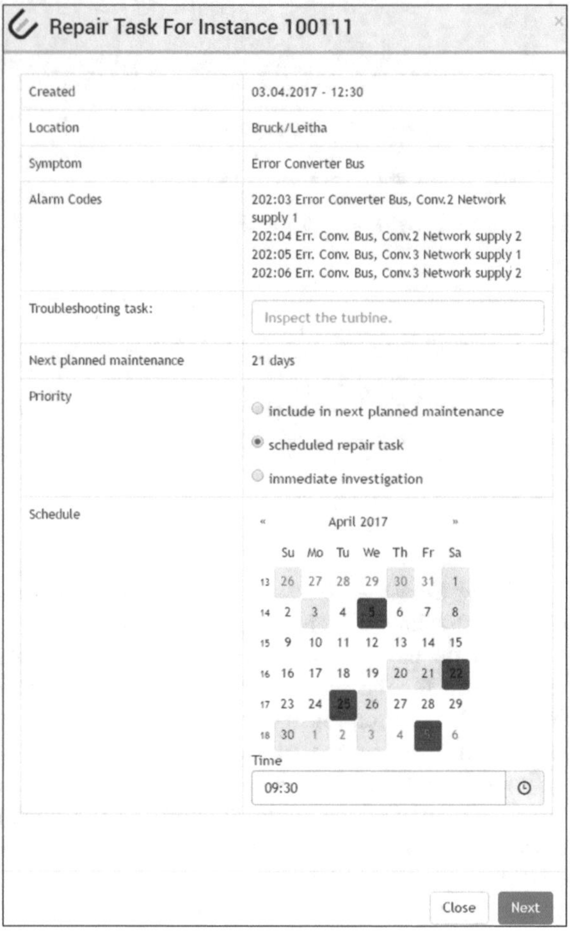

Fig. 2.4 Repair task screen

technicians. A simple navigation drawer is used to allow the user to switch quickly between the top-level sites. In Fig. 2.5a the *Home* screen of the mobile application is shown, where the technician can see all work assignments for the day as collapsible panels with additional information. Based on their tasks the technicians can obtain a list containing all necessary spare parts, tools, and safety equipment required for all maintenance activities planned on that day from the *Preparation* view depicted in Fig. 2.5b. The preparation would usually be performed at the service center, where the stockroom is also located (Step **5** in Fig. 2.2). Once at the turbine, an overview of the maintenance task for this particular instance and assignment is shown in the overview screen (see Fig. 2.5c), where an enforcement permit for the activity must be acquired.[5]

[5]A notification for the person responsible for the entire installation is automatically generated containing the request. Only after the permission has been granted, may the technicians perform the maintenance work.

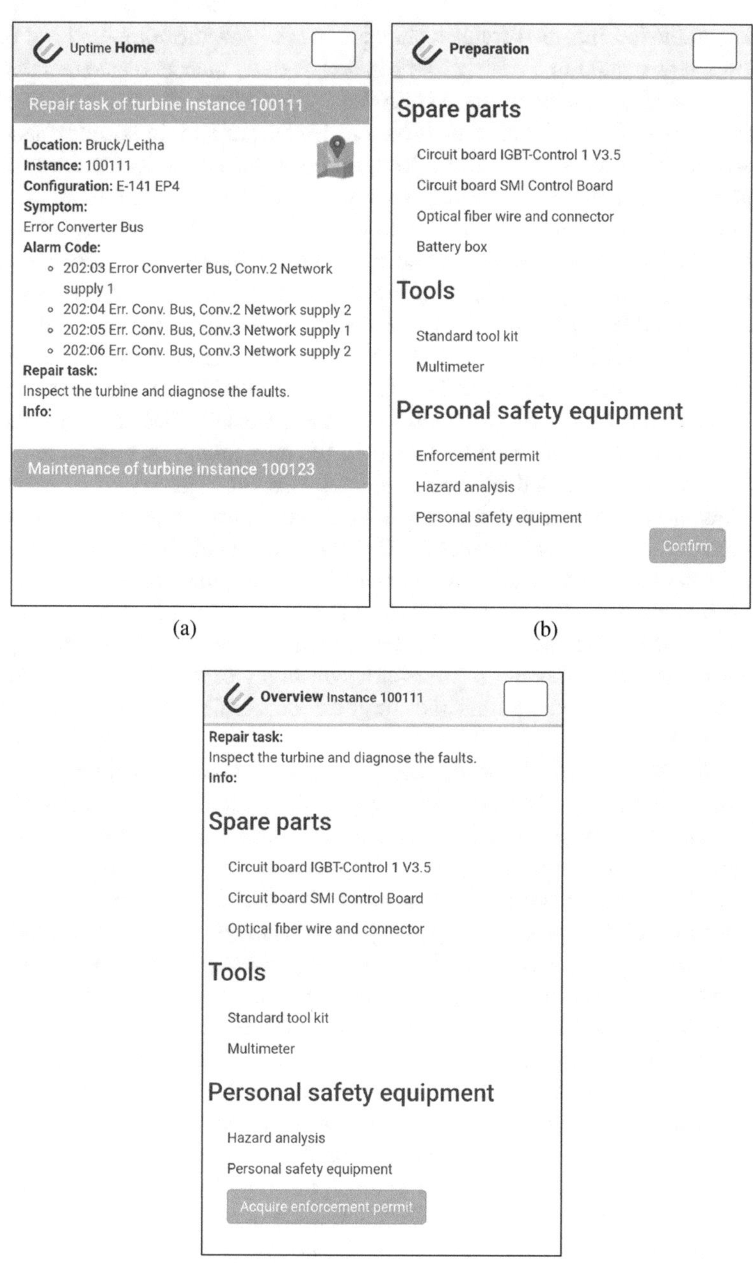

Fig. 2.5 Preparation and overview interface. (**a**) *Home* screen. (**b**) *Preparation* screen. (**c**) *Overview* screen

When at the turbine, the service technicians can review the computed and ranked diagnoses represented in a collapsible panel as well as their percentage value (see Fig. 2.6a). Expanding a diagnosis, a short description of the root cause of the failure is depicted as well as a **Repair task** button, leading to additional information on the required activities to restore a healthy turbine state, e.g., manufacturer manuals on repair and replacement of the faulty component. In cases where there is still no clear indication of the most likely diagnosis, the service technician can improve the initial results by supplying additional observations via **Improve diagnosis**. The next best probing points are determined via entropy as described in the previous section. To obtain the additional turbine state information the application asks the users to answer a set of questions as seen in Fig. 2.6b (Step **6a** in Fig. 2.2). Besides the textual representation, a picture is provided for each probe to facilitate visual identification. Each observation can either be confirmed (**Yes**), denied (**No**), or bypassed (**?**). The device camera can be used to document the measurements separately and the pictures are later appended to the maintenance report. The layout to enter new symptoms ensures efficient use of available screen space, logically structures the tasks, and requires minimal interaction. Once the additional observations have been made, the diagnoses are updated by manually restarting the computation given the new information from the device (Step **6b** in Fig. 2.2). Given the updated results, the probabilities and arrangements of the faults change accordingly in the *Diagnosis* screen with arrows indicating if the fault converges or not (see Fig. 2.6c). The diagnoses can be refined several times until an acceptable certainty for a fault has been reached.

Once the root cause of the detected anomaly has been repaired (Step **7** in Fig. 2.2), the service technicians must create a report of the activities (Step **8** in Fig. 2.2). This has previously been a tedious task and therefore an essential requirement by the users was that the application provides support for their reporting. Thus, the mobile application allows the user to notify the system of the final confirmed diagnosis as well as spare parts consumed and repair or replacement activities carried out (see Fig. 2.7a, b). The observations made and their visuals are automatically included in the report. Since it might be possible that a fault has not been considered within the failure assessment and thus abductive model, the interface provides a simple way for the service technician to reject all proposed diagnoses and add a custom fault, as shown in Fig. 2.7c. Based on the entered data the work assignment documentation is automatically generated and sent to the *Operations Center*. There it is either stored or post processed, depending on whether crucial information is missing from the report or a custom failure has been created, which needs to be analyzed and subsequently added to the failure assessment. It is currently planned to have the knowledge acquired from the custom failures inserted into the failure assessment manually after ensuring the information provided is sound and complete.

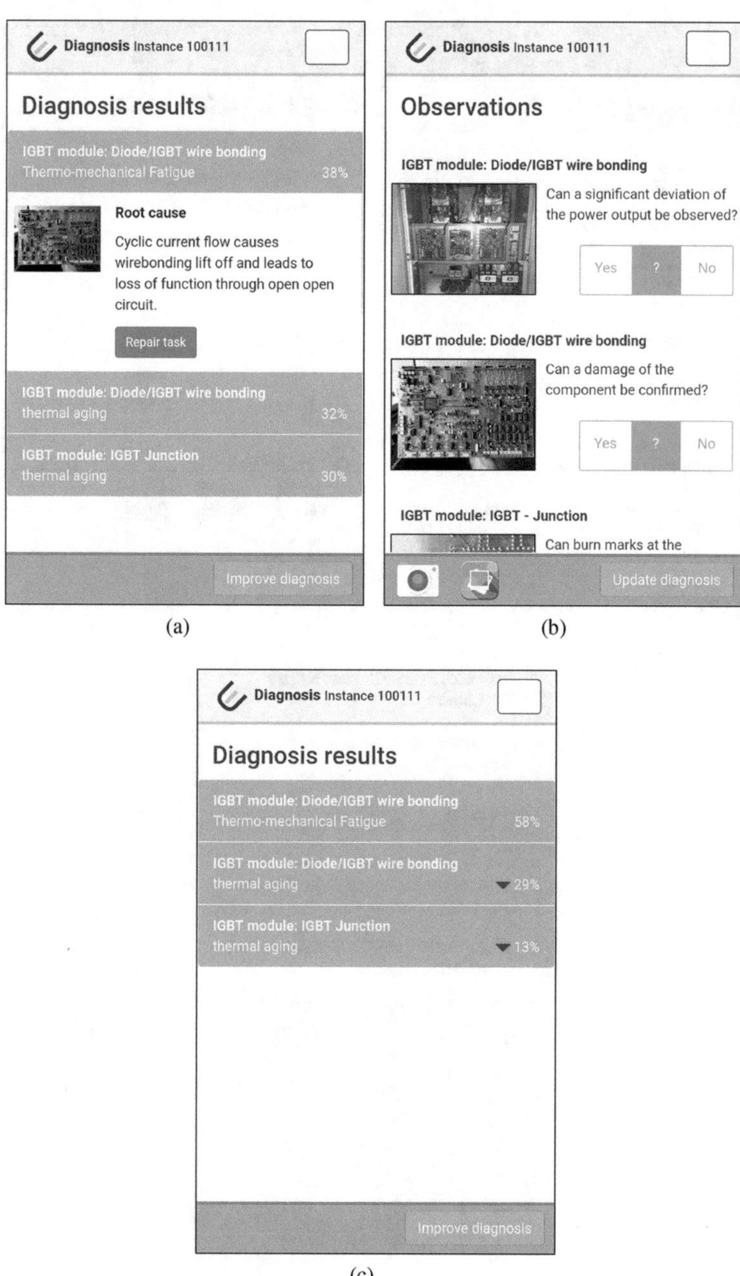

Fig. 2.6 Diagnosis and probing interface. (**a**) Initial *Diagnosis* screen. (**b**) Adding new observations. (**c**) *Diagnosis* screen after recomputation

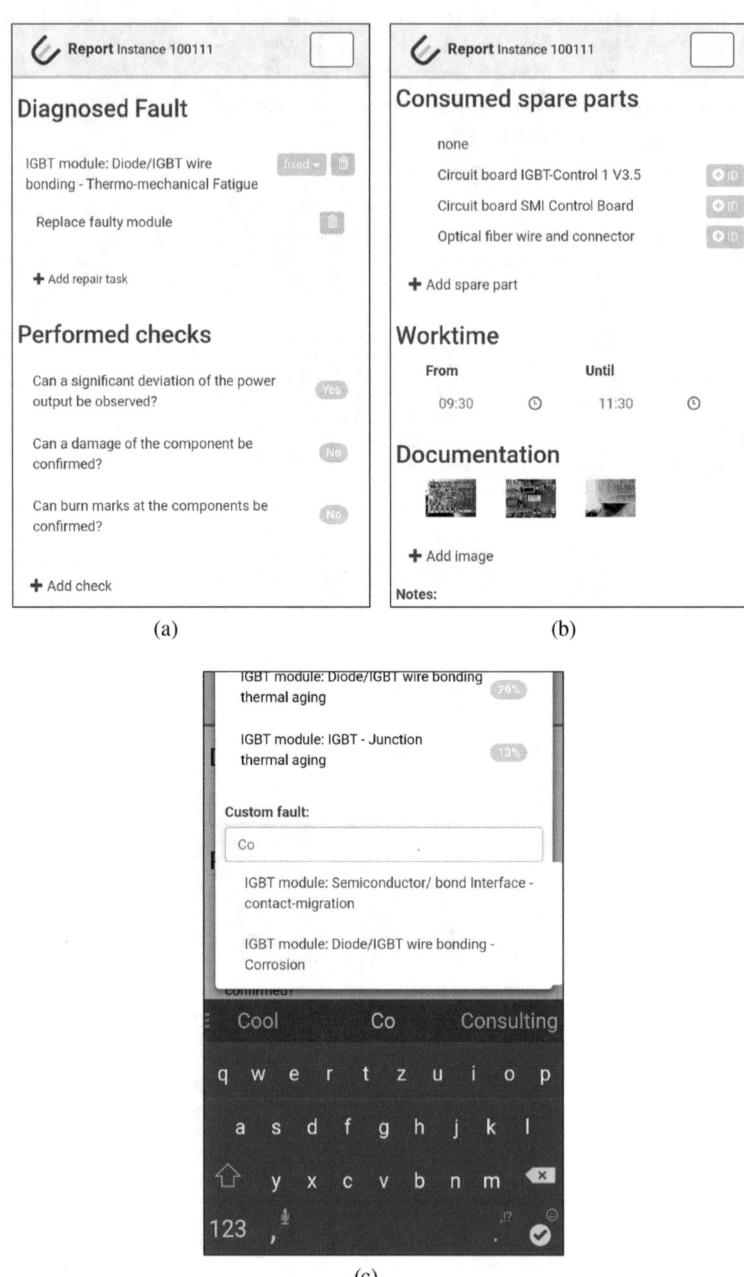

Fig. 2.7 Mobile reporting interface. (**a**) *Report* screen (I) [21]. (**b**) *Report* screen (II). (**c**) Adding a new custom fault

2.3.2 *Realization of the Diagnosis Engine*

The full integration of the diagnosis engine as well as the associated user interfaces into the existing Uptime Engineering condition monitoring tool was dealt with in three main project phases. Initially it was necessary to introduce a structure to manage the newly introduced system taxonomy. Next an interface was generated between the condition monitoring system and the MBD engine, via messaging bus. Finally, the user interface was adapted to incorporate the functionality as described in the previous section. Each of these project phases is now described in further detail.

The use of a strict taxonomy to describe individual system components (such as bearing, filter, fan) as well as sub-systems (such as gearbox, generator, pitch drive) within the turbine is critical for effective automation of the diagnostic process. The absence of such a system would lead to errors in the definition of observed system behavior, issues in regard to the model generation based on failure assessments, as also problems in interpretation of the recommendations. Furthermore, the naming of input signals (such as power, wind speed, oil temperature) and the behavior of such signals (e.g., "high power relative to the expectation, as defined by historical power performance") must be coded according to a strict and consistent taxonomy. Once established, this taxonomy must be applied accurately and consistently throughout the software environment. Further complication results from the need to develop a general solution that can be deployed across multiple organizations (e.g., several different energy utilities) and monitor technology from a variety of wind turbine manufacturers, each with its own internal preferences for system taxonomy. Therefore much attention was paid to the generation of a global, independent taxonomy specifically suited to the purpose of anomaly detection and fault diagnosis.

The MBD engine provided by Graz University of Technology is based on a Java implementation of the model mapping as also an ATMS for the fault identification portion and the entropy-based computation of the next best probing point, while Uptime Engineering's the condition monitoring system was developed using a range of technologies including .NET, MSSQL, Matlab, Java. The frameworks and languages have been selected based on their capabilities to perform signal analysis as well as their suitability for building commercial tools. An effective communication between these two systems was achieved using a Messaging Bus solution. Upon detection of abnormal system behavior, the condition monitoring software generates Events, with a specific content adhering to the taxonomy rules aforementioned. This Event is stored within the condition monitoring system and also converted into a message which is sent via the bus system to the reasoning engine. The message contains the latest observations for a specific instance (e.g., wind turbine) and the current failure assessment to be converted to an abductive KB. While the model generation can be performed offline, we decided to perform the transformation of the failure analysis whenever a diagnosis is triggered since the mapping implementation is very efficient and thus constitutes a negligible factor in the overall computation time. In addition, this means the engine can be memoryless from one diagnosis computation to the next and failure assessment updates are always considered during fault identification. The model-based reasoning engine then processes the information received and generates its own message, containing all possible root causes as well as recommendations for additional turbine state information that could be added as effective discriminators and hence improve the accuracy of the diagnosis. Note here that for each turbine instance a new MBD engine instance is created once an abnormal behavior has been encountered. Through this exchange of information, the capability of the condition monitoring system is significantly enhanced.

An intuitive and efficient user interface is seen as critical for acceptance of the system by its users, as described in Sect. 2.3.1.1 above. Since the software application is used both in the office environment (e.g., at the wind turbine service center) and on mobile applications (e.g., by service staff working on the wind turbines), it was important to develop a graphical interface that works effectively in both cases. In order to provide the user with a consistent solution and also to minimize development effort, it was decided to create "adaptive user interfaces", which recognize the device currently in use and automatically switch to the most suitable layout. The diagnosis GUI is currently in a prototype stage and still in development, while the back-end has already been fully developed and tested. Parts of the system, however, are already being evaluated by Uptime Engineering and the energy provider. Due to the lack of a user interface at this point in time, a broad and exhaustive acceptance study is planned for future work. The final product is expected to provide all relevant stakeholders with a comfortable user experience and improve and facilitate the fault diagnosis procedure of the wind turbine operator and its maintenance staff.

2.4 Conclusion

The theories and techniques of model-based reasoning have been applied to industrial problems, but the approach is not well represented as yet in practical every-day use. Computation of root causes relies on the presence of a suitable system description. From a practical point of view, however, due to the unavailability of tools that can be used by people who are not experts in logical-based modeling, model generation remains a hindering factor in the adoption of MBD in practice. As in the abductive variation of MBD the model characterizes the relations between faults and their manifestations. The proposed general process allows the utilizing of failure assessments commonly used in practice for computerized knowledge base generation. Automating this task should facilitate the introduction of MBD in industrial applications.

We applied this process to the domain of industrial wind turbines. Uptime Engineering maintains a comprehensive failure analysis of turbine faults and their effects, which can be used as input for the automatic diagnosis model generation. Furthermore, their condition monitoring software allows the detection of abnormal turbine behavior. Graz University of Technology implemented an MBD engine, which creates abductive knowledge bases from failure analyses, computes explanations based on the determined symptoms, and provides recommendations on additional observations to refine the diagnosis results based on entropy. Various aspects, e.g., as ensuring a suitable communication between existing software tools, need to be considered when deploying MBD to practical domains as the wind turbine industry. Moreover, (1) integrating the approach into already established work processes, (2) delivering a benefit to the users of the diagnosis software, and (3) providing intuitive and easy-to-use interfaces and interactions are key factors in promoting user acceptance. In this regard, a main goal was to facilitate not only the diagnosis portion of the work process, but also to simplify and improve the related tasks, such as the maintenance report creation. Working closely with the service staff and energy provider while creating and revising the user interface and interaction design as well as the application workflow ensures a usable template for the final software product. A prototype is currently being tested by Uptime Engineering and the energy provider to verify the feasibility of the approach.

Nevertheless open issues in the fault identification of wind turbines still remain. Even after performing a repair, a failure might prevail in the system and its manifestations will be visible for a certain period of time. In this case, the diagnosis must be repeated and the original failure report should be re-opened. In addition, the automatic acquisition of additional expert knowledge is challenging. Updating the failure assessment and thus the abductive model cannot solely be performed by the user due to quality concerns of the inserted data in regard to its technical soundness and the knowledge engineering capabilities of the service staff. Nevertheless, the question remains as to whether there is a feasible way to have the maintenance personnel make additions to the model. Further refinements to the diagnosis results based on knowledge about damage accumulation and load history could improve the fault identification process. This type of information can be exploited to determine the remaining life time of components which subsequently can be used to derive and update fault likelihoods. Other common considerations concern maintenance decisions based on the expected cost accounting for all expenditures associated with the diagnostic task, such as probing, repair, or replacement costs. In the future other application domains, such as truck fleets, should clarify whether abductive reasoning is a suitable diagnosis approach for industrial domains in general.

Acknowledgements The work presented in this paper has been supported by the FFG project Applied Model Based Reasoning (AMOR) under grant 842407 and the SFG project EXPERT. We would further like to express our gratitude to VERBUND Hydro Power GmbH.

References

1. B. Abichou, D. Flórez, M. Sayed-Mouchaweh, H. Toubakh, B. François, N. Girard, Fault diagnosis methods for wind turbines health monitoring: a review, in *European Conference of the Prognostics and Health Management Society* (2014)
2. R. Bakker, P. Van den Bempt, N.J. Mars, D.J. Out, D. van Soest, Issues in practical model-based diagnosis. Futur. Gener. Comput. Syst. **9**(4), 329–337 (1993)
3. C.S. Carlson, Understanding and applying the fundamentals of FMEAs, in *Annual Reliability and Maintainability Symposium* (2014)
4. L. Console, O. Dressler, Model-based diagnosis in the real world: lessons learned and challenges remaining, in *Proceedings of the Sixteenth International Joint Conferences on Artificial Intelligence* (1999), pp. 1393–1400
5. L. Console, D.T. Dupre, P. Torasso, On the relationship between abduction and deduction. J. Log. Comput. **1**(5), 661–690 (1991)
6. M.O. Cordier, P. Dague, F. Lévy, J. Montmain, M. Staroswiecki, L. Travé-Massuyès, Conflicts versus analytical redundancy relations: a comparative analysis of the model based diagnosis approach from the artificial intelligence and automatic control perspectives. IEEE Trans. Syst. Man Cybern. B Cybern. **34**(5), 2163–2177 (2004)
7. F.D. Davis, Perceived usefulness, perceived ease of use, and user acceptance of information technology. MIS Q. **13**, 319–340 (1989)
8. J. de Kleer, An assumption-based TMS. Artif. Intell. **28**(2), 127–162 (1986)
9. J. de Kleer, B.C. Williams, Diagnosing multiple faults. Artif. Intell. **32**(1), 97–130 (1987)
10. E. Echavarria, T. Tomiyama, W. Huberts, G.J. van Bussel, Fault diagnosis system for an offshore wind turbine using qualitative physics, in *Proceedings of EWEC* (2008)
11. T. Eiter, G. Gottlob, The complexity of logic-based abduction. J. ACM **42**(1), 3–42 (1995)
12. G. Fleischanderl, T. Havelka, H. Schreiner, M. Stumptner, F. Wotawa, DiKe-a model-based diagnosis kernel and its application, in *KI 2001: Advances in Artificial Intelligence* (Springer, Berlin, 2001)
13. G. Friedrich,G. Gottlob, W. Nejdl, Hypothesis classification, abductive diagnosis and therapy, in *Expert Systems in Engineering Principles and Applications* (Springer, Berlin, 1990), pp. 69–78
14. C.S. Gray, S.J. Watson, Physics of failure approach to wind turbine condition based maintenance. Wind Energy **13**(5), 395–405 (2010)
15. C. Gray, F. Langmayr, N. Haselgruber, S.J. Watson, A practical approach to the use of SCADA data for optimized wind turbine condition based maintenance, EWEA Offshore Wind Amsterdam (2011)
16. Z. Hameed, Y. Hong, Y. Cho, S. Ahn, C. Song, Condition monitoring and fault detection of wind turbines and related algorithms: a review. Renew. Sust. Energ. Rev. **13**(1), 1–39 (2009)
17. R. Koitz, F. Wotawa, Finding explanations: an empirical evaluation of abductive diagnosis algorithms, in *Proceedings of the International Conference on Defeasible and Ampliative Reasoning*, CEUR-WS.org (2015), pp. 36–42
18. R. Koitz, F. Wotawa, From theory to practice: model-based diagnosis in industrial applications (2015), in *Proceedings of the Annual Conference of the Prognostics and Health Management Society* (2015)
19. R. Koitz, F. Wotawa, On the feasibility of abductive diagnosis for practical applications, in *Proceedings of the 9th IFAC symposium on fault detection, supervision and safety of technical processes* (2015)
20. R. Koitz, F. Wotawa, Integration of failure assessments into the diagnostic process, in *Proceedings of the Annual Conference of the Prognostics and Health Management Society* (2016)

21. R. Koitz, J. Lüftenegger, F. Wotawa, Model-based diagnosis in practice: interaction design of an integrated diagnosis application for industrial wind turbines, in *Proceedings of the 30th International Conference on Industrial Engineering and Other Applications of Applied Intelligent Systems* (2017)
22. H.J. Levesque, A knowledge-level account of abduction, in *Proceedings of the Eleventh International Joint Conferences on Artificial Intelligence* (1989), pp. 1061–1067
23. B. Lu, Y. Li, X. Wu, Z. Yang, A review of recent advances in wind turbine condition monitoring and fault diagnosis, in *Power Electronics and Machines in Wind Applications, 2009. PEMWA 2009. IEEE* (IEEE, Piscataway, 2009), pp. 1–7
24. H. Milde, T. Guckenbiehl, A. Malik, B. Neumann, P. Struss, Integrating model-based diagnosis techniques into current work processes–three case studies from the INDIA project. AI Commun. **13**(2), 99–123 (2000)
25. J. Nielsen, Iterative user-interface design. Computer **26**(11), 32–41 (1993)
26. D. Poole, R. Goebel, R. Aleliunas, Theorist: a logical reasoning system for defaults and diagnosis, in *The Knowledge Frontier* (Springer, New York, 1987), pp. 331–352
27. L. Rademakers, A. Seebregts, B. van Den Horn, Reliability analysis in wind engineering. Netherlands Energy Research Foundation ECN (1993)
28. R. Reiter, A theory of diagnosis from first principles. Artif. Intell. **32**(1), 57–95 (1987)
29. M. Schlechtingen, I.F. Santos, S. Achiche, Wind turbine condition monitoring based on scada data using normal behavior models. Part 1: system description. Appl. Soft Comput. **13**(1), 259–270 (2013)
30. L. Simon, A. Del Val, Efficient consequence finding, in *Proceedings of the Seventeenth International Joint Conference on Artificial intelligence* (2001), pp. 359–365
31. P. Struss, Model-based problem solving, in *Handbook of Knowledge Representation*, Chap. 10, ed. by F. van Harmelen, V. Lifschitz, B. Porter (Elsevier Science, Oxford, 2008), pp. 395–465
32. P. Struss, A. Malik, M. Sachenbacher, Case studies in model-based diagnosis and fault analysis of car-subsystems, in *Proceedings of the 1st International Workshop Model-Based Systems and Qualitative Reasoning* (1996), pp. 17–25
33. P. Struss, M. Sachenbacher, C. Carlén, Insights from building a prototype for model-based on-board diagnosis of automotive systems, in *Proceedings of the International Workshop on Principles of Diagnosis* (2000)
34. Y. Sun, D.S. Weld, A framework for model-based repair, in *Proceedings of the Eleventh National Conference on Artificial Intelligence* (1993), pp. 182–187
35. N. Tazi, E. Châtelet, Y. Bouzidi, Using a hybrid cost-FMEA analysis for wind turbine reliability analysis. Energies **10**(3), 276 (2017)
36. M. Wilkinson, B. Hendriks, F. Spinato, K. Harman, E. Gomez, H. Bulacio, J. Roca, P. Tavner, Y. Feng, H. Long, Methodology and results of the reliawind reliability field study, in *European Wind Energy Conference and Exhibition 2010, EWEC 2010*, Sheffield, vol. 3 (2010), pp. 1984–2004
37. F. Wotawa, On the use of abduction as an alternative to decision trees in environmental decision support systems, in *International Conference on Complex, Intelligent and Software Intensive Systems, 2009. CISIS'09.* (IEEE, Piscataway, 2009), pp. 1160–1165
38. F. Wotawa, Failure mode and effect analysis for abductive diagnosis, in *Proceedings of the International Workshop on Defeasible and Ampliative Reasoning* (2014), pp. 1–13
39. F. Wotawa, I. Rodriguez-Roda, J. Comas, Environmental decision support systems based on models and model-based reasoning. Environ. Eng. Manag. J. **9**(2), 189–195 (2010)
40. W. Yang, J. Jiang, Wind turbine condition monitoring and reliability analysis by SCADA information, in *Second International Conference on Mechanic Automation and Control Engineering (MACE), 2011.* (IEEE, Piscataway, 2011), pp. 1872–1875
41. A.S. Zaher, S. McArthur, A multi-agent fault detection system for wind turbine defect recognition and diagnosis, in *2007 IEEE Lausanne Power Tech* (2007), pp. 22–27

Chapter 3
Fault Detection and Localization Using Modelica and Abductive Reasoning

Ingo Pill and Franz Wotawa

3.1 Introduction

On an abstract level, fault diagnosis encompasses two tasks; that of fault detection where we decide whether there is a fault, and that of fault isolation where we are interested in identifying the cause of some unexpected behavior. This book is a perfect showcase that there are many solutions to these two non-trivial tasks, where in practice the available resources, the application domain, the availability of detailed system models, and many other aspects limit our choices for a specific scenario.

In this chapter we take a focus on diagnosis scenarios where we have some Modelica system model and would like to detect and isolate occurring faults. As will be discussed in Sect. 3.2, Modelica [12] is an intuitive to use programming language that allows us to directly execute our models in simulations. Due to features like the availability of libraries for digital circuits, fluids, and mechanics, Modelica is an attractive means for modeling hybrid cyber-physical systems. The scenario we are facing in this chapter is now one where we do have a Modelica system model, and observe some actual behavior that we would like to evaluate. In particular, we would like to determine whether there is a fault in the system, and if there is a fault, then we would like to derive diagnoses explaining the encountered error—exploiting the Modelica system model in this process.

A key element of any model-based diagnosis approach is the system model that allows us to reason about the correctness of individual system components [6, 9, 25]. Unfortunately, the task of coming up with an effective diagnosis model has been turning out to be a stumbling block, prohibiting a wide-spread adoption of

I. Pill (✉) · F. Wotawa
Institute for Software Technology, Graz University of Technology, Graz, Austria
e-mail: ipill@ist.tugraz.at; wotawa@ist.tugraz.at

M. Sayed-Mouchaweh (ed.), *Diagnosability, Security and Safety of Hybrid Dynamic and Cyber-Physical Systems*, https://doi.org/10.1007/978-3-319-74962-4_3

model-based diagnosis in practice. In particular, there is a gap between development models we create, e.g., for some simulation, and the models we require for diagnosing the deployed system. This gap originates in the different objectives behind the models. So, for instance, a simulation-oriented model usually has to be as detailed and as close to the system's real-world behavior as is required by the simulation objectives. In contrast, diagnosis models usually rely on qualitative system descriptions that merge behaviors we do not need to distinguish for diagnostic purposes into qualitative (in other words symbolic) ones with the aim of restricting the search space. So far, the expertise and additional resources needed for coming up with a well-defined diagnosis model have been countermanding the advantages of model-based approaches in many cases. A prominent aspect here is that we often lack options for integrating the modeling process for development purposes and the process of modeling for diagnosis purposes.

In this chapter, and in particular in Sect. 3.4.2 we thus discuss a solution for automatically compiling Modelica development models into ones that we can use for diagnosis. In the literature, there have been several approaches dealing with this idea, e.g., [21, 22, 28] (see Sect. 3.6). In contrast to earlier work, we are interested in providing models to be used for abductive diagnosis. This means that we extract cause-effect rules from Modelica models, where such rules are intuitive to designers familiar with failure mode and effect analysis (FMEA) [4, 15]. This familiarity makes the approach quite attractive for practical purposes. The basic idea for the extraction is to compare simulations of the *correct* model with simulation results for *faulty* variants that we create via fault injection [30]. In case of detected deviations, we extract a rule stating that the introduced fault leads to the observed deviations.

Before triggering a diagnostic process, we have to determine whether the observed behavior is faulty in the first place. Bearing noisy signals, minimal parameter deviations for component instances, and physical component degradation over time on our minds, we present in Sect. 3.3 three approaches for comparing observed with simulated behavior. That is, based on average values and tolerances, temporal band sequences, or the Pearson correlation coefficient, we aim to detect the presence of a fault when comparing simulated with observed signal behavior.

As is illustrated by the examples spread throughout the chapter, the proposed combination of technologies provides an intuitive approach that minimizes the additional resources needed for diagnosis purposes, and which is based on concepts like simulation and cause-effect-rules that engineers are, in principle, familiar with—certainly a plus when it comes to consider the attractiveness of deploying a new technology to enrich an already installed and proven development process. This is also highlighted in our case studies in Sect. 3.5, where our corresponding observations are one pillar of the summary that we provide in Sect. 3.7.

3.2 Using Modelica for Describing a System's Behavior

There are many languages that we can use to model physical and hybrid systems, including Modelica [12] which is an object-oriented, open, and multi-domain language. Using Modelica, we define models via sets of equations that can range from simple algebraic equations to complex differential ones. When using Modelica, a designer has the luxury of being able to draw on a huge variety of available libraries targeting, e.g., digital circuits, electronics, mechanics, or fluids. All these characteristics distinguish Modelica from other modeling languages, making it very flexible, easy to employ, and an attractive means for modeling hybrid systems. In Fig. 3.1 you can see simple example code for describing a pin in an electronic circuit, a grounded pin, and an abstract component with two pins.

An attractive feature for established development workflows is that Modelica is optimized for simulation. Consequently, we have that any Modelica model has to ensure that its corresponding equations allow for computing exactly one solution, i.e., an assignment of variable values that solve all the equations at any point in time. Otherwise, an error message is raised. This means, however, that we cannot use Modelica for diagnosis directly, due to the lacking capabilities for dealing with unknown values or sets of values to be assigned to a variable.

The main question now is how we *can* use Modelica for diagnostic purposes. While we introduce the formal background of our corresponding concept in the following sections, let us at this point briefly introduce the underlying ideas using the voltage-divider circuit shown in Fig. 3.2. It contains two resistors `R1` and `R2` (100 and 50 Ω) and a battery `BAT` with a nominal voltage of 12 V. If every component works as intended, we see voltage drops of $V1 = 8$ V at `R1` and $V2 = 4$ V at `R2`. If now the battery would be completely empty (or one of its pins

Fig. 3.1 Simple Modelica code for a pin, a "ground pin," and a component with two pins

```
connector MyPin
  Real v;
  flow Real i;
end MyPin;

model MyGround
  MyPin p;
equation
  p.v = 0.0;
end MyGround;

partial model MyComponentSimple
  MyPin p,m;
  Real v;
  Real i;
equation
  v = p.v - m.v;
  i = p.i;
  0.0 = p.i + m.i;
end MyComponentSimple;
```

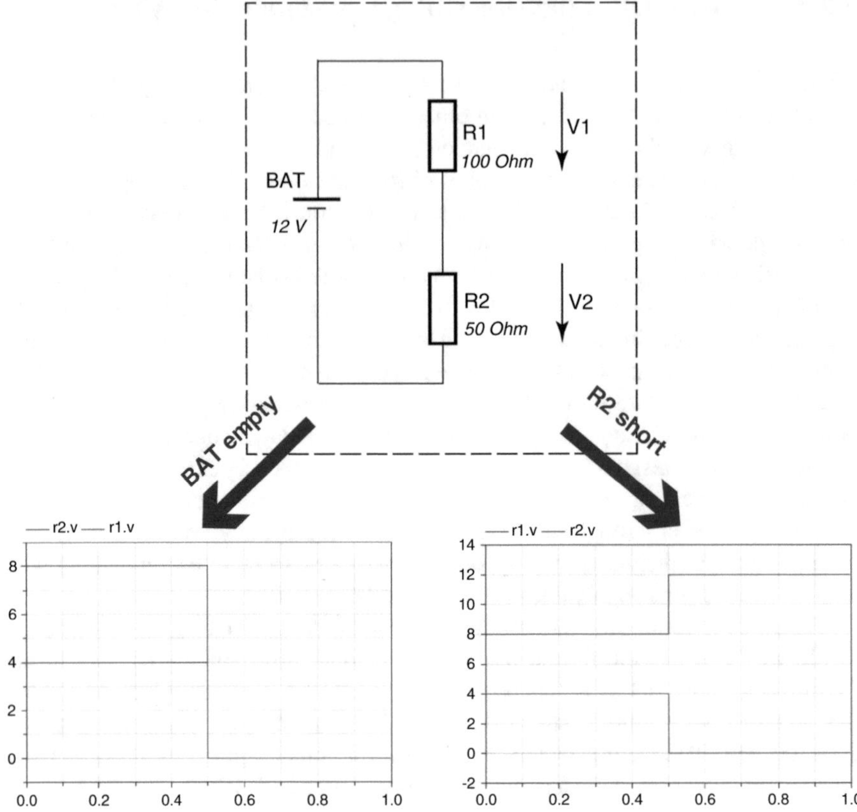

Fig. 3.2 A voltage divider circuit

is broken) the voltage drops $V1$ and $V2$ would decline to 0 V. On the other hand, if there is a short at R2, $V2$ would become 0 V and $V1$ would rise to 12 V (see Fig. 3.2 for corresponding diagrams, when assuming that the faults appear individually after half a second).

Such knowledge can be used to create an abductive diagnosis model. There we would state, for instance, that an empty battery causes both $V1$ and $V2$ to be 0 V, and/or that the voltage drops on both resistors are smaller than expected then:

$$emptyBat \rightarrow (val(v1, 0) \wedge smaller(v1))$$
$$emptyBat \rightarrow (val(v2, 0) \wedge smaller(v2))$$

A similar model for a short resistor R2 could look as follows, where other faults like broken pins can be handled in a similar way:

$$short(\texttt{R2}) \rightarrow (val(v1, 12) \wedge higher(v1))$$
$$short(\texttt{R2}) \rightarrow (val(v2, 0) \wedge smaller(v2))$$

Fig. 3.3 A MyComponent model augmented with a health state, and a derived resistor's model

```
type FaultType = enumeration(
     ok,broken,short,empty);

partial model MyComponent
  MyPin p,m;
  Real v;
  Real i;
  FaultType state(start = FaultType.ok);
equation
  v = p.v - m.v;
  i = p.i;
  0.0 = p.i + m.i;
end MyComponent;

model MyResistor
  extends MyComponent;
  parameter Real r;
equation
  if state == FaultType.ok then
    v = r * i;
  elseif state == FaultType.short then
    v = 0;
  else
    i = 0;
  end if;
end MyResistor;
```

Now, if we would have models for simulating the individual faults' effects, we could automate this rule extraction idea. Ideally, we would have a single simulation model, where we only need to activate/deactivate the individual faults to be considered in a simulation (switching between equations when doing so). Then, comparing a simulation where no fault is active with simulations where we activate individual faults would enable us to obtain data about an individual fault's effects and in the end to come up with the desired abductive diagnosis model.

Before offering the details of our concept for creating such an augmented simulation model in Sect. 3.2.1, let us briefly illustrate the needed steps for some Modelica code related to the voltage divider example. That is, as a first step, we add a system state *mode* of the newly introduced type *FaultType* to the abstract component model from Fig. 3.1, which results in the model *MyComponent* as shown in Fig. 3.3. We use this state variable to select between equations for the nominal and individual fault behavior. For a resistor's model *MyResistor* derived from *MyComponent* and also shown in Fig. 3.3, we define three such modes, and depending on the fault state the corresponding equations are used for our simulations.

If we have a closer look at *MyResistor*, we can see that our changes lead to a hybrid model, even if the core model is an analog one. That is, we do have a discrete valued system state for the component's health, and the continuous system signals for the resistor's voltage and current. Consequently, this shows that our concept works for hybrid models and we can use this simple scenario for illustration purposes. Please note that we briefly discuss extending this circuit to include a bulb, a capacitor, and a switch for controlling the lighting situation in Sect. 3.5—which is closer to the expectations about a hybrid system (especially if the switch is controlled by a digital pulse-width controller).

3.2.1 A Modelica Simulation Model with Fault Modes

As depicted in the introduction, at the beginning of our process we have a system's Modelica model that was developed for simulation purposes. Aiming at an automated extraction of the desired cause-effect rules, we need also characterizations of fault effects. Our basis for obtaining such fault effect data is that of using fault injection [30] and simulating the faulty behavior. To this end, we aim at a system model where the individual components may have several modes describing their nominal and faulty behavior. As suggested above, we furthermore assume that the Modelica model offers the means to enable or disable individual fault modes defined for the individual components. Every such mode of a particular component can be a *hypothesis* explaining some observed faulty behavior, i.e., the observed *symptoms*, in our abductive diagnostic reasoning (see Sect. 3.4.1). Please note that we also distinguish input from output variables (signals). That is, input variables are those for specifying a certain desired input scenario, and the outputs are those variables that we can observe and which provide the symptoms.

Definition 1 (System Model) A *system model* $\mathbf{M} = (COMP, MODES, \mu, I, O, \mathbf{P})$ is a tuple that comprises a set of components *COMP*, a set of modes *MODES* that has at least the correct mode *ok* as element, a function $\mu : COMP \mapsto MODES$ mapping components to their featured modes, a set I of variables considered as inputs, a set O of variables considered as outputs, and a Modelica model $\mathbf{P}$ that allows us to set an individual mode $m \in \mu(c)$ for each individual component $c \in COMP$.

Example 1 In Fig. 3.4, we show the Modelica source code for our voltage-divider circuit example from Fig. 3.2, reusing components from Figs. 3.1 and 3.3. `MyBattery` and `MyResistor` (Fig. 3.3) are sub-classes of `MyComponent` (Fig. 3.3), inheriting its equation, but extending it to capture the more specific behavior of the component type. The union of all fault modes for the individual components (type `FaultType` in Fig. 3.3) contains `ok` for the ordinary nominal behavior, `broken` and `short` for a resistor's fault modes, and `empty` for a faulty battery. Hence, a *system model* $(COMP, MODES, \mu, I, O, \mathbf{P})$ for our voltage divider circuit would comprise the following elements:

$$
\begin{aligned}
&COMP = \{\texttt{bat}, \texttt{r1}, \texttt{r2}\}\\
&MODES = \{ok, broken, short, empty\}\\
&\mu(\texttt{bat}) = \{ok, empty\}\\
&\mu(\texttt{r1}) = \mu(\texttt{r2}) = \{ok, broken, short\}\\
&I = \emptyset\\
&O = \{\texttt{r1.v}, \texttt{r2.v}\}
\end{aligned}
$$

Fig. 3.4 The Modelica program implementing the voltage divider circuit

```
model MyBattery
  extends MyComponent;
  parameter Real vn;
equation
  if state == FaultType.ok then
    v = vn;
  else
    v = 0.0;
  end if;
end MyBattery;

model MySimpleCircuit
  MyResistor r1(r=100);
  MyResistor r2(r=50);
  MyBattery bat(vn=12);
  MyGround gnd;
equation
  connect(bat.p,r1.p);
  connect(r1.m,r2.p);
  connect(bat.m,r2.m);
  connect(bat.m,gnd.p);
end MySimpleCircuit;
```

■

In addition to a *system model*, we need to introduce the concept of simulation. That is, the computation of values for a system's variables over time, given the system model and an input scenario. To this end, we first introduce the concept of *mode assignments*, i.e., assignments of modes to components at points in time.

Definition 2 (Mode Assignment) Let *TIME* be a finite set of time points. A mode assignment Δ is a set of functions $\delta_i : COMP \times TIME \mapsto MODES$ that assign to each component $c \in COMP$ for each time point $t \in TIME$ a mode $m \in \mu(c)$, i.e., $\Delta = \{\delta_1, \ldots, \delta_{|TIME|}\}$ where $\forall i \in \{1, \ldots, |TIME|\} : \forall t \in TIME : \forall c \in COMP : \delta_i(c, t) \in \mu(c)$.

During simulation, a *mode assignment* allows us to change a component's behavior and in turn change that of the entire system—even dynamically if desired. Thus also in our definition of a simulation function, we use a mode assignment.

Definition 3 (Simulation) Let us assume that we have a *system model* **M** as of Definition 1, a test bench **T** specifying the desired system inputs over time, a mode assignment Δ, and an end time t_e. A *simulation* function sim is a function that computes via **P** the values of all variables over time between 0 and t_e, considering (a) test bench **T** for inputs *I* and (b) the mode assignment Δ.

We can easily implement such a simulation function sim using a Modelica simulator. To this end, we construct a new test bench **T**′ from **T** and Δ. A typical test bench **T** for a Modelica circuit SUT would follow, e.g., the following structure if we desire the inputs to change over time:

```
model Testbench
   SUT sys;
equation
  if (time < t1) then
    ....  // First inputs
  elsif (time >= t1 and time < t2) then
    .... // Next inputs
  elsif
    ....
  else
    ....
  end if;
end Testbench;
```

When taking Δ into account, we can easily extend $\mathbf{T}$ to derive $\mathbf{T}'$ that contains also the mode assignments for the individual components.

Example 2 Let us continue Example 1, considering the mode assignment $\delta(\texttt{bat}, ok, 0)$, $\delta(\texttt{r1}, ok, 0)$, $\delta(\texttt{r2}, ok, 0)$, $\delta(\texttt{bat}, ok, 0.5)$, $\delta(\texttt{r1}, ok, 0.5)$, $\delta(\texttt{r2}, short, 0.5)$. Due to the fact that the voltage divider has no input values, we can easily obtain test bench $\mathbf{T}'$ by considering only Δ.

```
model TestbenchPrime
   MySimpleCircuit sut;
equation
  if (time < 0.5) then
    sut.r1.state = FaultType.ok;
    sut.r2.state = FaultType.ok;
    sut.bat.state = FaultType.ok;
  else
    sut.r1.state = FaultType.ok;
    sut.r2.state = FaultType.short;
    sut.bat.state = FaultType.ok;
  end if;end Testbench;
```

■

Using a Modelica simulator and the extended test bench $\mathbf{T}'$, the simulation function sim can be defined as a call to this simulator using $\mathbf{P} \cup \mathbf{T}'$ and end time t_e as parameters. All values for outputs $o \in O$ in $\mathbf{M}$ will be computed during the simulation, and we assume that they are returned as a set of tuples (o, v) for t where $o \in O$ and v gives o's value.

3.3 Comparing Signals for Detecting the Presence of Faults

Before starting a diagnostic process, we would like to determine whether the system exhibits some unexpected behavior in the first place. To this end, we would compare expected with observed signal behavior in order to classify the observed one to be faulty or within expected limits. In this section, we compare several techniques for such a comparison. All of them rely on some assumptions and corresponding definitions. That is, we assume that signals change their value over time. Second, we assume that there is a fixed finite set *TIME* of points in time where we consider the signals' values. Note that for a hybrid system, the sampling frequency will most likely be determined by the analog part, rather than the clocked digital one. And third, we assume that we know the values for "reference" signals for *each* $t \in TIME$.

Given such data for two signals s_1 and s_2, we are then interested in assessing whether their temporal behavior is equivalent with respect to a chosen equivalence operator. It is important to note that such an equivalence operator might take certain tolerance values (to cover for expected system parameter deviations and expected exogenous influences like noise and temperature effects) into account. Formally, we express the equivalence problem for two signals as follows:

Definition 4 (Equivalence Problem) Given two signals s_1 and s_2 from a set of signals *SIGS*, a set of time points *TIME*, and a function $f : SIGS \times TIME \mapsto VALUE$ mapping signals to their corresponding values at a certain point $t \in TIME$. We say that s_1 is equivalent to s_2, i.e., $s_1 = s_2$, if and only if $\forall t \in TIME : f(s_1, t) =_C f(s_2, t)$ for a certain equivalence operator $=_C$. If s_1 is not equivalent to s_2, we write $s_1 \neq_C s_2$. In particular, if there exists a point $t \in TIME$ such that $f(s_1, t) =_C f(s_2, t)$ does not hold, we write $f(s_1, t) \neq_C f(s_2, t)$.

When comparing two signals for diagnostic purposes, one signal is the reference describing the expected behavior, and the other one is the observed signal. For such a diagnostic scenario, we are not only interested in whether we indeed have $s_1 = s_2$, but also in knowing the first timestamp where the equivalence would be violated. In the following definition we capture this situation:

Definition 5 Given all prerequisites of Definition 4. If $s_1 \neq_C s_2$, we define t_D as the first point in time where the signal values fail to be equivalent with respect to $=_C$, i.e., $f(s_1, t_D) \neq_C f(s_2, t_D) \wedge \forall t \in TIMEsuchthatt < t_D : f(s_1, t) =_C f(s_2, t)$.

Intuitively, the point in time when we are able to detect a signal deviation should be as close as possible in time to the faults' occurrence. For given observable signals, the delay between the time of fault detection and the time where the isolated fault occurred is a good measure for the quality of the operator $=_C$. Other measures are the capability to detect all faults, and the rate of false alarms (such that a certain behavior is identified as being faulty when it is not). In the following, we discuss several different definitions of comparison operators $=_C$, starting with intuitive comparisons using tolerance values, and continuing with temporal band sequences and the Pearson correlation coefficient.

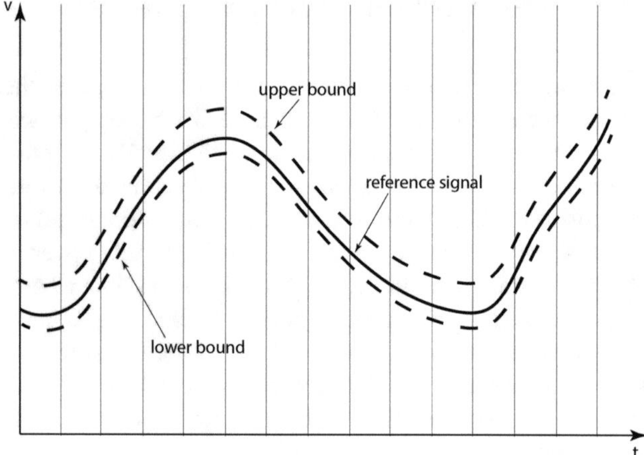

Fig. 3.5 Defining the bounds for the expected signal corridor using tolerances in the value

3.3.1 Tolerance Values

For the tolerance value approach illustrated in Fig. 3.5, we intuitively take the reference signal and derive upper and lower signal bounds for the expected signal corridor by considering two individual values for the respectively allowed distance to the reference values. Formally, we define the equivalence operator $=_T^{(v_l,v_u)}$, such that v_l represents the lower bound and v_u the upper bound tolerance value.

Definition 6 (Tolerance Value Comparison Operator) In accordance with Definition 4, we define the comparison operator $=_T^{(v_l,v_u)}$ as a variant of $=_C$ as follows, where parameters $v_l, v_u \geq 0$ define the desired tolerance values for the lower and upper bounds, and we assume that s_1 is the reference signal:

$$f(s_1,t) =_T^{(v_l,v_u)} f(s_2,t) \equiv_{DEF} (f(s_1,t) - v_l) \leq f(s_2,t) \leq (f(s_1,t) + v_u)$$

Following the definition, we can easily see that with the static definitions of v_u and v_l, we have that the *normal* distances of the upper and lower bounds to s_1 vary over time, in contrast to the constant vertical distances. Thus, while it allows for possible shifts in the amplitude of the signal, slight shifts of signal parts in time (relative to each other) are not considered systematically. In order to address this to some extent, we can define tolerance also using the normal distance from the reference signal as is illustrated in Fig. 3.6. That is, for every $t \in \mathit{TIME}$ we can calculate an upper and a lower point using the normal on the tangent passing through the point, in order to obtain functions for the upper and lower bounds.

Definition 7 (Alternative Tolerance Value Comparison Operator) As a variant of Definition 6, we define the comparison operator $=_{T_F}^{(v_l,v_u)}$ as follows:

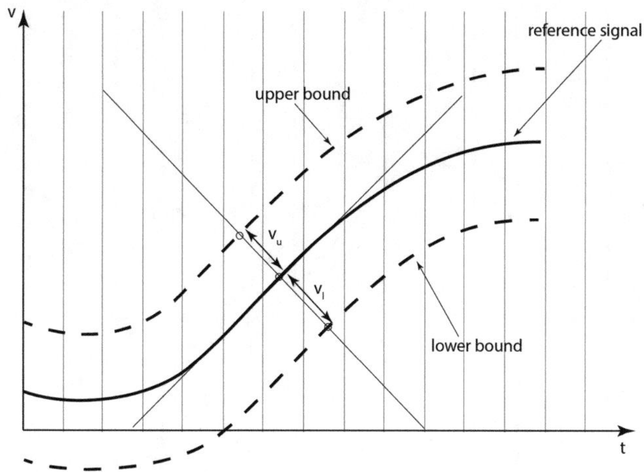

Fig. 3.6 An alternative way of defining tolerance

$$f(s_1, t) =_{T_F}^{(v_l, v_u)} f(s_2, t) \equiv_{DEF} f_l(s_1, t) \leq f(s_2, t) \leq f_u(s_1, t)$$

Let Δ be the first derivative of function f (s.t. $\Delta = f'(s_1, t)$), $\delta s_u = \frac{v_u}{\sqrt{1+\Delta^2}}$, $\delta t_u = \delta s_u * \Delta$, $s_l = \frac{v_l}{\sqrt{1+\Delta^2}}$, and $\delta t_l = \delta s_l * \Delta$. Then the upper and lower bound functions f_u and f_l are computed as follows:

$$\forall f(s_1, t) : f_l(s_1, t + \delta t_l) = f(s_1, t) - \delta s_l$$

$$\forall f(s_1, t) : f_u(s_1, t - \delta t_u) = f(s_1, t) + \delta s_u$$

With this definition we have to assume in principle a differentiable function f. Thus, when considering this method for a digitized signal with a finite set *TIME* of timestamps, one has to take care that two segments with individual (and possibly different) steepnesses Δ_{t-} and Δ_{t+} would meet at such a point $t \in TIME$.

An approximation for such a scenario would then be to compute *two* points with the individual steepnesses Δ_{t-} and Δ_{t+} of the adjacent segments for the convex side (upper or lower bound), and for the concave side one point using the average of the two steepnesses. Interpolations between those computed points then represent the functions f_l and f_u. While this takes care of the situation from an abstract point of view, still, for high values for v_l / v_u together with rapidly changing signals, this can lead to issues. That is, let us assume that we have a signal that has a steep rising edge immediately followed (i.e., within a millisecond) by a steep falling edge. If we now assume a value of 0.1 for v_l and compute the lower bound for some timestep in the rising edge, then, if the angle of the curve is steep enough on both flanks, it can occur that with the falling edge, the signal would fall below this value fast enough so that even the reference signal would violate this lower bound value.

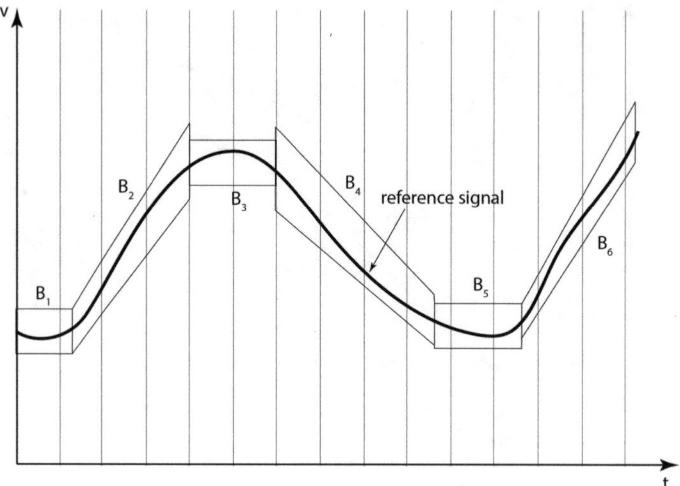

Fig. 3.7 Reference signal covered using a temporal band sequence $\langle B_1, B_2, B_3, B_4, B_5, B_6 \rangle$

A more complex solution addressing also these scenarios would be to compute circles with diameters v_u and v_l around all signal values (the areas within the circles—or respective circle segments (divided by s_1) if v_u and v_l differ—describe all acceptable signal values), and f_l and f_u could be defined using tangents to these circles for approximating the lower and upper hulls. While this might be intuitive for a human designer, please note that this certainly would be quite complex computation-wise.

3.3.2 Temporal Band Sequences

Loiez et al. [19] introduced temporal band sequences for diagnostic purposes. In particular they provided the foundations for using temporal band sequences directly for model-based analog circuit diagnosis. Later on, Alaoui and colleagues [1] presented the results of experiments comparing temporal band diagnosis with statistical diagnosis approaches based on the false positive rate of the approaches. In the context of fault detection, temporal band sequences can be seen as an extension of tolerance approaches. In Fig. 3.7, we show an illustration of a reference signal being covered by a sequence of temporal bands, each individual band defining part of the corridor we expect the signal to adhere to.

According to Loiez et al. [19], a temporal band is an area between two points in time, where the upper and lower bounds are given using boundary functions. The authors defined functions f_l, f_u as polynomials, e.g., $\sum_{i=0}^{n} a_i \cdot t^i$ where the a_is are the parameters. Formally, we can define a temporal band as follows:

Definition 8 (Temporal Band) Given a signal s, a temporal band B is a tuple $((t_i, t_e), (f_l, f_u))$ where

- t_i and t_e are time points $\in$ *TIME* with $t_i < t_e$.
- f_l and f_u are functions defining the lower and upper bounds of the band such that $\forall t \in [t_i, t_e[: f_l(s,t) \leq f_u(s,t)$.

Definition 9 (Temporal Band Sequence) A sequence of temporal bands $TBS = \langle B_1, \ldots, B_k \rangle$ is a temporal band sequence for *TIME*, iff for all $j = 1, \ldots, k-1$ with $B_j = ((t_i^j, t_e^j), (f_l^j(s,t), f_u^j(s,t)))$ and $B_{j+1} = ((t_i^{j+1}, t_e^{j+1}), (f_l^{j+1}(s,t), f_u^{j+1}(s,t)))$ it holds that $t_e^j = t_i^{j+1}$, and for all $t \in TIME$ it holds that $t_i^0 \leq t < t_i^k$.

A temporal band sequence *TBS* connects several individual temporal bands, where we have to ensure that *TBS* defines a temporal band for all $t \in TIME$, and that the sequence is ordered in respect of time. We say that a temporal band sequence is defined for a reference signal s, if for each function value $f(s,t)$ there exists a temporal band $((t_i, t_e), (f_l(s,t), f_u(s,t))) \in TBS$ with $t_i \leq t < t_e$ so that $f_l(s,t) \leq f(s,t) \leq f_u(s,t)$. From these definitions, we can easily define the equivalence of signals with respect to temporal bands:

Definition 10 (Temporal Band Sequence Equivalence) Given signals s_1, s_2 with a function and a temporal bands sequence $TBS = \langle B_1, \ldots, B_k \rangle$ for the reference signal s_1. Then we have $f(s_2, t) =_{TBS} f(s_1, t)$ if and only if

$$\exists((t_i, t_e), (f_l(s_1,t), f_u(s_1,t))) \in TBS : (t_i \leq t < t_e) \wedge (f_l(s_1,t) \leq f(s_2,t) \leq f_u(s_1,t))$$

It is important to note that such bands could be in principle defined without referring to a reference signal (s_1 in the definition), so that functions f_l and f_u of a band are just functions of time t. So, we could define them also using "lines" as upper and lower bound functions (see Fig. 3.7). Furthermore it is not necessary that the boundary values for two connected temporal bands match each other.

Since we noted that temporal band sequences can be seen as an extension of tolerance approaches, immediate questions arising are those regarding the differences and whether there is an advantage to be gained with using temporal band sequences. As we can gather from Definition 6, we have that the tolerance values v_l and v_u are constants for the tolerance value approach, and thus define a validity corridor of a certain vertical width $(v_l + v_u)$ with the reference signal defining the vertical placement. Temporal bands as of Definition 8 allow us to define tolerance as a dynamic function of the reference signal and time. This allows us to take temporal signal segments where we expect more noise into account. That is, since tolerance can be defined via a dynamic function, we can define a higher tolerance for this segment without sacrificing preciseness for segments where we expect almost no deviation from the "ideal" reference signal. Temporal band sequences as of Definition 9 then add the comfort that we can use more than one function for defining tolerance for the signal's whole duration, i.e., by defining individual temporal bands with individual tolerance functions for the signal's individual segments in time. Thus, especially if the tolerance function would consider time only as mentioned before, the concept of having a sequence of segments adds further flexibility and room for optimization.

3.3.3 Pearson Correlation Coefficient

In his Master's thesis and the resulting book [14], Schneider used the Pearson correlation coefficient for comparing a measured signal with a reference. In the system testing framework he considered, using the Pearson correlation coefficient offered improvements on previous comparison algorithms used for classifying signals as being equivalent or different. Thus it seems an attractive variant also for our comparison of a reference signal with observed behavior. Some advantage of this coefficient is that it is well known to be invariant to linear transformations of either signal. Depending on the actual scenario and fault effect, this can also be an issue though, as you can see in our experiments in Sect. 3.5. In our context, we can formalize the Pearson correlation coefficient $r(s_1, s_2)$ for signals s_1 and s_2 as follows:

Definition 11 (Pearson Correlation Coefficient) For signals s_1 and s_2 over *TIME*, we define the Pearson correlation coefficient r as follows, where m_1 and m_2 are the mean values of signals s_1 and s_2 respectively:

$$r = \frac{\sum_{t \in TIME}((f(s_1, t) - m_1) \cdot (f(s_2, t) - m_2))}{\sqrt{(\sum_{t \in TIME}(f(s_1, t) - m_1)^2) \cdot (\sum_{t \in TIME}(f(s_2, t) - m_2)^2)}}$$

In the literature we can find the following interpretation of the result:

$0.0 \leq r \leq 0.2$	Too weak a linear connection
$0.2 < r \leq 0.5$	Weak to moderate linear connection
$0.5 < r \leq 0.7$	Significant linear connection
$0.7 < r \leq 0.1$	High to perfect linear connection

When defining equivalence based on the Pearson correlation coefficient for our context, we have to take such values into account.

Definition 12 (Pearson Correlation Coefficient Equivalence) Given two signals s_1 and s_2 together with their behavioral functions over time *TIME* and their Pearson correlation coefficient r as of Definition 11. We say that s_1 and s_2 are Pearson correlation coefficient equivalent (or PCC equivalent for short—$s_1 =^{r_b}_{PCC} s_2$) if and only if $r \geq r_b$ for a decision value r_b.

In this definition, obviously r_b is not fixed but has to be obtained from experiments with the problem domain in order to maximize the fault detection rate and minimize the false positive rate. An educated guess could be, for instance, $=^{0.75}_{PCC}$ dividing the best correlation range given in the above table into halves. Please note in this respect also our corresponding experiments in Sect. 3.5.

3.4 Abductive Diagnosis for Modelica Models

In this section we show how to automatically derive a knowledge base for abductive diagnosis using the Modelica system model and simulation concept discussed in Sect. 3.2. Before we disclose our rule extraction process in Sect. 3.4.2, we briefly recapitulate the basic definitions of abductive diagnosis in Sect. 3.4.1.

3.4.1 Abductive Diagnosis

Let us first introduce the concept of a *knowledge base*. A knowledge base comprises a set of horn clause rules *HC* over propositional variables *PROPS*. When diagnosing engineered systems, such propositional variables state, for example, some components' particular mode or a certain value. Considering our voltage-divider example from Fig. 3.2, we might use, e.g., proposition $\mathit{short}(\texttt{R2})$ for stating that the resistor R2 has a short. A proposition $\mathit{nok}(\texttt{R2}.v)$ might be used to indicate that the voltage at resistor R2 deviates from the expected voltage. For stating a behavior, e.g., saying that a short of R2 leads to an unexpected voltage drop at the same resistor, we use the horn clause $\mathit{short}(\texttt{R2}) \rightarrow \mathit{nok}(\texttt{R2}.v)$. Using the definitions of [11], we define a knowledge bases formally as follows:

Definition 13 (Knowledge Base (KB)) A *knowledge base* (KB) is a tuple $(A, \mathit{Hyp}, \mathit{Th})$ where $A \subseteq \mathit{PROPS}$ denotes a set of propositional variables, $\mathit{Hyp} \subseteq A$ a set of hypotheses, and $\mathit{Th} \subseteq \mathit{HC}$ a set of horn clause sentences over A.

In the context of our work, hypotheses correspond directly to causes, i.e., faults. Thus, in the following, we use the terms hypothesis and cause interchangeably.

Example 3 A partial KB for our voltage divider example looks like:

$$\begin{pmatrix} \{\mathit{short}(\texttt{R2}), \mathit{nok}(\texttt{R2}.v)\}, \\ \{\mathit{short}(\texttt{R2})\}, \\ \{\mathit{short}(\texttt{R2}) \rightarrow \mathit{nok}(\texttt{R2}.v)\} \end{pmatrix}$$

■

In the next step we define a *propositional horn clause abduction problem*.

Definition 14 (PHCAP) Given a knowledge base (A, Hyp, Th) and a set of observations $Obs \subseteq A$, the tuple (A, Hyp, Th, Obs) forms a propositional horn clause abduction problem (PHCAP).

A solution for some PHCAP is a set of hypotheses that allows deriving the given observations or symptoms. The following definition from [11] states this formally.

Definition 15 (Diagnosis; Solution of a PHCAP) Given a PHCAP (A, Hyp, Th, Obs), a set $\Delta \subseteq Hyp$ is a solution if and only if $\Delta \cup Th \models Obs$ and $\Delta \cup Th \not\models \bot$. A solution Δ is *parsimonious* or *minimal* if and only if no set $\Delta' \subset \Delta$ is a solution.

A solution Δ for some PHCAP is an *explanation* for the given observations. Thus we refer to Δ as *abductive diagnosis* (or diagnosis for short). In Definition 15, diagnoses do not need to be minimal or parsimonious. In most practical cases, however, only minimal diagnoses or minimal explanations for given effects are of interest. Hence, from here on, we assume that all diagnoses are minimal ones, if not specified explicitly otherwise.

Example 4 Let us continue Example 3 and add the observation $nok(\mathtt{R2}.v)$ to the KB to form a PHCAP. The only solution for this problem is $\{short(\mathtt{R2})\}$. If we assume an observation $\neg nok(\mathtt{R2}.v)$, then there is no solution given the partial KB of Example 3. ■

Finding minimal diagnoses for a given PHCAP is an NP-complete problem (see [11]). However, computing all parsimonious solutions can be done easily and efficiently in cases where the number of hypotheses is not too big. An algorithm for computing abductive solutions might use De Kleer's Assumption-based Truth Maintenance System (ATMS) [7], where we refer the interested reader to [8] for an ATMS algorithm. When using an ATMS for abductive diagnosis, we only need to encode observations as a single rule, i.e., for observations $Obs = \{o_1, \ldots, o_k\}$, we generate a new proposition σ and add $o_1 \wedge \ldots \wedge o_k \rightarrow \sigma$ to the theory Th that is passed to the ATMS. The label of the corresponding node of σ is an abductive diagnosis for *Obs*. Due to the rules for the node labels, which only comprise hypotheses, it is ensured that the solution is minimal, sound, complete, and consistent. For more technical details, including computing distinguishing diagnoses and characterizing knowledge bases according to their capability of distinguishing diagnoses, we refer the interested reader to [31]. For a discussion on whether abductive reasoning can be used in practice, we recommend reading [16].

3.4.2 Automated Rule Extraction

As briefly depicted in the introduction, we extract the desired cause-effect rules from a system model **M** via comparing the outcome of two simulation runs. That is, one run where no fault is enabled for any component, and one with exactly one fault mode enabled for some individual component. The difference between

Algorithm 1 Rule extraction from Modelica models

Require: System model $(COMP, MODES, \mu, I, O, \mathbf{P})$, a test bench $\mathbf{T}$, a time t_f where a fault should be injected, and an end time t_e
Ensure: A KB (A, Hyp, Th)
1: Let I_p be the propositional representation of I.
2: Let A be I_p and let Hyp, and Th be empty sets.
3: **for** c in $COMP$ **do**
4:     **for** m in $\mu(c) \setminus \{ok\}$ **do**
5:         Let $\Delta_1(c, 0) = ok$
6:         Let $\Delta_2(c, 0) = ok$ and $\Delta_2(c, t_f) = m$
7:         **for** c' in $COMP \setminus \{c\}$ **do**
8:             $\Delta_1(c', 0) = ok$
9:             $\Delta_2(c', 0) = ok$ and $\Delta_2(c', t_f) = ok$
10:         **end for**
11:         Let $\mathsf{sim}(\mathbf{P}, \mathbf{T}, \Delta_1, t_e)$ be $\mathbf{B}_{corr}$.
12:         Let $\mathsf{sim}(\mathbf{P}, \mathbf{T}, \Delta_2, t_e)$ be $\mathbf{B}_{faulty}$.
13:         Let D be the result of $\mathsf{diff}(\mathbf{B}_{corr}, \mathbf{B}_{faulty})$ considering variables in O only.
14:         Add all elements of D to A.
15:         Add the proposition $m(c)$ to A and Hyp.
16:         **for** d in D **do**
17:             Add the rule $m(c) \wedge (\bigwedge_{p \in I_p} p) \rightarrow d$ to Th
18:         **end for**
19:     **end for**
20: **end for**
21: **return** (A, Hyp, Th)

the two behaviors is mapped to a proposition that we can use as an observation in a PHCAP (see Definition 14). Also the inputs are mapped to propositions, as are the assigned modes of components $c_i \in COMP$ that are not *ok*. The latter are furthermore considered as hypotheses for diagnosis purposes.

Algorithm 1 formalizes the individual steps for our rule extraction concept. In lines 1–2, we initialize the individual sets of the knowledge base (see Definition 13) used in the computation. That is, the set of propositions A, the set of hypotheses Hyp, and Th—the horn clauses over A. We assume here that we have input values over time in I, as is necessary for stimulating the system under consideration. We further assume that this information can be mapped to a set of propositional variables and that it is also represented in the given test bench $\mathbf{T}$. A simple mapping would state that each variable occurring in I over time is ok, i.e., $ok(v, t)$ such that v is a variable and $t \in TIME$ a point in time where a value is set.Alternatively, we might state a precise certain value, which has to be provided in order to be able to use the extracted rules for diagnosis purposes, i.e., $value(v, t, x)$ such that v is the name of the variable, $t \in TIME$ the considered timestamp, and x the value to be set for v at t.

Lines 3–20 implement the core of our rule extraction process. For all components and their modes, we iterate over lines 5–17 in lines 3–4. In the first part of the inner loop's body (lines 5–10), we generate the mode assignments for the correct simulation run (Δ_1) and for the faulty one (Δ_2): With the exception of the currently "active" component c (chosen in line 3), all components are assigned mode *ok*. For component c though, we assign mode m (selected in line 4) for Δ_2, and *ok* for Δ_1. Note that we implement a "simple" temporal mode assignment with only *one permanent change* at t_f. Considering variants more complex in a temporal sense would require many more simulation runs which would affect computational complexity. Thus, for our purpose of isolating fault effects at specific points in time, this restriction seems to be appropriate. Note that t_f should be selected such that the system's initialization is finished so that we can observe the ordinary expected behavior.

In lines 11–13, we start the simulation runs and compare the observed results with function diff. In the simulation function sim, we assume that the test bench $\mathbf{T}$ is merged with the mode assignment Δ ($\mathbf{T} \rightarrow \mathbf{T}'$) as discussed in Sect. 3.2.1 (see Example 2). Please note that the set *TIME* is indirectly defined by the sampling frequency of these simulation runs and t_e—and is thus not a parameter of Algorithm 1.

The function diff deserves special attention in that there are two potential concepts for implementing it. Either, diff returns a propositional representation of values for all output variables from O if there is at least one output where the behaviors differ. Or, we have that diff reports the *deviation* of values between $\mathbf{B}_{corr}$ and $\mathbf{B}_{faulty}$, stating, e.g., that the value caused by the fault for some observed variable $o \in O$ is smaller than the nominal value determined via the correct simulation run $\mathbf{B}_{corr}$. Please note that also Struss used the latter variant for diagnosis purposes [29].

The former mapping option requires a propositional representation for translating the real simulation values to a qualitative domain. For example, we might consider only values like 0, v_{max}, or v as qualitative values for a variable, such that 0 represents the zero value, v_{max} the maximum value that can be reached, and v any value between 0 and v_{max}. Obviously, the diagnosis capabilities might vary, depending on the chosen qualitative representation and the encoded information. In [26, 27], the authors discuss this issue, i.e., the task of finding an appropriate task dependent qualitative abstraction, and also show how to automate this abstraction. Now let us formalize these two approaches at implementing diff:

Qualitative representations: For a qualitative representation, we assume a quantitative domain D and its qualitative representation D_Q, together with a mapping function $\rho : D \mapsto D_Q$. Then, we define diff as:

$$\mathsf{diff}(B_1, B_2) = \begin{cases} \emptyset & \text{if } \nexists (x, v) \in B_1, (x, v') \in B_2 : v \neq v' \\ \{val(x, \rho(v)) | (x, v) \in B_2\} & \text{otherwise} \end{cases}$$

Deviation models: In case we prefer a deviation model, we are "only" interested in whether the value for a variable is, e.g., smaller, equal, or larger than its expected value. Hence, we define diff straightforwardly as follows:

$$\mathsf{diff}(B_1, B_2) = \{o(x) | (x, v) \in B_1, (x, v'') \in B_2 \wedge v' \; o^* \; v\}$$

In this definition, o and o^* represent the relational operator on the side of the qualitative and quantitative domain, respectively. For example, the operator $o \in \{smaller, equal, larger\}$ in the qualitative domain corresponds to $o^* \in \{<, =, >\}$ in the quantitative (integer, real) domain. While we showed the obvious relational operators that we might want to consider, for some projects, we might also want to add, e.g, ones indicating that there is a huge difference in the values (i.e., a deviation above a certain threshold).

In the last part of the rule extraction algorithm's core functionality (i.e., lines 14–18), we add the obtained propositions, hypotheses, and horn clause rules to the respective sets. Finally, we return the knowledge base in line 21, which we can then use to specify and solve diagnosis problems as of a PHCAP (see Definition 14).

It is easy to see that, by construction, our rule extraction algorithm works as expected. If we assume that sim and diff terminate, then termination is ensured since we only have a finite number of components and modes. In respect of time, the complexity of Algorithm 1 is bounded by $O(|COMP|^2 \cdot |MODES|)$ when assuming that sim and diff run in unit time. Note that in practice simulation is very much likely to be responsible for most of the experienced run-time. However, since we can automate all parts of the algorithm and can execute them before deploying the system (and in turn before deploying the diagnosis engine), the time complexity of the rule conversion process seems to be negligible since we do not encounter it when running the diagnosis engine itself.

3.5 Case Studies

In order to show the viability of our proposed method, we carried out two case studies. So let us start with the *first case study* showing the results we obtained when using Algorithm 1 for our voltage-divider example. Please note that we focus first on the model extracting step, and only for the second case study we illustrate also the aspects of the fault detection step as discussed in Sect. 3.3. The corresponding source code for the first Modelica model is shown in Fig. 3.4, and the list of components *COMP*, the list of behavioral modes *MODES*, etc. were derived in Examples 1 and 2. When applying Algorithm 1 considering $COMP = \{\texttt{bat}, \texttt{r1}, \texttt{r2}\}$, $MODES = \{ok, broken, short, empty\}$, and assuming t_f to be set to 0.5 s like for Example 2, we would obtain the following results when simulating the resulting faulty behavior.

Component	Mode	v_1	v_2	e_1	e_2
BAT	*empty*	0	0	Smaller	Smaller
R1	*short*	0	12	Smaller	Larger
R1	*broken*	12	0	Larger	Smaller
R2	*short*	12	0	Larger	Smaller
R2	*broken*	0	12	Smaller	Larger

In this table, we also state the deviations (or effects e_1, e_2) for variables v_1 and v_2 and the individual fault modes (let us remind you that the nominal values should be 8V for v_1 and 4V for v_2). What we see also from our example is that the values obtained when simulating the Modelica program do not necessarily guarantee that we are able to distinguish between all diagnoses. For example, both a *broken* R1 and a *short* R2 would produce the same values for v_1 and v_2, so that we do not end up with a single explanation/diagnosis in general. Let us now use this table to generate the qualitative and the deviation model (concerning diff in Algorithm 1) for the voltage divider. For both models we have the same set of hypotheses:

$$\left\{ \begin{array}{c} \mathit{empty}(\mathtt{BAT}), \mathit{short}(\mathtt{R1}), \mathit{broken}(\mathtt{R1}), \\ \mathit{short}(\mathtt{R2}), \mathit{broken}(\mathtt{R2}) \end{array} \right\}$$

Qualitative model: For this kind of model we assume a qualitative domain of $\{0, (0, 4), 4, (4, 8), 8, (8, 12), 12\}$ such that we consider all voltage values occurring in the correct model. The open interval (x, y) stands for any value larger than x and smaller than y. For this representation, the algorithm would return the following rules:

$$\begin{array}{c} \mathit{empty}(\mathtt{BAT}) \rightarrow \mathit{val}(v1, 0) \\ \mathit{empty}(\mathtt{BAT}) \rightarrow \mathit{val}(v2, 0) \\ \mathit{short}(\mathtt{R1}) \rightarrow \mathit{val}(v1, 0) \\ \mathit{short}(\mathtt{R1}) \rightarrow \mathit{val}(v2, 12) \\ \mathit{broken}(\mathtt{R1}) \rightarrow \mathit{val}(v1, 12) \\ \mathit{broken}(\mathtt{R1}) \rightarrow \mathit{val}(v2, 0) \\ \mathit{short}(\mathtt{R2}) \rightarrow \mathit{val}(v1, 12) \\ \mathit{short}(\mathtt{R2}) \rightarrow \mathit{val}(v2, 0) \\ \mathit{broken}(\mathtt{R2}) \rightarrow \mathit{val}(v1, 0) \\ \mathit{broken}(\mathtt{R2}) \rightarrow \mathit{val}(v2, 12) \end{array}$$

For this case, the set of propositions includes the hypotheses (like *empty*(BAT) and the elements in $\{\mathit{val}(v1, 0), \mathit{val}(v1, 12), \mathit{val}(v2, 0), \mathit{val}(v2, 12)\}$. No other qualitative values are necessary for this example.

Deviation model: The example's deviation model comprises the following rules:

$$\begin{array}{c} \mathit{empty}(\mathtt{BAT}) \rightarrow \mathit{smaller}(v1) \\ \mathit{empty}(\mathtt{BAT}) \rightarrow \mathit{smaller}(v2) \\ \mathit{short}(\mathtt{R1}) \rightarrow \mathit{smaller}(v1) \\ \mathit{short}(\mathtt{R1}) \rightarrow \mathit{larger}(v2) \\ \mathit{broken}(\mathtt{R1}) \rightarrow \mathit{larger}(v1) \\ \mathit{broken}(\mathtt{R1}) \rightarrow \mathit{smaller}(v2) \\ \mathit{short}(\mathtt{R2}) \rightarrow \mathit{larger}(v1) \\ \mathit{short}(\mathtt{R2}) \rightarrow \mathit{smaller}(v2) \\ \mathit{broken}(\mathtt{R2}) \rightarrow \mathit{smaller}(v1) \\ \mathit{broken}(\mathtt{R2}) \rightarrow \mathit{larger}(v2) \end{array}$$

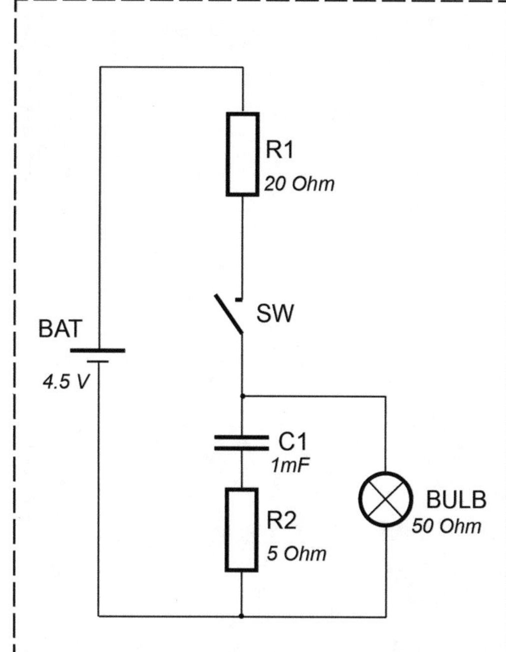

Fig. 3.8 A switch circuit example

In this case, the set of propositions is formed by the hypotheses together with propositions $\{smaller(v1), larger(v1), smaller(v2), larger(v2)\}$.

Both kinds of model represent the obtained information in a qualitative way. It is worth mentioning that, in addition, we might also want to encode in a knowledge base that some values cannot occur at the same time. For example, a voltage drop cannot be larger *and* smaller than some value at the same time. Such knowledge can be added easily via stating $smaller(v1) \wedge larger(v1) \rightarrow \bot$ and $smaller(v2) \wedge larger(v2) \rightarrow \bot$. Of course, we can also automate the process of adding such mutual exclusiveness data.

The purpose of the *second case study* is to show that our approach can be applied also to hybrid/analog circuits comprising capacitors and switches such that the behavior over time is more complicated. In Fig. 3.8, we show such a circuit, where we use a switch `SW` for turning a bulb `BULB` on or off. The purpose of capacitor `C1` is such that the bulb stays on (red line in Fig. 3.9) for a short while after switching it off (see blue line in Fig. 3.9), drawing from the energy stored while loading `C1`—see the green line for the current in the capacitor). Thus, we have one input, i.e., `SW`, and one output, i.e., `BULB` transmitting light or not. Appropriate sets of behavioral modes would be $\{ok, empty\}$ for the battery, and $\{ok, short, broken\}$ for the other components. Via corresponding simulations, we obtained the results shown in the following table, where we focus on the first difference between the expected value of *light* and the observed one at a particular point in time. In the table we see also the value of the input variable *on* stating the status of `SW`.

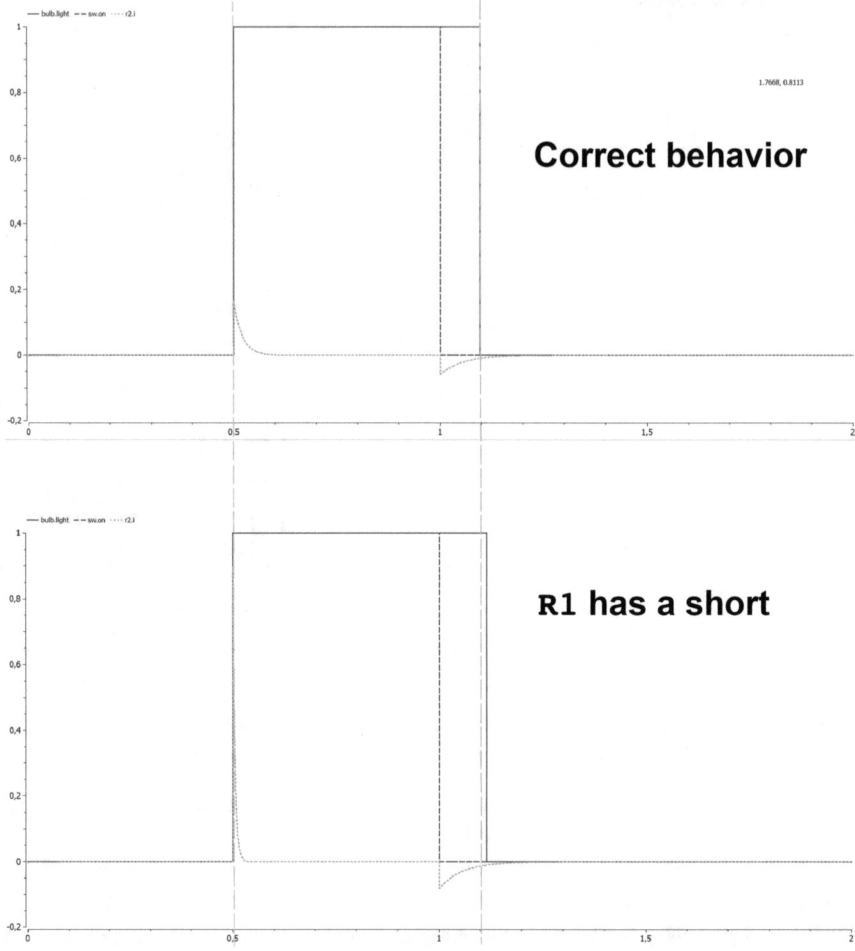

Fig. 3.9 The correct behavior of the switch circuit behavior (picture at the top) versus the behavior for the case that R1 has a short (picture at the bottom)

Component	Mode	*On*	*Light*	*Time*
BAT	*empty*	On	Off	0.5
R1	*short*	Off	On	1.13
R1	*broken*	On	Off	0.5
SW	*short*	Off	On	0.0
SW	*broken*	On	Off	0.5
C1	*short*	Off	Off	1.0
C1	*broken*	Off	Off	1.0
R2	*short*	Off	Off	1.09
R2	*broken*	Off	Off	1.0
BULB	*short*	On	Off	0.5
BULB	*broken*	On	Off	0.5

Again, it is evident from the table that we cannot distinguish between all faults via the obtained deviations. We also see from Fig. 3.9 that for a short in `R1`, the output might have the same shape, but would show a slightly different timing. Using the obtained table of behavioral deviations, computing the knowledge base for abductive diagnosis as described in this paper is straightforward. However, for more detailed results, we should probably introduce the means for reasoning about differences in the timing, e.g., stating that `BULB` goes off too late. Thus, we see that our approach for extracting a PHCAP can be used for many systems, and that the detail level of the obtained knowledge base depends on the chosen deviation description as expected and discussed.

The slight timing deviations for the faulty case are also something that affects fault detection as described in Sect. 3.3. That is, let us assume that we can observe the bulb's current I, its voltage V, and also its lighting status *light*. In particular, it is more likely that we have sensors there, rather than for the current of the capacitor in Fig. 3.9 (but which is be equal to the current of the bulb if the switch is off).

To this end, let us assume that we have $v_l = 0.01/v_u = 0.02$ for both tolerance approaches, and let us neglect the more flexible TBS approach at this point since we will see that already the basic tolerance approaches work well. Of course, we will also have a look at the PCC. Considering the faulty scenario, we have that the bulb's current I is completely different in value when we turn the switch on, since then it is only determined by the bulb's resistance and the battery voltage (`R1` has a short). Assuming that we have simulations with a 1 kHz sampling rate, both tolerance approaches can recognize the fault at simulation time 501 ms when focusing on the bulb's current or voltage. This means right for the first sample when the fault in `R1` starts affecting the circuit after the switch is turned on. If we, however, consider the light status *light* instead, we can detect the fault's presence only at simulation times 1.098 s and 1.101 s respectively (tolerance/alternative tolerance). This means about 100 samples after turning the switch off, and quite some time later than when considering I or V. This was to be expected though, since the fault effects would not be visible earlier for this "signal." Consequently, the observed signals have to be chosen wisely.

As we wrote above, while there are differences in the timing, the "shape" of the signals is quite similar for the scenario. That is, while there is a difference also in the amplitude (at 0.65 s we have $I = 90\,\text{mA}/V = 4.5\,\text{V}$ for the faulty scenario, and $I = 64.3\,\text{mA}/B = 3.21\,\text{V}$ for the correct scenario), the overall shape is still similar. Due to the specific values in the scenario, this results in the PCC values to be 0.992 for the current, 0.992 for the voltage, and 0.975 for the lighting status. So, while we would have violations with the tolerance approaches (and thus also if we defined a TBS), the PCC method is not ideal for this scenario despite its advantages that we discussed in Sect. 3.3. Consequently it is not only the observed signals that have to be chosen wisely, but also the fault detection method(s). In this respect, we would like to point out that a portfolio approach would be of advantage in practice.

3.6 Related Research and Discussion

For model-based diagnosis approaches, e.g. [9, 25], we can use the diagnostic model for fault localization via checking whether the observations are consistent with the behavior predicted by the model for the considered stimuli. In Sect. 3.3, we considered several approaches that allow us to detect the presence of some fault(s) via considering the output and some tolerance specified one or the other way. That is, without the use of a model, but using a "specification" in the form of an expected behavior corridor or our expectations regarding a statistic correlation to reference signals. As discussed in that section, these concepts have been considered in fault detection and diagnosis before [14, 19], and perfectly fit our scenario of Modelica models of hybrid systems where we can simulate the expected behavior.

We proposed in Sect. 3.4 an automated extraction of a diagnostic model that would support abductive diagnostic reasoning for identifying the cause(s) of some unexpected behavior as briefly outlined in [23]. In the literature, there have been three major approaches to using Modelica models for diagnosis. The first one implements the basic idea of applying those changes necessary for computing diagnoses to the language Modelica itself. This includes adaptions for handling unknown behavior, allowing us to come up with simulations where no single value can be determined anymore, and the introduction of corresponding fault modes and their behavior. In [20], Lunde presented such an approach leading to the language Rodelica as used in the model-based diagnosis system RODON [3]. In contrast to [20], we neither change the modeling language, nor do we rely on specific simulation engines. Rather, we use Modelica and its available simulation infrastructure. Regarding available data, we also assume that the components' specific fault modes are known and that we can activate and deactivate them individually during simulation.

The second approach augments Modelica models with fault modes and their behavior. In [10, 22] the authors correspondingly suggested to automatically augment Modelica models, and to use them for diagnosis as follows. When simulating the model with faults turned on or off, the outcome is compared with the expected fault free behavior. In case of differences that are considered to be large enough, the activated fault mode can be given back as result. The interesting idea behind [22] is that the authors suggested a Bayesian approach for checking similarity. This work is close to ours, with an important difference. That is, we use an augmented Modelica model for creating a knowledge-base for *abductive* reasoning, which can be used later on, i.e., after deployment, for diagnosis purposes.

The third approach to using Modelica for diagnosis (see [28]) uses corresponding models for computing a system's expected behavior to be compared with the actually observed and measured one. In case of a deviation, a model-based diagnosis engine is then used for computing explanations, i.e., diagnoses. The required model is extracted from the Modelica model such that basic components are replaced with qualitative models in order to come up with a component-connection model. Correspondingly, we can use this concept for all cases where there are qualitative models available for all the library components contained in a model. An advantage of this approach is that we can use Modelica for obtaining a system's structure and for checking deviations between the expected and observed system output. A significant drawback is, however, that somebody has to develop the qualitative models for library components used in the system model.

There is further work dealing with diagnosis in the context of Modelica. For example, Lee et al. [18] presented a very different and interesting approach. Also relying on component fault models, they use machine learning for identifying the root cause of an issue, rather than using a logic-based reasoning. In particular, the idea is to use simulation results from models where a component is assumed to be faulty for training a belief network. Once a certain behavior is observed, the network can then be used to isolate the corresponding faulty component. A related approach was presented in [21] where the authors used partial models of a railway switch to learn diagnosis classifiers with a random-forest algorithm.

In contrast to Lee and colleagues, we rely on model-based diagnosis and in particular on abductive diagnosis. We furthermore do not rely on machine learning and we are also able to derive all diagnoses for a certain behavior, which is usually not possible when relying on belief networks.

The concept discussed in this chapter extends previous work in the domain of abductive diagnosis [13, 31] where we used FMEA like tables for extracting the cause-effect rules. In those papers, the authors introduced an algorithm that converts available tabular data on the available components, potential faults occurring for the individual components and the resulting effects, as well as supporting conditions into horn clauses for abductive diagnosis. The advantage of that approach is that it draws on information that is available in practice. Hence, an easy integration into existing processes for diagnosis and monitoring can be assured. On the downside, the tabular information has to be provided manually. The concept discussed in this chapter improves on this situation, in that we suggest to use popular modeling languages like Modelica for generating the desired abductive diagnosis models.

While the proposed concept has several advantages, our current variant of Algorithm 1 has a limitation that is of interest for practical applications. That is, when gathering the data for creating the knowledge base KB, we limit the simulations to the exploration of single fault behavior in order to limit the algorithm's complexity. If we assume that individual faults are indeed independent, this still allows for a correct isolation of diagnoses for multi fault scenarios. If faults interact with each other though, KB does not contain any data about effects resulting from such interactions, which will add some impreciseness regarding multi-fault diagnosis results.

If KB shall rely also on data about possible fault interactions, we need to conduct also simulations where multiple faults are active simultaneously. This would certainly increase the algorithm's complexity though. For example, if assuming that the mutual exclusiveness of fault modes for a single component still holds, we would need then $|MODES|^{|COMP|}$ simulations for investigating the entire behavior for some input stimulus, compared to $|COMP|.|MODES|$ simulations for the current version. For a mere 20 components with 5 fault modes each, this would result in

5^{20} simulations compared to 100 simulations for the current algorithm version. A combinatorial exploration like adopted for combinatorial testing [2, 5, 17] could be a promising technique to tackle this problem via exhaustively considering *all* locally possible fault interactions for *all* component subsets of a certain size $x < |COMP|$ (x is then referred to as the *strength* of the combinatorial approach). The advantage then would be that a single simulation with several activated faults would cover more than one such local combination of various component subsets. For considering the effect, let us assume a simple scenario where we have two modes OK and F_1 and three components $\{c_1, c_2, c_3\}$ that can all feature either of the two modes. Furthermore let us assume that the faults are active for the whole simulation duration. For an exhaustive approach we would need $|MODES|^{|COMP|} = 2^3 = 8$ simulations, while for a combinatorial strength of $x = 2$, we could use the following four simulations shown in the table below. We can easily see that for any component subset of size two, and for any possible mode assignment to the components in this subset, there is at least one simulation that features exactly this mode assignment.

	$\Delta(c_1)$	$\Delta(c_2)$	$\Delta(c_3)$
Simulation 1	OK	OK	OK
Simulation 2	OK	F_1	F_1
Simulation 3	F_1	OK	F_1
Simulation 4	F_1	F_1	OK

In [24] we outlined a preliminary concept for such an extension of Algorithm 1, and discussed several questions that immediately arise (like that of exploiting structural dependency information in the combinatorial exploration) and should be investigated as well as solved in order to gain enough confidence in the effectiveness of such a concept for practical purposes. In the context of combinatorial testing, Kuhn et al. showed, for instance, that a strength of $x = 6$ should suffice in practice, but it is an open question whether this would suffice for our purposes. Since we usually have a large number of components (e.g., 100), requiring a strength of $6 << |COMP|$ would allow a significant reduction in the number of required simulations though. Still, future research will have to show the effectiveness of adopting such a concept.

3.7 Summary

When facing the decision of whether to employ some model-based diagnosis approach for a project, more often than not the resources and knowledge to create the needed model prohibit a positive decision. With the concept discussed in this chapter, we address some of the issues that make the required modeling step so demanding. That is, we show for a popular modeling language how to *automatically* derive a model containing cause-effect rules that we then can use for abductive diagnosis. We use Modelica's simulation engine to compute these rules, and the only data a designer has to provide are data that she would consider during FMEA anyway. This data—concerning component fault models and input vectors for triggering them—can, which is important, be defined in Modelica. In particular, a designer can add multiple behaviors for a component in Modelica, and our approach will enable them individually in order to simulate the corresponding behavior for deriving the desired diagnosis model.

The derived abductive diagnosis model can then be used to solve propositional horn clause abduction problems which basically consist of the data we derived when creating our cause-effect rules plus the observed symptoms. An advantage of this type of model-based reasoning is that it is quite intuitive to what a maintenance expert would think about when considering FMEA data, but offers the advantage of a formal background and thus is amenable to automated reasoning. For fault detection, we discussed three different approaches to trigger the diagnosis process by considering tolerances or statistic correlation in the context of the observed signals.

So far, we have been limiting our algorithm to single fault simulations. For future work, we intend to extend our concept to include also simulations of scenarios with multiple faults using a combinatorial exploration strategy as outlined in Sect. 3.6. Like we mentioned, we plan to extend the algorithm also to incorporate a more general temporal fault activation as hinted at in our definitions. Last but not least, while we showed the viability of our approach for some examples, we need also industrial sized case studies as showcases. Hopefully corresponding results can contribute to increasing the deployment of model-based diagnosis in practice.

References

1. R.M. Alaoui, B.O. Bouamama, P. Taillibert, Diagnosis based on temporal band sequences - a empirical comparison to statistical approaches, in *Proceedings of the Automation Congress*, vol. 17 (2004), pp. 435–440
2. M.N. Borazjany, L. Yu, Y. Lei, R. Kacker, R. Kuhn, Combinatorial testing of ACTS: a case study, in *2012 IEEE Fifth International Conference on Software Testing, Verification and Validation* (2012), pp. 591–600. https://doi.org/10.1109/ICST.2012.146
3. P. Bunus, O. Isaksson, B. Frey, B. Münker, Rodon - a model-based diagnosis approach for the DX diagnostic competition, in *International Workshop on Principles of Diagnosis (DX)* (2009)
4. M. Catelani, L. Ciani, V. Luongo, The FMEDA approach to improve the safety assessment according to the IEC61508. Microelectron. Reliab. **50**, 1230–1235 (2010)
5. D.M. Cohen, S.R. Dalal, M.L. Fredman, G.C. Patton, The AETG system: an approach to testing based on combinatorial design. IEEE Trans. Softw. Eng. **23**(7), 437–444 (1997). https://doi.org/10.1109/32.605761
6. R. Davis, Diagnostic reasoning based on structure and behavior. Artif. Intell. **24**, 347–410 (1984)
7. J. de Kleer, An assumption-based TMS. Artif. Intell. **28**, 127–162 (1986)
8. J. de Kleer, A general labeling algorithm for assumption-based truth maintenance, in *Proceedings AAAI* (1988), pp. 188–192
9. J. de Kleer, B.C. Williams, Diagnosing multiple faults. Artif. Intell. **32**(1), 97–130 (1987)

10. J. de Kleer, B. Janssen, D.G. Bobrow, T. Kurtoglu, B. Saha, N.R. Moore, S. Sutharshana, Fault augmented Modelica models, in *24th International Workshop on Principles of Diagnosis (DX)* (2013), pp. 71–78
11. G. Friedrich, G. Gottlob, W. Nejdl, Hypothesis classification, abductive diagnosis and therapy, in *International Workshop on Expert Systems in Engineering* (1990)
12. P. Fritzson, *Object-Oriented Modeling and Simulation with Modelica 3.3 – A Cyber-Physical Approach*, 2nd edn. (Wiley-IEEE Press, New York, 2014)
13. C.S. Gray, R. Koitz, S. Psutka, F. Wotawa, An abductive diagnosis and modeling concept for wind power plants, in *9th IFAC Symposium on Fault Detection, Supervision and Safety of Technical Processes* (2015)
14. S. Hannes, *Automated Measurement Evaluation for Combustion Analysis* (AV Akademikerverlag, Riga, 2016). https://www.akademikerverlag.de/catalog/details//store/de/book/978-3-639-67777-5/automated-measurement-evaluation-for-combustion-analysis; https://dl.acm.org/citation.cfm?id=3002640
15. P.G. Hawkins, D.J. Woollons, Failure modes and effects analysis of complex engineering systems using functional models. Artif. Intell. Eng. **12**, 375–397 (1998)
16. R. Koitz, F. Wotawa, On the feasibility of abductive diagnosis for practical applications, in *9th IFAC Symposium on Fault Detection, Supervision and Safety of Technical Processes* (2015)
17. D.R. Kuhn, R.N. Kacker, Y. Lei, Sp 800-142. Practical combinatorial testing. Tech. rep. (2010). http://nvlpubs.nist.gov/nistpubs/Legacy/SP/nistspecialpublication800-142.pdf
18. D. Lee, B. Lee, J.W. Shin, Fault detection and diagnosis with modelica language using deep belief network, in *Proceedings of the 11th International Modelica Conference*, Versailles (2015), pp. 615–623. https://doi.org/10.3384/ecp15118615
19. E. Loiez, P. Taillibert, Polynomial temporal band sequences for analog diagnosis, in *15th International Joint Conference on Artificial Intelligence, IJCAI'97* (1997), pp. 474–479
20. K. Lunde, Object oriented modeling in model based diagnosis, in *Modelica Workshop 2000 Proceedings* (2000), pp. 111–118
21. I. Matei, A. Ganguli, T. Honda, J. de Kleer, The case for a hybrid approach to diagnosis: a railway switch, in *26th International Workshop on Principles of Diagnosis*, (2015), pp. 225–234. http://ceur-ws.org/Vol-1507/dx15paper29.pdf
22. R. Minhas, J. de Kleer, I. Matei, B. Saha, Using fault augmented modelica models for diagnostics, in *10th International Modelica Conference* (2014)
23. B. Peischl, I. Pill, F. Wotawa, Abductive diagnosis based on modelica models, in *27th International Workshop on Principles of Diagnosis (DX)* (2016). No archival proceedings
24. I. Pill, F. Wotawa, Model-based diagnosis meets combinatorial testing for generating an abductive diagnosis model, in *28th International Workshop on Principles of Diagnosis (DX'17)*, ed. by M. Zanella, I. Pill, A. Cimatti, Kalpa Publications in Computing, vol. 4 (EasyChair, 2018), pp. 248–263. https://easychair.org/publications/paper/7t86
25. R. Reiter, A theory of diagnosis from first principles. Artif. Intell. **32**(1), 57–95 (1987)
26. M. Sachenbacher, P. Struss, Automated qualitative domain abstraction, in *International Joint Conference on Artificial Intelligence* (2003), pp. 382–387
27. M. Sachenbacher, P. Struss, Task-dependent qualitative domain abstraction. Artif. Intell. **162**(1–2), 121–143 (2005). https://doi.org/10.1016/j.artint.2004.01.005
28. R. Sterling, P. Struss, J. Febres, U. Sabir, M.M. Keane, From modelica models to fault diagnosis in air handling units, in *Proceedings of the 10th International Modelica Conference*, Lund (2014)
29. P. Struss, Deviation models revisited, in *Working Papers of the 15th International Workshop on Principles of Diagnosis (DX-04)* (2004)
30. J. Voas, G. McGraw, Software fault injection: inoculating programs against errors. Softw. Test. Verif. Reliab. **9**(1), 75–76 (1999)
31. F. Wotawa, Failure mode and effect analysis for abductive diagnosis, in *Proceedings of International Workshop on Defeasible and Ampliative Reasoning (DARe-14)* (2014)

Chapter 4
Robust Data-Driven Fault Detection in Dynamic Process Environments Using Discrete Event Systems

Edwin Lughofer

4.1 Introduction

In today's industrial systems with increasing complexity, the supervision of (on-line) processes [3] is of utmost importance in order to guarantee a failure-free continuation of their life-cycles and thus to reduce the likelihood of waste during production. This also leads to a decrease of wrong feedbacks to operators (half on-line systems) or to system/control components (within full in-line installations), which may guide the whole process into a "wrong direction." Process or system failures could even lead to risks for operators in certain constellations (e.g., consider a leaky pipe transferring emission gases) [11] or to significant down-times of system components, which would make the process non-operable. This may even lead to halts of important machines. Production waste should be ideally kept at a low level in order not to significantly increase the production costs or not to disappoint customers due to the delivery of faulty, defective or incomplete items [37].

Therefore, the concept of *fault detection*, as the business of detecting the occurrence of a fault in a system, has been seen as one of the major priorities in several companies and (applied) research programmes across Europe: for instance, within the Horizon 2020 programme, there exist several objectives related to the Factories of the Future (FoF) track[1] which particularly demand for developments within the field of fault detection and diagnosis as well as predictive maintenance. The concept of fault detection was formally defined by IFAC Technical Committee

[1] http://ec.europa.eu/research/industrial_technologies/factories-of-the-future_en.html.

E. Lughofer (✉)
Department of Knowledge-Based Mathematical Systems, Johannes Kepler University Linz, Linz, Austria
e-mail: edwin.lughofer@jku.at

M. Sayed-Mouchaweh (ed.), *Diagnosability, Security and Safety of Hybrid Dynamic and Cyber-Physical Systems*, https://doi.org/10.1007/978-3-319-74962-4_4

SAFEPROCESS as the "Determination of faults present in a system and the time of detection." They also defined a fault as an "Unpermitted deviation of at least one characteristic property or variable of the system from acceptable/usual/standard behaviour" [58].

The techniques for fault detection existing in literature can be loosely classified into [108]:

1. differential equations, modeling physical laws within the systems [33, 57]
2. fault models based on expert knowledge from experienced operators of the system [31], and
3. fault detection based on data-driven models serving as a fault-free reference of interrelations and dependencies among process variables [11, 106].

The first two techniques, even when successful, are mainly bounded by their (high) development effort, which may last over several months and thus induce significant costs for companies and/or research facilities. In order to reduce the efforts, automatic extraction of these dependencies in multi-sensor systems [42] using either data-driven methods, machine learning methods and/or fusion methods has been emerged during the last years. This approach led to the so-called *data-driven System Identification based fault detection (FD)*, where the learnt models are known as System Identification (SysID) models [75, 94, 112] and reflect higher-dimensional relations between the considered variables. The models are then used as monitoring reference for the nominal, fault-free situation of the system. In contrast to other data-driven (more univariate) FD techniques that rely on (1) the supervision of abnormal behavior based on the recorded measurements [25], (2) frequency-based analysis of measurement signals [20, 98], or (3) autoregressive moving average models [122], SysID models do not require any re-occurring (typical) patterns representing the fault-free cases in the signals and thus are applicable to a wider range of measurement signals.

A particular challenge in fault detection systems is the detection of faults at an early stage, ideally as early as possible before more severe failures or downtrends in the quality of components or products can happen at all. This often makes the application of a-posteriori checks by surface inspection [21, 52] or structural health monitoring [45] inefficient and unpracticable. Therefore, the usage of process/system variables which are able to track the state of the system over time, thus to contain information about the system state at discrete time points already at 'processing stage', is of much more interest. An example from a real-world process (rolling mills at cold rolling plants) is provided in Fig. 4.1, which shows the development of a particular process variable (its values measured by a sensor) over a certain time frame. Obviously, the sensor signal of the process variable starts to significantly drift at around Sample #450 (exceeding a tolerance band shown as solid line), but the real failure became visible at the (surface of the) steel sheet product much later, namely at around Sample #600, which is a time gap of 150 min = 2 1/2 h in this case. Hence, in this case an adequate warning to operators at the early stage would have omitted a bad product and thus would have reduced the waste and costs significantly.

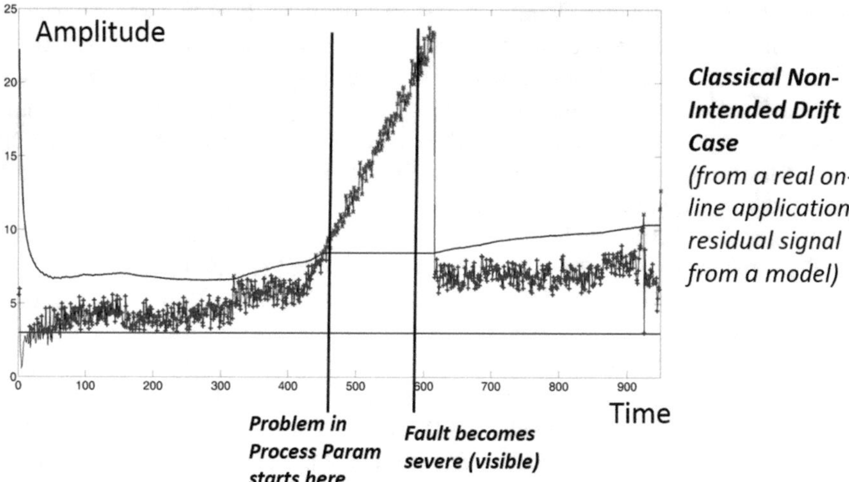

Fig. 4.1 Problem starts already much earlier in the process value than the real fault becomes visible and recognizable from outside—a real-world example from condition monitoring at rolling mills [105]

In this chapter, we therefore concentrate on the fully automated early detection of faults by means of supervising the process variables using two particular types of data-driven Sys-ID models, namely (1) causal relation networks and (2) distribution-based reference models. We thereby assume that process variables are recorded regularly with a certain frequency by multiple sensors, which can be typically located among different spatial sites in order to collect data from different positions within the system, and fuse them adequately [42]. We will address several well-known state-of-the-art techniques as well as recently introduced approaches for the two types of data-driven FD models (Sects. 4.3 and 4.4). The basic FD idea in all of them is that new incoming states (reflected in newly recorded samples) are checked in terms of their "deviation degree" to the fault-free reference models, and upon violation a certain tolerance level, a fault can be alarmed.

A particular challenge to the approaches will arise when significant dynamics in the system is present. Then, the reference models require to be (regularly) updated in order to include the latest trend(s) of the system. This omits high false positives (leading to wrong alarms) and false negatives (leading to misses) due to wrong predictions and quantifications on new samples. Techniques from the fields of incremental learning of data-driven models [28, 66] as well as evolving systems [12, 62, 77] can be used for establishing model updates with a reasonable stability-plasticity tradeoff and a good parameter convergence (as will be described in detail throughout Sect. 4.5). However, the major problem especially in an FD system is how to distinguish between intended changes (=regular system dynamics, drifts, etc.) [64, 67] and non-intended changes. The latter typically points to upcoming problems/faults in the system which should therefore not be integrated into the model updates (but better alarmed to system operators). In both cases, an exceed of a signal tolerance band as shown in Fig. 4.1 is highly likely. Thus, an automatic distinction without operator intervention is not reliably possible. At the end of this chapter (Sect. 4.6) we will discuss and present a concept how discrete event signals [102] can be used in combination with predictive mappings and change isolation (as a form of fault isolation [65]) for a fully automatic distinction. To our best knowledge, such a distinction and such a combined approach forming a kind of *hybrid dynamic FD system* has been not addressed in literature so far.

The organization of this chapter is thus arranged in the following way: first, we define in more detail the problem statement when conducting FD in dynamic multi-sensor networks (Sect. 4.2), then we describe various approaches for establishing causal relation networks and distribution-based reference models from multiple channels recorded within sensor networks and how fault and anomaly detection can be achieved based on such model types (Sects. 4.3 and 4.4). In Sect. 4.5, we extend the description to the case when significance dynamics is present in the system by demonstrating ways how the reference models can be updated and evolved over time. Finally, we address a concept for the automatic distinction between intended and non-intended changes (Sect. 4.6), which can be used (1) for more reliable fault detection and (2) for more stable model updates.

4.2 Problem Statement

Within the context of an FD approach relying on data-driven SysID models, typically a large amount of data is gathered from multi-sensor networks, where various sensors may contain different channels and where these may be also located among spatial (local) sites. A typical sensor network example is shown in Fig. 4.2, containing five sensors located at five different sites; each of the sensors contains one or more channels, as indicated by the number of strokes going into them. We assume that the channels are dynamically and continuously measured over time and recorded through the central database sink (bottom box). Sensors #2, #3, and #4 are connected, thus could exchange the channel data and perform a partial modeling on their channel views. The data is fused and stored at a central database server (bottom box) which is connected to all sensors (and included channel), thus is able to process all 16 channels in parallel together with the event signals coming from a process control system. Synchronization and fusion of the data is a sophisticated challenge, especially in Big Data applications, but not the focus of this chapter—we thus refer to [2, 63]. We assume that the data-driven modeling phase can be performed on the central database server.

In principle the modeling and also the whole FD can be conducted in each sensor individually, which is, for instance, established in so-called smart or intelligent sensors embedding some artificial intelligence on chip devices [18, 41]. However, this typically leads to a smaller partial view of the whole possible interrelations between channels in the system than when performing the modeling at the central

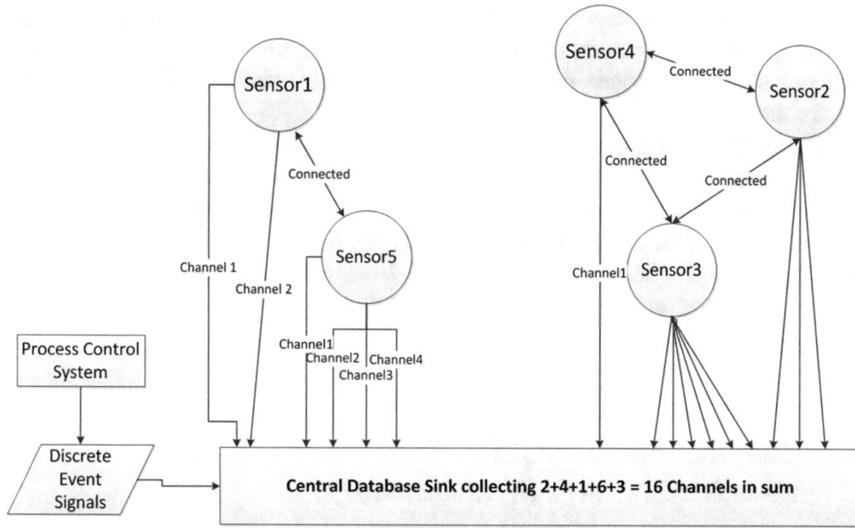

Fig. 4.2 Example of a multi-sensor network with five spatially distributed sensor sites, each sensor measuring one or more channels, which are all collected in a central data sink; discrete event signals from the process control system also may flow into the data sink

site. In particular, some important channels embedding high FD capabilities may be excluded, and thus an increase of misses of particular faults may be caused, especially of those ones which crystallize much better in higher-dimensional channel spaces than in more univariate views [26, 108]. On the other hand, the isolation of faults is much easier or even for free.

The aim of a data-driven modelling approach is to establish as many relations between all the M available channels as possible solely based on the data, and this is fully automatically by appropriate learning procedures. This ideally includes most of the channels in order to approach an "all-coverage" situation of so-called reference models, which decreases the likelihood of overseeing any potential faults. In order to address system dynamics properly and to handle any regular changes (such as system drifts [64] or non-stationary environmental conditions [101]), the established reference models should be updateable over time with high robustness. The whole model may also include some additional statistical help measures, which needs to be updated synchronously to the "real" model.

The selection of the measurements with which the model should be updated is indispensable once intended and non-intended changes occur, both typically leading to violations of the model f (see Sects. 4.3 and 4.4 below). This is because only in the former case the model should be updated to reflect new operations modes, etc., as otherwise faults/failures/problems in the system would be wrongfully learnt into the models. Therefore, we propose a new concept in Sect. 4.6 how to distinguish between intended changes and non-intended ones. It is based on a combination of change isolation and influence analysis between discrete event signals and process variables. This combination leads to a *hybrid dynamic system*, where the event signals (reflecting, e.g., crisp switches or parameter settings) play an essential role because they often induce intended changes in the system and in the continuous sensor measurement flow.

Ideally, fault and anomaly detection and the concepts for distinguishing intended and non-intended changes (drifts) are done on *instance-based level*, i.e. for each newly recorded sample instance it can be decided whether it stems from a faulty or abnormal system situation. This abandons any delays which would occur when using past data chucks, sliding windows, etc. In the following Sects. 4.3 and 4.4, the concentration lies on fault detection concepts based on data-driven reference models which are able to fully operate on *instance-based level*.

4.3 Residual-Based Fault Detection Based on Causal Relation Networks (Model-Based)

This approach builds upon linear or non-linear system identification from (measurement) data [94]. The basic intrinsic motivation is the intention to find causal relations and dependencies between certain system variables, basically among *channels* (also termed *process values*) which are characterizing the system states and are being permanently recorded over time by one or several sensors (see Fig. 4.2). The essential point is that this can be done in a fully *unsupervised manner*, which means that no quality information about the current process/system/product states needs to be available. Therefore,

1. The automatization capability of such an approach is expected to be very high.
2. Annotation effort in terms of labeling costs for historic data samples [87] can be completely avoided. Opposed to classification approaches for fault detection [34, 74], where ML classifiers are trained based on pre-labelled data, there is no necessity to collect data in advance and to divide it into faulty and non-faulty phases.
3. It is possible to detect problems at an early stage, as the inter-relations between variables are expected to be violated *before* the fault is recognized on the industrial plant or on the production items themselves. This could be, e.g., verified before in [92, 105, 108] for several industrial applications, and is especially true when mappings are achieved with a high predictive quality and a high confidence in their output predictions.

These circumstances trigger an applicability to a wide range of (on-line) quality control systems, where quality information and/or analytical models cannot be provided at all [117].

The basic motivation for using such an unsupervised, automated approach within the context of quality control is to represent various states in the system for the nominal, fault-free case (=the regular process case), and to check new on-line data

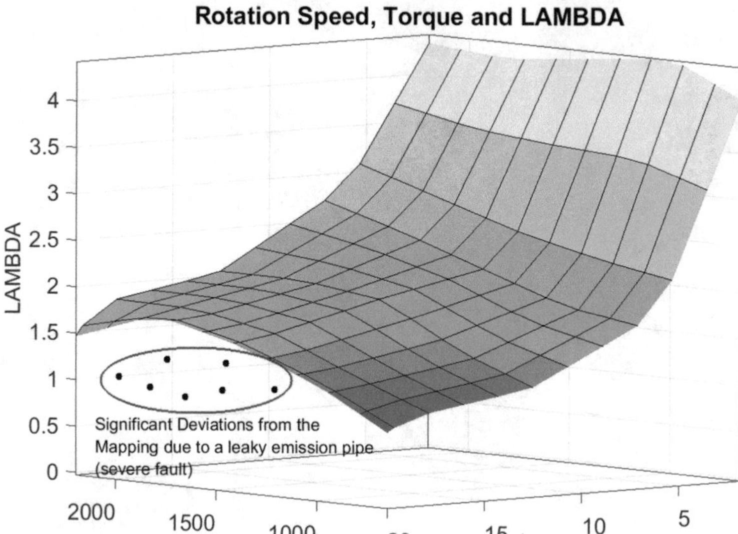

Fig. 4.3 Characteristics engine map (colored surface) as a model for representing the causal relation between torque, rotation speed and Lambda; data points, which have been caused by a fault in the system, significantly deviate from the map, indicating potential fault candidates

from the production how well they fit into the models [84], often represented by mappings and approximation surfaces. An example of a faulty situation is shown in Fig. 4.3 for a three-dimensional characteristic map established at an engine test bench [11]. New on-line data samples are shown as dotted points and significantly deviate from the real functional trend (surface of the engine map), thus can be correctly detected as faults when including an appropriate analysis of the residuals, see below (in this case shown in Fig. 4.3, these have been in fact caused by a leaky pipe).

Usually, there are several such causal relation mappings within a complete causal relation network, which then can be checked for violations and deviations of new samples altogether (for details see Sect. 4.3.2)—an example of such a network from a micro-fluidic chip production process is provided in Fig. 4.4, where the nodes represent various channels = process values in this case (from sensors). Dark (red) highlighted nodes occur as targets and white ones only as inputs in the partial causal relation mappings; the input/output relations are given by the arrows which automatically define all these mappings (e.g., target sink $S0$ can be represented by a causal mapping using $T11$, $T21$, $T144$, $P125$, $P126$, and $T143$ as input). Such causal relation networks may also be used by operators and experts for gaining insights into the system process and interpretability purposes—e.g., the network shown in Fig. 4.4 contains a clear clustered structure of dependencies between measurement channels, which, e.g., could be used for parameter fine-tuning of the chip production process.

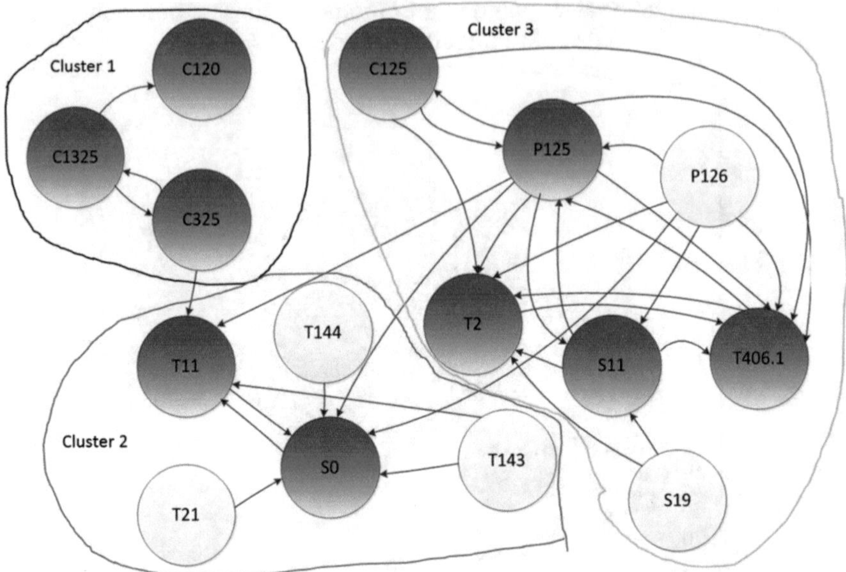

Fig. 4.4 A typical causal relation network from an industrial production system (of micro-fluidic chips); dark (red) nodes denote the target channels in *high quality relations*, all inflowing vertices indicate input channels to these

4.3.1 Establishment of Causal Relation Networks

One central goal of the whole modeling framework is to establish as many high-qualitative mappings between the channels as possible, in order to assure a nice coverage of the parameter space. This in turn decreases the likelihood to oversee any upcoming problems respectively potential faults in the system: the more (distributed) channels are involved, the higher the detection capabilities of the system. Separate validation data is important in order to judge the model quality in a fair manner and to avoid over-fitting effects [51].

A typical modeling framework for causal relation networks employing data-driven system identification is shown in Fig. 4.5, whose components will be explained in the itemization below. The big block into which the data set is fed is carried out for each channel, meaning that each channel is fixed as target (sink) node in the network once and a causal mapping is tried to be established based on the remaining channels—whereas we assume to have M channels available from the multi-sensor network (compare with Fig. 4.2). Ideally, it contains various system states and modes—how to best establish a "rich" data set in this direction is typically application-dependent.

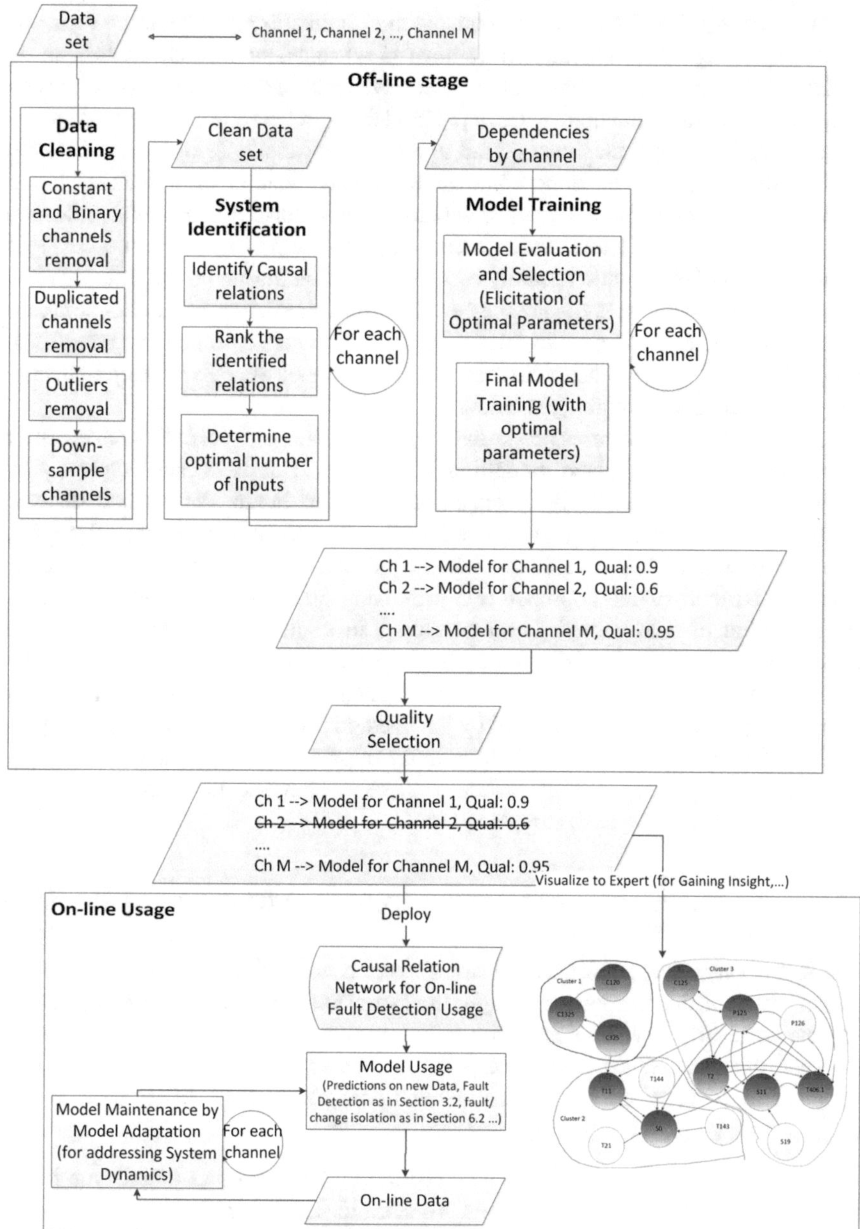

Fig. 4.5 Typical data-driven system identification framework aiming for an "all-coverage" approach to establish causal mappings between the M channels recorded from sensors (recording stored in a data set)

Data Cleaning The first two components are application-dependent and include issues regarding which channels to remove in advance, such as channels with no information content regarding faults in the systems, channels that appear pretty constant (they would not contribute in the modeling process), or channels which are simply duplicated and thus would lead to dummy correlations. Here, it is possible to include some specific expert knowledge about channel pre-selection, based on her/his expertise about knowing which channels have a substantial influence on the system process and which ones not. This helps to reduce the curse of dimensionality effect, which usually causes over-fitting during model training from data [51]. Alternatively, someone may even exploit here the spatial sensor structure as indicated in Fig. 4.2, e.g., to take only those channels as input for data-driven model building whose sensors lie close to the sensor of the current target channel for which a model is sought (see below).

Outliers removal concerns pre-processing on the sample level and is intrinsically important to assure a robust modelling, afterwards. This is because typically an outlier is characterized by an unexpected deviation in the data which denotes an erroneous situation (e.g., as arised during the data recordings). Typically, the distribution of the data streams from the channels is not known advance. Thus, a distribution-free approach is a promising option which loosens the strict interpretation of data densities and envelopes in a statistical sense. This can be accomplished by characterizing an outlier to have an untypical occurrence in the feature space in terms of a distance from the main trend in the training data [125]. The main trend can be characterized by the average distance av_{dist} between samples pairs $dist(\mathbf{x}_i, \mathbf{x}_j)$. Then, the idea is to record samples which have an extraordinary above-average number of high distances to other samples, i.e. a kind of outlier degree is calculated for each sample $\mathbf{x}_i$ by:

$$Deg_i = \frac{card(Cand_i)}{N-1} \qquad card(Cand_i) = |\{\mathbf{x}_j | i \neq j, dist(\mathbf{x}_i, \mathbf{x}_j) \geq av_{dist}\}|, \quad (4.1)$$

which measures the proportion of samples which have an above-average distance to sample $\mathbf{x}_i$. If this proportion is high, the sample can be recognized as an outlier sample as clearly deviating from the main average trend of the training data in-between distances. Once having extracted all outlier degrees $Deg_i, i = 1, \ldots, N$, a threshold is extracted based on the functional trend shape of its ordered list $Deg_{i*}, i* \in \{1, \ldots, N\}$ with $Deg_{i*} \geq Deg_{(i-1)*}$.

System Identification For each single channel (chosen once as target), the inter-relations with the remaining channels are elicited. Each iteration takes a channel $ch_1, \ldots, ch_m$ as the target of the relation to be identified (=the "cause") and searches through the remaining ones the best possible combination to "explain" the target (="the effect of the cause"). This can formally be expressed by:

$$ch_i(t) = f_i(ch_{i1}, \ldots, ch_{iL}) \qquad \forall i = 1, \ldots, M \quad (4.2)$$

Usually, a subset of the remaining channels is used, i.e. $L < M$ is aimed to be as small as possible, in order to reduce the curse of dimensionality effect [51]; possible methods for determining the optimal number of inputs L are variable ranking methods [48] or feature space transformation method extracting the most essential components such as partial least squares analysis coming with different variants [7], curvilinear component analysis [32], stochastic neighbor embedding [54], or factor analysis [15].

The choice of the function f in (4.2) depends strongly on the current data set and contained learning problem. For instance, if a significant non-linearity can be expected to be present in the system, linear models are not an adequate choice; or, if some interpretation requirements are given by the experts (so they want to check and inspect the models), black-box models such as neural networks or support vector machines should be avoided. Fuzzy systems with interpretability assurance techniques embedded [80] or symbolic formulas (as obtained through genetic programming (GP) [1] or fast function extraction (FFX) for compact symbolic formula representations [90]) are much better alternatives then. A promising way to establish f is given by interpreting the modeling problem as a regression problem. This has the advantageous side-effect to act as a kind of filter in cases when the channel signals are non-smooth, showing abrupt changes, which, however, do not indicate faults, but are basically due to the nature of the process. In order to underline this visually, an example of the regression process and its advantageous impact on fault detection is shown Fig. 4.6.

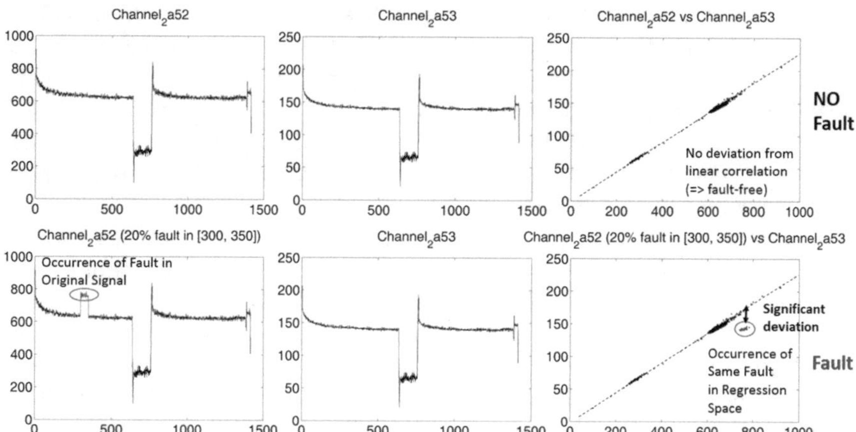

Fig. 4.6 Upper row: original correlated measurement channels with significant discontinuities (abrupt changes in the signals), which vanishes in the joint product-amplitude space when establishing a smooth regression model between them (right image); lower row: influence of a fault in the first channel (indicated by ellipsis): its pattern (between samples# 350 and 400) cannot be safely discriminated to an abnormal, but fault-free sudden change in the original signal (samples# 600–700), however in the product-amplitude space (right image) a clear deviation of these faulty samples to the regression trend as indicated by the line shows up

Equation (4.2) represents the classical static case leading to causal relation mappings. Often, dynamic causal relations are present in the system which needs the integration of the time components for establishing adequate prediction horizons between causes and effects. This leads to the functional definition:

$$ch_i(t) \leftarrow f_i(ch_{i1}(t-n_1), ch_{i1}(t-n_1-1), \ldots, ch_{i1}(t-n_2), \ldots, ch_{iL}(t-n_1), \ldots, \\ ch_{iL}(t-n_2)) \qquad \forall i = 1, \ldots, M \tag{4.3}$$

In [106], a system identification approach for extracting dynamic relations by employing channel space transformation is suggested, whose components (including orthogonal transformations and channel expansions due to lags) are nicely visualized in an overall workflow framework.

Checking the Model-Ability of the Process Once the inputs are identified for each target channel, the model-ability for each sub-problem can be justified with the so-called Gamma Test [38]; the aim is to check how the mean squared error (MSE) can be minimized, independently from the chosen model structure/architecture. As a side output of the Gamma test, the gradient of regression fit provides a useful information on the complexity of the process under study [104].

Model Training and Evaluation Once the inputs are identified for all target $ch_1, \ldots, ch_M$ (and the model-ability is checked), the models are trained, i.e. their structures and parameters are extracted from the training data, whose concrete algorithms depend on the chosen functional mapping structure f (see discussion above). The evaluation approaches often use an N-fold cross validation (CV) strategy [114] or repeated bootstrapping [36] to elicit the optimal parameter setting(s) over a pre-defined grid of learning parameters. This relies on model selection techniques, because for each setting an own model is constructed, which are based on the tradeoff between model accuracy and model complexity—models with lower complexities, but similar accuracies are usually preferred, as being less prone to over-fitting on new samples [27].

A straightforward technique of model selection is simply to use the minimal error as achieved from the statistical evaluation procedure (e.g., the minimal CV error) when conducting it with different parameter settings, typically arranged in a grid-like manner. This technique, however, does not take into account the complexity or sensitivity of models. Especially, lower complex models with a similar or even the same accuracy as higher complex ones should be tendentially preferred during model selection. An often applied possibility is to integrate also the model sensitivity regarding the training data selection, which can be measured by the variance of models errors obtained over different (CV) folds. In this way, a prominent selection in literature [19] is the model having lowest complexity but still being not significantly worse than the model with lowest error, i.e. its error lying within the one-sigma sensitivity band of the model with the lowest error. Figure 4.7

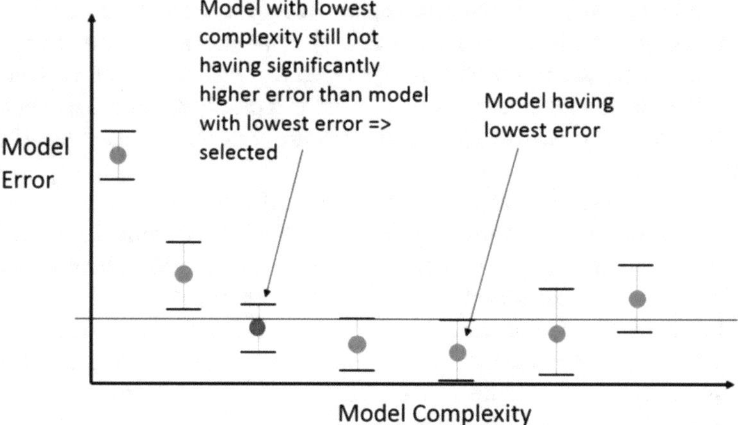

Fig. 4.7 Enhanced model selection example, also respecting the model sensitivity (indicated by vertical markers around the dots) when being trained from different data set folds/bootstraps

shows exemplarily how model selection is done in this way. Other techniques include the direct punishment of model complexity in the error calculation such as AIC (Akaike's information criterion), BIC (Bayesian information criterion) and extensions, see [27].

Model Assessment The purpose is to check the expected model quality on new unseen (on-line) data. Only the models with sufficient quality together with their uncertainty information (confidence intervals) are then further processed for fault detection (see subsequent section). This is because for low quality models typically many false alarms and misses can be expected due to the delivery of inaccurate predictions.

Sufficient quality Q_i can be understood as a combination of high expected predictive quality PQ_i and high output certainty $Cert_i$, which can be, for instance, combined in a weighted convex combination when normalizing the certainty outputs to [0, 1]:

$$Q_i = \alpha PQ_i + (1 - \alpha) \cdot Cert_i > threshold \tag{4.4}$$

If (4.4) holds, the model where channel ch_i appears as target can be used for residual analysis and is thus also established in the final causal relation network. PQ_i denotes the expected predictive quality of the model, which can be, for instance, measured during the training process by the K-fold cross-validated adjusted R^2 measure, which is given by:

$$CV\left(\bar{R}^2\right) = \frac{1}{K}\sum_{i=1}^{K}\left(1 - \frac{\frac{1}{N/K-p-1}\sum_{k=(i-1)(N/K)+1}^{i(N/K)}\left(y_k - \hat{y}_k\right)^2}{\frac{1}{N/K-1}\sum_{k=(i-1)(N/K)+1}^{i(N/K)}\left(y_k - \bar{y}\right)^2}\right) \tag{4.5}$$

with p the complexity of the model, typically comprised by the number of inputs (+ number of structural components in case of non-linear models), and $\bar{y}$ the mean value of the target y. Obviously, higher p's lead to a punishment of more complex models. $Cert_i$ denotes the output certainty and is typically measured by confidence intervals, error bars, sensitivity over the training folds/bootstraps, etc. [30], see subsequent section.

Another viewpoint of model assessment is the reliability of the established causal relation network and embedded mappings to be able to recognize fault and failure modes in the system. Due to its visual and eye-catching transparency, experts can easily check whether the structure is reliable and essential from "systems point of view," i.e. in a medical, physical, biological sense. This includes the check whether important variables are included at all or not, whether the relations are plausible, etc.: if so, we can assume some meaningfulness of the data with respect to potential failure modes, as these usually affect (one or more) important variables and violate valid relations among these. The detection of these violations will be studied in the subsequent section.

4.3.2 Advanced Residual Generation and Analysis

Assuming to have $m \leq M$ high quality mappings within the causal relation networks available. Now, when a new on-line sample $\mathbf{x} = (ch_1(t), \ldots, ch_M(t))$ at time instance t is recorded, which contains the values from all M channels. Then, first the corresponding subset of this value is elicited for each target channel ch_i, i.e. $\mathbf{x}_{\text{sub}} = (ch_{i1}(t), \ldots, ch_{iL}(t))$, which in the static case can then be sent into the model as defined in (4.2) for receiving a prediction of channel i, $\overline{ch}_i$. In case of a high model quality, it is expected that the prediction is close to the measured value, thus the residual

$$res_i(t) = \frac{ch_i(t) - \overline{ch}_i(t)}{cert_i} \tag{4.6}$$

is small. In the dynamic case, the past recordings for all channels $(ch_{i1}(t-1), \ldots, ch_{iL}(t-1), ch_{i1}(t-n_2), \ldots, ch_{iL}(t-n_2))$ need to be stored (for instance, in a ring-buffer containing n_2 samples). These then can be processed through (4.3) for obtaining $\overline{ch}_i(t)$, the prediction in the current time instance t from $n_2 - n_1$ past values of channels $ch_{i1}, \ldots, ch_{iL}$ which are used in (4.3), namely from $(ch_{i1}(t-n_1), \ldots, ch_{iL}(t-n_1), ch_{i1}(t-n_2), \ldots, ch_{iL}(t-n_2))$.

In (4.6), $cert_i$ denotes a normalization factor which is important in order to adjust the residual to the model output uncertainty. The higher the output uncertainty for a current prediction is, the less trustful the residual becomes. In such cases, a high (untypical) residual may thus not necessarily indicate an anomaly. A possible and likely false alarm is therefore prevented by normalization. Or in other words, for models where the predictions are highly uncertain, thus allowing a large bandwidth

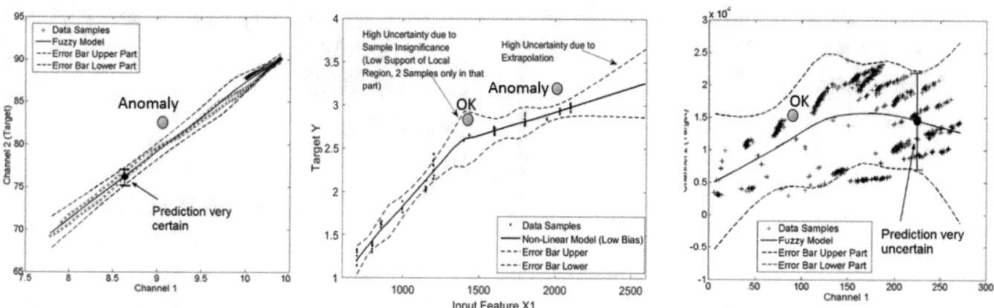

Fig. 4.8 Three cases of model uncertainty shown as dotted lines around the solid model trends; the grey big dot denotes a new incoming sample and has the same deviation to the predicted value (=vertical distance to the model in sold line) in all cases. However, it is obviously more likely to be an anomaly in case of narrower bands falling out of these (as in the left case and in the middle right case) than when occurring significantly inside these bands (as in the middle left and the right case)

of prediction values, samples significantly deviating from the mapping may not be assessed as anomalies. On the other hand, for models producing very certain predictions, tiny deviations from the real mapping trend can be already classified as anomalies with high likelihood. Figure 4.8 shows three examples of error bars (with different widths) for two-dimensional examples as dotted lines surrounding the solid lines.

The model uncertainty can be estimated during the training process and measured in various forms. The most convenient way is to estimate it through the expected error on separate validation data, as can be, for instance, approximated with the usage of the cross-validation error [114] or bootstrapping [36] methods, see Chapter 7 in [51] for a comparison on error estimation methods.

A more accurate representation of the model uncertainty can be expressed by so-called *local error bars* or *confidence intervals* [49]. These respect the change in uncertainty over local parts of the input space, so in some parts the model outputs may be more certain than in others—as is shown in the examples in Fig. 4.8. Such error bars are easy to compute for linear models (usually based on the Fisher information matrix reflecting internal parameter uncertainty) [94], but are more sophisticated in case of non-linear models, as requiring specific developments and derivations with respect to the chosen model architecture and internal model structure [81, 97].

The magnitude of the residuals defined in (4.6), however, may be not sufficient to discriminate between normal and anomal system behavior. It is more the trend of the residual signals over time which makes the appearance of current residuals atypical. Moreover, it is very difficult or even impossible to set a fixed, reliable threshold on the magnitude in advance (upon exceed of which, an alarm is triggered), which discriminates well between normal and real abnormal behaviors. Examples of residual signals and an inappropriate setting of the magnitude threshold (from a real-world application case) are provided in Fig. 4.9, the left case shows the detection

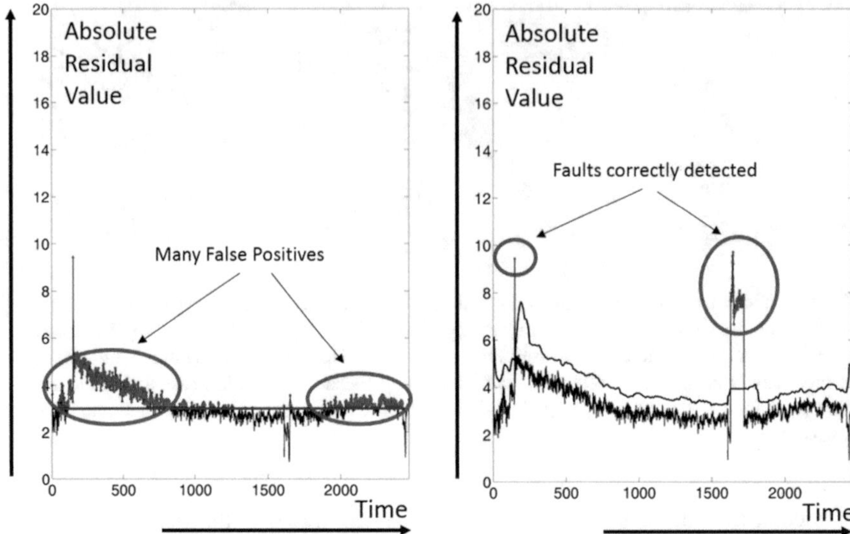

Fig. 4.9 Left: residual signal trend line and a fixed threshold of 3 (as tuned on some initial data set), leading to many false alarms indicated by small red dots; right: the same with the dynamic tolerance band, avoiding false alarms, but still being exceeded by (and thus being able to detect) the one sample outlier fault at the beginning and the significant jump towards the end

outcomes in case of a fixed threshold (a value of 3), which has been a good default value on first initial trial-and-error runs within the same application. Obviously, it leads to too many false alarms, as not being able to follow the natural (fault-free) dynamic trend of the residual signal.

Thus, in order to abandon a fixed threshold and to make the warning level dynamically adaptive, it is necessary to analyze the residual signals over time how they develop, evolve. In [106], a statistical-oriented tolerance band is introduced which checks the hypothesis whether a new residual falls out of the univariate Gaussian process model. Therefore, the tolerance band is defined as:

$$tolband_i(t) = \mu_i(t) + n \cdot \sigma_i(t) \tag{4.7}$$

with n a factor typically set to (inbetween) 2 or 3 in order to retrieve a 2-sigma or 3-sigma area covering 95.9–99.6% of the regular cases. Thus, the fixed magnitude threshold $]0, \infty[$ is shrinked to a relatively local threshold inbetween 2 or 3. The mean value $\mu_i(t)$ and the standard deviation $\sigma_i(t)$ over past values can be updated in an incremental, decremental way over consecutive sliding windows (thus, no re-estimation is necessary [106]):

$$\mu_i(t) = \frac{N_1 \mu_i(t-1) + res_i(t) - res_i(t-T)}{N_2} \tag{4.8}$$

$$\sigma_i(t) = \frac{1}{N_2}(N_1\sigma_i(N-1) + N_2\Delta\mu_i(t)^2 + (\mu_i(t) - res_i(k))^2 - N_2\Delta\mu_i(t-T)^2 - (\mu_i(t-T) - res_i(t-T))^2) \quad (4.9)$$

whereas $res_i(t-T) = 0, \mu_i(t-T) = 0, N_1 = t-1, N_2 = t$ for all $t < T$ and $N_1 = N_2 = T$ for $t \geq T$, and $\Delta\mu_i(t) = \mu_i(t) - \mu_i(t-1)$.

This then leads to the dynamically adaptive threshold as shown in the right image in Fig. 4.9: false alarms are completely avoided, but the tiny peak-based (outlier-type) fault at the beginning is still detected, so is the longer, severe abrupt fault towards the end of the stream. Please note that the update of the μ and σ and thus the tolerance threshold is only triggered when the signal stays within the tolerance band. In case of detected faults, μ and σ are kept constant such that they are not wrongly adjusted to include faulty situations.

4.3.3 Enhanced Residual Analysis

The strategy aforementioned in the previous section may get biased towards significant false alarms in case when there is significant high noise level in the data, which leads to a (mostly slight) exceed of the tolerance band, typically occurring in some regular intervals over time.

In order to overcome such unpleasant situation, on-line filtering techniques were suggested in [108] which can be applied within incremental adaptive learning scenarios from streams over time based on sliding windows concepts. This approach employs different types of filters, namely modified moving average filters (adaptable in single-pass manner), median average filters (adaptable due to re-estimation over sliding-windows), and Gaussian filters with a parametrizable number of bells (adaptable in single-pass manner). The results on four real-world application scenarios showed remarkable performance boosts of the ROCs (Receiver Operating Curves): these curves visualize the false positive rates (x-axis) versus the true positives rates (y-axis). The faster the curve rises and thus the larger the area under it becomes, the better the performance of the method is. A performance example of enhanced residual analysis employing filters based on a causal relation network for engine test benches, where each mapping is realized by a sparse fuzzy inference learning scheme (termed as *SparseFIS* [82]), is shown in Fig. 4.10. The ROC curves for filtered and non-filtered residuals (in two line styles and colors) are drawn. These curves are obtained when using different threshold levels in the tolerance band, whose values are steered by the factor n in (4.7) (here using $n = 1, 2, \ldots, 8$), and plotting the false positive (FP) and true positive (TP) rates obtained from an (on-line) test set (including known faults) as points (FP,TP) in the coordinate system. Higher AUCs (areas under the curve) point to better methods. Thus, obviously, the performance is significantly improved with the moving median filter.

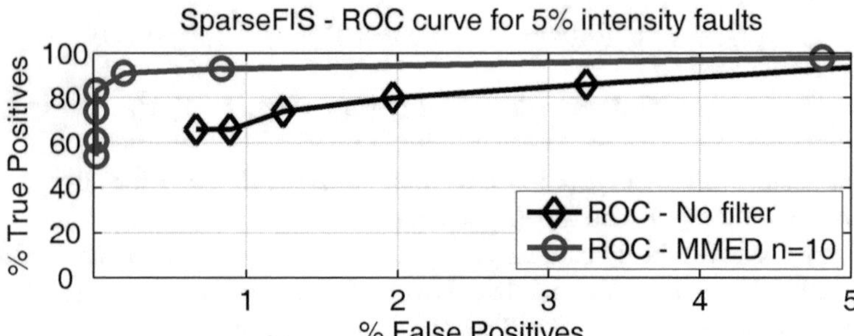

Fig. 4.10 Significant improvement in the fault detection performance when applying moving median filter (MMED) in the residual analysis obtained from fuzzy mappings within the causal relation networks (the mappings are achieved through SparseFIS method [82])

Another enhancement concerns a multi-variate analysis of residual signals extracted from more than one model, i.e. from several mappings within a causal relation network. Assuming to have m high quality models available, this would lead to an m-dimensional space of residual signals which can be imagined (plotted) sample-wise as stationary points in this space, as it is assumed that all channels are consecutively measured and collected in parallel at a central sink (Sect. 4.2, Fig. 4.2). An exemplary realization of such a residual signal space is shown in Fig. 4.11 for a two-dimensional case. In this joint feature space, atypical occurrences can then be characterized by significant deviations to density distributions reflecting the nominal, fault-free case [16]. Thereby, it is assumed that these distributions are estimated from the first on-line phase, i.e. from initial signal values retrieved from the causal relation models. Various methods exist in literature for characterizing and establishing the distribution of samples in the feature space together with the recognition of deviations and novelty content in new data chunks, see [26] and [88]. These will be discussed in more detail in Sect. 4.4 below.

An alternative how to treat residual signals in a univariate way is demonstrated in [81]. This is a more generic form of the approach discussed in Sect. 4.3.2, which does not assume that residuals are approximately normally distributed during the nominal, fault-free case. It therefore uses the concept of trend analysis in univariate signals with the usage of regression fits and its surrounding confidence bands. Once a new sample (or a bunch of new samples) falls out of this band, which characterizes the most recent trend of the signal, a problem/fault may be indicated. An example is provided in Fig. 4.12 below. The newly loaded point denotes an anomaly due to a significant jump of more than double of the most recent past values. Such trend fits can be incrementally sample-wise updated with the usage of recursive (weighted) least squares (R(W)LS) approach [75], which is a modified form of the Kalman filter [61] including the Kalman gain for achieving fast updates of inverse Hessian matrices (second order information to assure robust convergence). The inclusion of a smooth forgetting factor mechanism [113] is important in order to always represent the most recent local trend of the signal.

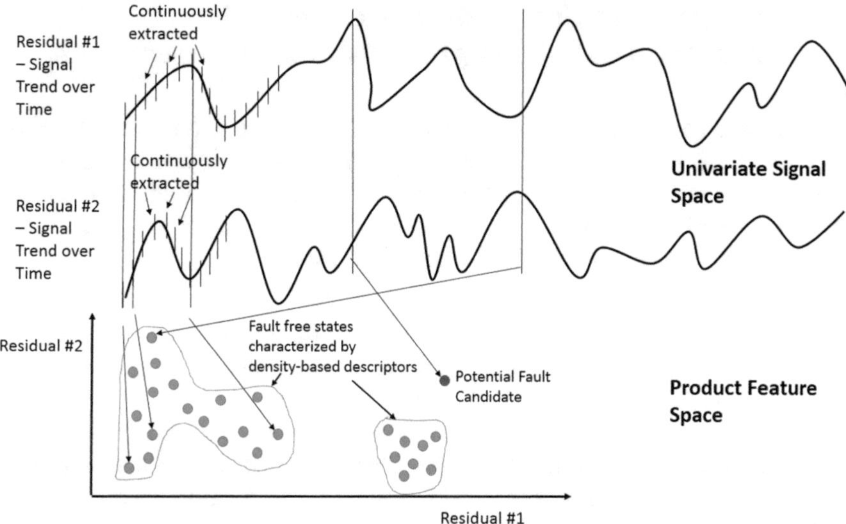

Fig. 4.11 Data is transformed from the time-series based residual signals (upper rows) to the two-dimensional product feature space (lower row); the fault-free states characterized by density-based descriptors are highlighted by surrounding hulls, a new sample (in red) seems to fall significantly outside these regions, thus marking a potential fault candidate

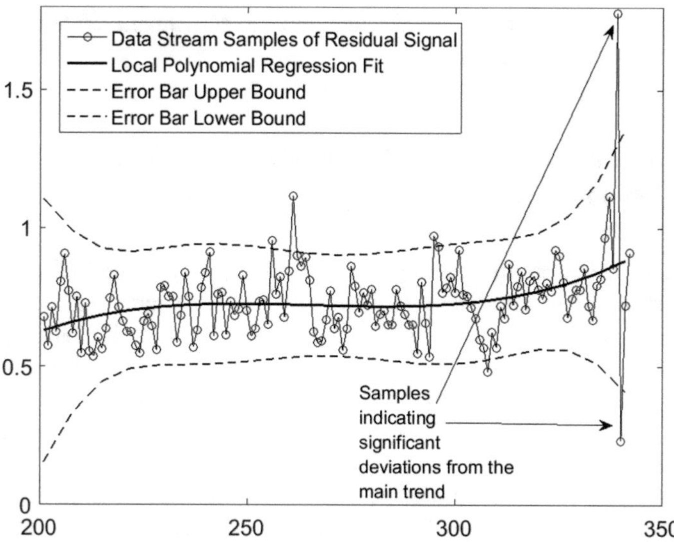

Fig. 4.12 A typical tend of a residual signal (its absolute value) over time with some slight, regular fluctuations from Sample 200 to 340, but suddenly jumping to an untypical values (indicated by arrow markers), the error bars marking the confidence region of the local trend are shown as dashed lines

4.4 Recognition of Untypical Occurrences in the Unsupervised Feature Space (Distribution-Based)

This is the second major research line in today's data-driven fault detection approaches and usually applied when no model(s) describing causal relations or dependencies within the system is(are) available. In this sense, residuals as defined in (4.6) cannot be explicitly calculated and thus their signals not analyzed/tracked over time with any of the techniques described in Sect. 4.3.

There may be several cases why causal relations are not available:

- No significant implicit relations or dependencies between system variables exist, such that models with sufficient quality can be extracted; i.e., the system is not "modellable."
- The extraction of (complex, highly non-linear) causal relation models takes significant amount of time, especially when the number of channels is very broad and/or a detailed optimization of parameters (within a statistical evaluation procedure) is too time intensive.
- The data has not been sufficiently recorded such that high-dimensional causal relation models cannot be reliably built without significant over-fitting.
- A fast model update with new on-line data should be carried out (ideally in real-time) to account for system dynamics, which may not be possible in case of huge causal relation networks.

In such occasions, the options discussed in the subsequent subsections can be considered as reliable alternatives.

4.4.1 Deviation Analysis in the (Global) Principal Component Space

The basic principle of this concept, as, e.g., used in [96] or [115] for on-line fault detection, is to elicit how well a new incoming data sample $\mathbf{x}$ fits into the *principal component space* of the data, represented by loadings $p_1, \ldots, p_M$ extracted from the original $N \times M$ training data matrix X (comprising N samples as rows and all M sensor channels as columns). The principal component technique treats all process values equally and searches for successive orthogonal directions in the multi-dimensional space which best explain the variance contained in the data [60]. The directions are linear combinations of the original features and thus represent rotations of the original main coordinate system. This results in an optimal representation of the (fault-free) data with respect to its variance elongations along certain directions in the high-dimensional feature space. It does not require that concrete dependencies, relations between variables are present in the system, so this restriction of causal relation networks is cancelled.

In order to reduce the dimensionality, usually a subset of k loadings (corresponding to the k largest eigenvalues of the covariance matrix $\Sigma = X^T X/(N-1)$ is used, where k is elicited through a criterion reflecting the "optimal capture of the variations in the data." Typically, the criterion is a saturation threshold (such as 95 or 98% of the data), or it can be elicited dynamically through the knee point in the accumulated variance explained curve [60]. The basic idea is now to use the reduced principal component space and to check how a new sample fits into it with respect to two criteria:

- Its deviation to the center of the data along the projected component space—termed as T^2 *statistics.*
- Its orthogonal deviation to the principal component space—termed as *Q-statistics.*

Both deviation concepts are visualized in Fig. 4.13 below, assuming a two-dimensional (reduced) principal component space (spanned by the two orthogonal axes) and a three-dimensional original data space: the sample (marked by a bold fat cross) appears in the three-dimensional space and its deviation degree to the principal component space (sometimes also called energy [96]) is measured by the Q-statistics:

$$Q = e * e^T \tag{4.10}$$

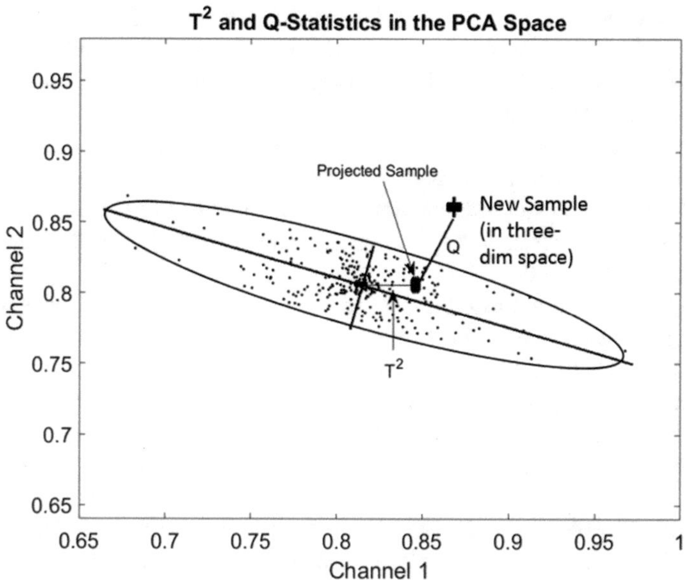

Fig. 4.13 Hotelling (T^2) and Q-statistics for a data set example; the new sample is shown as bold fat cross and appears in the three-dimensional space, the ellipsoid span the reduced PCA space in the two-dimensional projected plane

with $e = \mathbf{x} - \mathbf{x}P_kP_k^T$, with P_k containing the k first (main) loadings as column vectors. It can be matched against a statistical, dynamic threshold, directly extracted from the data according to a certain significance level, thus no tuning is required, see [24]. Its projected deviation to the center of the PCA space is calculated by the T^2 statistics:

$$T^2 = \sum_{i=1}^{k} \left(\frac{t_i}{\lambda_i} \right)^2 \tag{4.11}$$

with t_i the score of the ith component in the current data sample and λ_i the eigenvalue of the ith component. It can be fundamentally compared with the critical value of the Fisher-Snedecor distribution, see [24].

4.4.2 Analysis in the Partial Local Principal Component Space

The aforementioned approach is able to model the data variance behavior in a global way, as it extracts its covariance matrix for all samples over the whole input space, which then defines an ellipsoidal influence region of the data—as drawn in Fig. 4.13 by a dark, black solid line. However, in practical settings the data may be spread up in different local regions which are not necessarily connected and which do not follow the same trend; e.g., there could be various operation modes or transient conditions included—this is even the ideal case for retrieving a "rich" data set covering the whole usage space (if it is not the case, operation modes may dynamically show up and should be integrated on the fly, again requiring local structures, see subsequent Sect. 4.5).

An example is visualized in Fig. 4.14. There, the different local clouds are represented in different point markers and colors (green, blue, red). When being joined together, a global representation of the first two principal components would lead to the axes represented by the two orthogonal (dotted) straight lines and the big, inaccurate ellipsoid.

In such cases, a pre-clustering step would help to recognize the actual number of data clouds and to be able to perform a partial principal component analysis per cloud. Possible (prototype-based) clustering techniques to be applied in batch, off-line mode are Gustafson-Kessel clustering [47] or a generalized variant of vector quantization [79]. Both use the Mahalanobis distance measure to update the inverses of all the cluster covariance matrices within multiple iterations over the complete data set. Then again, as in the global case, any significant deviation from the recognized ellipsoids (one per cluster), as measured in terms of local Q-statistics or T^2-statistics, may point to anomalies or other upcoming problems.

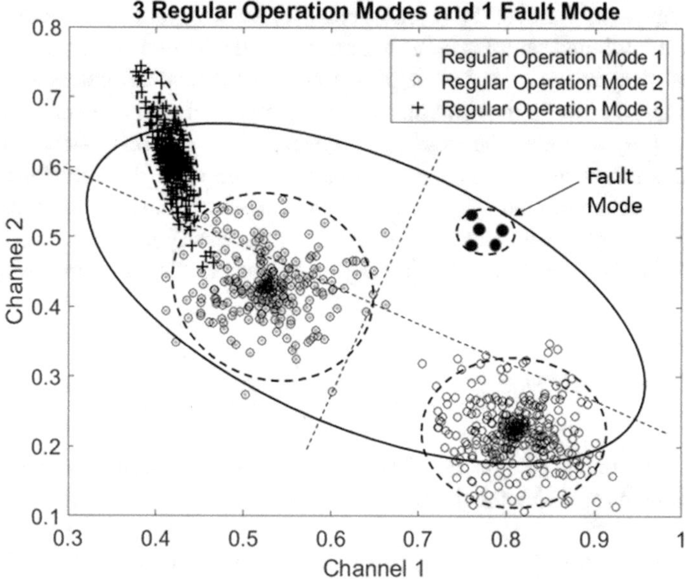

Fig. 4.14 Different local data clouds indicated by different point markers and representing different (normal, fault-free) system states, the big dark dots denote new samples belonging to an anomal behavior (not seen before), the big solid ellipsoid would be the shape obtained when extracting the (inexact) covariance matrix from all samples together—the fault mode could not be detected from the inexact global representation, but from the more accurate local one

4.4.3 Advanced Possibilities Based on Mixture Models and (Generalized) Data Clouds

A possible significant extension has been presented in [68], which expands the usage of classical principal component directions in fault detection to the usage of probabilistic principal components (PPCA) [116], which embed a local dimensionality reduction and thus acts on a smaller sample covariance matrix (with $p \times q$ entries = parameters instead of $p(p+1)/2$ parameters, with typically $q << p$). In this way, it suffers less from curse of dimensionality than using the full sample covariance matrix.

A further extension is provided in [93], where a mixture of probabilistic PCAs is employed for representing one single operation mode/condition (instead of one single PPCA). This provides more freedom for well representing a normal operation mode behavior appearing in sample clouds with arbitrary shapes. Interestingly, a mixture of PPCA is a special case of a class of Parsimonious Gaussian Mixture Models (PGMM), as demonstrated in [91].

Although a mixture of PCA can better represent operation modes inducing arbitrarily distributed samples in the feature space, it still embeds a(n) (partial local) ellipsoidal description with the help of a strict mathematical construct. In order to relax this strict definition and underlying formulation of mode components, the authors in [29] exploited the more loose concept of *data cloud*—according to the original definitions in [9]—to be used for fault detection and identification purposes. A cloud is just represented by its center μ_i, calculated through the mean of the neighboring points being close to the cloud, and a density measure. The latter can be calculated for each sample differently belonging to the cloud in the following way:

$$D_i(\mathbf{x}) = \frac{1}{1 + \|\mathbf{x} - \mu_i\| + \frac{1}{N_i}\sum_{k=1}^{N_i} \mathbf{x}_k - \|\mu_i\|^2} \tag{4.12}$$

With N_i the number of data samples seen so far and belonging to cloud i, $\mathbf{x}$ the current sample for which the density is computed and μ_i the mean value over all N_i samples.

In [9], it has been shown that this is equivalent to the classical definition of the Cauchy distribution, but can be exactly updated in recursive manner. Hence, a cloud is loosely represented by a sample density region without requiring a fixed, predefined shape, and thus is more generically applicable than covariance matrix estimates and partial local (P)PCAs. A comparison between conventional clustering-based and cloud-based partitioning of the feature space is shown in Fig. 4.15.

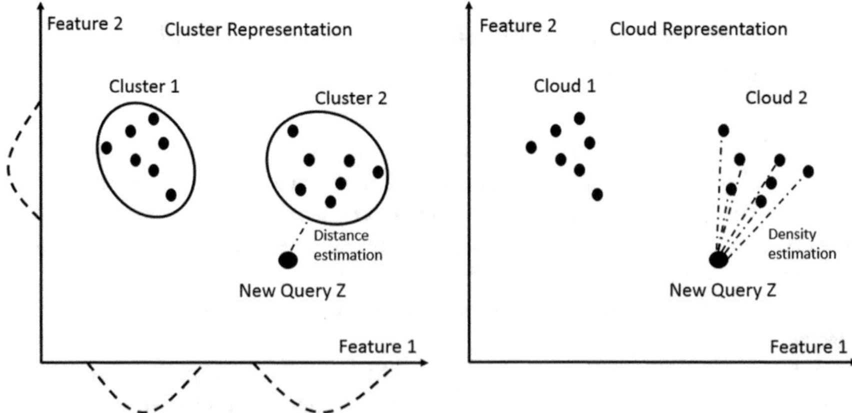

Fig. 4.15 Left: clusters representing operation modes defined in conventional ellipsoidal shapes (as in the previous approaches above), right: clouds are just loose representations of sample densities—a new sample Z is then associated to belong to the nearest cloud according to its density [9]

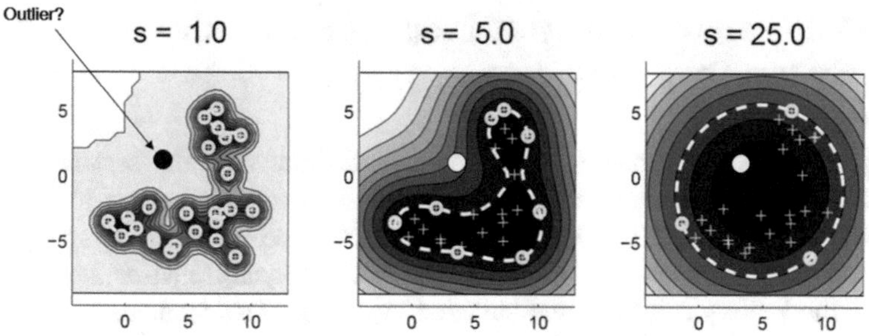

Fig. 4.16 Different parametrizations of the one-Class SVM classifier lead to different complexities of the hulls

4.4.4 One-Class Classifiers for Non-linear Hull Representations

A further generalization of the cloud-based approach can be achieved through a one-class classifier representation of the normal operation mode. The basic idea is to characterize fault-free data by a non-linear and, if necessary, a complex hull which internally embeds the data. The hull then defines the border of the decision between fault and non-fault. There are different possibilities for establishing complex linear boundaries, by using different types of one-Class classifiers. One of the most convenient options is the so-called one-Class SVM classifier [89], which is able to model data clouds with (a) very complex shapes and (b) appearing wide-spread over partial local regions (i.e., partial sample-embedding clouds which are not necessarily connected). The most essential parameter is the kernel width size which steers the degree of non-linearity and can be optimized through the reverse cross-validation approach [126], which is not requiring any supervised samples (with target values available). An example for different parameterizations of this width is provided in Fig. 4.16.

The structural elements are support vectors, which are sample points lying on or close to the border of the hull. In case when the kernel width is low ($s = 1$), a highly non-linear shape of the hull is induced and the sample marked by a dark dot is recognized as an outlier, which may represent a new state (upon confirmation by increased sample significance). In case when the width is high ($s = 25$), an almost spherical characterization of the fault-free training samples is induced, leading to a more generous surrounding of the sample cloud. One-class SVMs have been successfully applied for anomaly and fault detection in [26, 40, 88].

A final, remarkable note goes to the applicability range of all unsupervised approaches: they can be applied to original sample signals (raw process values), features extracted from original sample signals as well as to residual signals in the same way. It is basically always possible to transfer the multi-variate input signals into a multi-dimensional feature (sample) space as shown in Fig. 4.11.

4.5 Self-adaptive Reference Models for Handling System Dynamics

The aforementioned modeling approaches work fine when the system dynamics is low and the environmental influences are stationary or known to have little effects on the process. However, in case of systems/processes embedding significant dynamics over time as well as in case of non-stationary environments, off-line trained models typically become outdated over time. This leads to a successively increasing deterioration of their predictive performance [101], which in turn also decreases their fault detection capabilities. For instance, a big error in predictions of a target in a causal relation network also causes a big error in the residuals according to (4.6). This means that the residuals are not representative and the likelihood of frequent false alarms becomes pretty high. Furthermore, upcoming new process states, operation modes [69] or system behaviors or simply changes in the process settings (leading to so-called drifts) [64] which were not included in the historical data are often not adequately represented by the model. This typically leads to severe extrapolation cases during model inference processing, which are known to be risky and to lead to erroneous model behavior [77, Chapter 4].

In order to resolve such system dynamics, it is necessary to adapt and mostly also to evolve the models on the fly, ideally in single-pass on-line manner [22] based on new incoming recordings from the sensor(s) (networks). *Single-pass* means that a data stream is sequentially passed through the model update engine and older samples do not need to be re-entried [44]: once a sample or a whole chunk of samples is processed through the update engine, it is discarded forever. This makes it very efficient for fast on-line (or even real-time) training purposes [22]. *On-line* means that the update engine is able to work fully autonomously without any user intervention or without any (time-intensive) batch re-design phases requiring model and/or process experts [17, 50]. This makes it very efficient in terms of efforts and man-power for servicing and maintenance.

Basically there can be three concepts when updating reference models for fault detection:

- Incremental adaptation of parameters [109]: this concerns the update of the model parameters estimated during the off-line stage. This accounts for dynamic adjustment of the models to process changes and for increasing parameter significance (refinement). Often, this is conducted in recursive manner, which induces an exact convergence to the (hypothetical) batch off-line solution, which would have been obtained when being fed with all the data seen so far at once. Thereby, an objective function with a clear optimization goal can be formulated which is solved in incremental, single-pass manner [55, 118].
- Changes in the model structure by evolving and pruning model components (neurons, fuzzy rules, support vectors, etc.) with incremental learning methods on the fly [81]: this accounts for knowledge expansion and contraction in order to be able to integrate new modes and states and to delete obsolete ones. It is important for assuring compactness of the models and for preventing an ever-growing

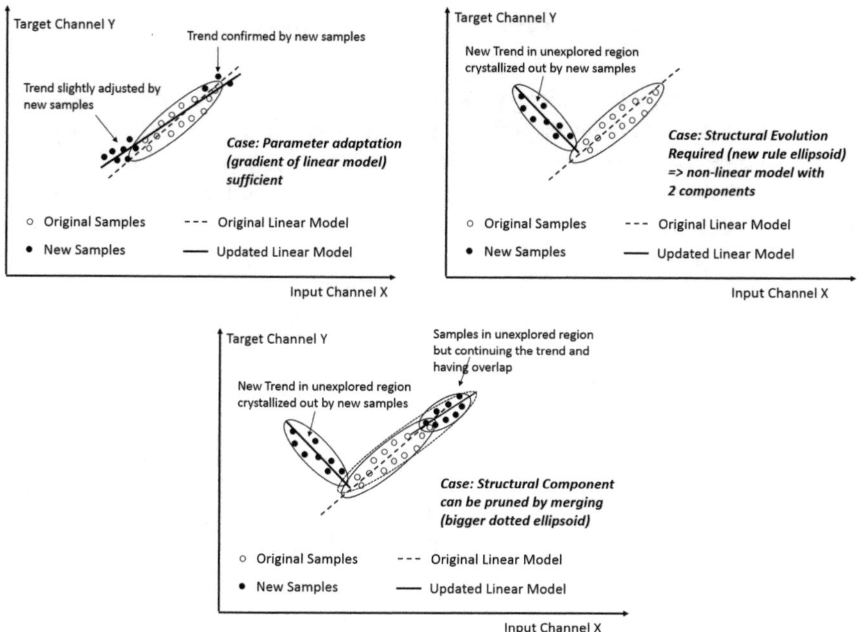

Fig. 4.17 Upper left: case where parameter adaptation (gradient of linear model) is sufficient to follow the slightly updated trend in the lower region; upper right: case where component evolution is required in order to sufficiently resolve the arising non-linear trend due to new samples → non-linear model with two components; lower: case where structural pruning (merging) can be performed to reduce complexity of the model (as there are two clusters alongside representing the same functional trend)

structure, which would cause significant computational burden. Techniques from the field of evolving (intelligent) systems are appropriate choices to address these tasks [12, 62, 77].

- Changes in the input structure of the models by updating the influence of input features in the original or transformed feature space on the fly [71]; this accounts for significant changes in the relationships contained in the systems (between variables/channels)—e.g., when introducing new educts [23] or product types [59] which often change the constellation in the production process. In the original space, it is addressed by on-line feature selection [119] and weighting mechanisms [85]. The latter acts in a continuous, smooth manner without explicit structural changes. This induces higher stability than a crisp removal or adjoinment of features. In the transformed space, an incremental update of the principal components in unsupervised (PCA) [120] and supervised form (PLS) [123] may be conducted.

Figure 4.17 demonstrates a two-dimensional regression modeling example (input channel X is regressed on Y) showing in which cases only parameters are required to be updated (left), structural components need to be evolved (middle), and structural components should be merged (right).

4.5.1 Self-adaptive Causal Relation Networks

In the realm of causal relation networks, the update may take place within the particular partial mappings as well as in changes of the whole network structure by the addition or deletion of arrows between the nodes (compare with Fig. 4.4).

The partial mappings typically appear in form of regression models, i.e. having continuous targets and some input channels (original or transformed ones). Thus, in case of *linear parameters*, the recursive update is very often accomplished with the recursive (weighted) least squares (R(W)LS) estimator due to favorable convergence properties, see [77, Chapter 2]. It is based on the (weighted) least squares functional and defined as:

$$\hat{\mathbf{w}}(k+1) = \hat{\mathbf{w}}(k) + \gamma(k)(y(k+1) - \mathbf{r}^T(k+1)\hat{\mathbf{w}}(k)) \tag{4.13}$$

$$\gamma(k) = \frac{P(k)\mathbf{r}(k+1)}{\frac{\Lambda}{\Psi(\mathbf{x}(k+1))} + \mathbf{r}^T(k+1)P(k)\mathbf{r}(k+1)} \tag{4.14}$$

$$P(k+1) = \frac{1}{\Lambda}(I - \gamma(k)\mathbf{r}^T(k+1))P(k) \tag{4.15}$$

with $\mathbf{w}$ the linear parameter vector, $P(k) = (R(k)^T Q(k) R(k) + \lambda I)^{-1}$ the (ridge regularized) inverse weighted Hessian matrix (with λ optimized during batch learning stage) and $\mathbf{r}(k+1) = [1\ x_1(k+1)\ x_2(k+1)\ \ldots\ x_p(k+1)]^T$ the regressor values of the $k+1$th data sample containing p input variables. Λ denotes a forgetting factor, which steers the degree of exponential forgetting (=outweighing older samples more and more over time), with default value equal to 1 (no forgetting).

Forgetting may be important to increase flexibility in a very dynamic process, thus to represent the latest trend more accurately; however, a too fast down-weighting of older samples may lead to catastrophic forgetting and thus to significant deterioration of model performance [4]. In this sense, an appropriate setting of the forgetting factor is a challenging issue. Self-adaptive approaches based on tracked change intensity levels have been proposed and discussed in [110, 113]. The symbol Ψ denotes a weight which can be given individually to the actual sample $\mathbf{x}(k+1)$ (apart from a regular pre-defined forgetting weight). Thus, different samples may receive different "importance" degrees during the update process. Whenever a sample receives a low weight, the Kalman gain $\gamma(k)$ in (4.14) becomes a value close to 0 and the update of P and $\mathbf{w}$ is marginal. This update mechanism has the great advantage to converge in each iteration step, as the optimization function is a (convex) parabola and the update step is equivalent to the Gauss-Newton iteration step [13]. On the other hand, a disadvantage is that it may over-fit in case of complex model structures, as it does not integrate any punishment or regularization term. Therefore, extensions of RLS such as generalized RLS (GRLS) [121], sparse RLS [14], or kernel-based RLS with extensions [127] may be more robust choices in case of higher-dimensional inputs.

For the *non-linear parameters* (defining sizes, orientations and shapes of structural components in non-linear models), advanced incremental optimization techniques such as the incremental Levenberg-Marquardt or the incremental Gauss-Newton algorithm can be applied, see [95] for a collection of possible algorithms. More complex incremental optimization algorithms are needed when extending the conventional least-squares problems with punishment terms regarding model complexity and feature space coverage [86] in order to reduce the risk of over-fitting. A collections of possible algorithms are presented in [12, 77] and [95].

Often, the predictive mappings are non-linear machine learning or soft computing (regression) models containing several so-called *structural components* ((fuzzy) rules, neurons, terms in symbolic formulas, etc.), which model different parts of the system. This gives them the opportunity (1) to model any degree of non-linearity contained in the system with sufficient accuracy (universal approximation theorem [124]) and (2) to expand and shrink on the fly based on new data, usually with little "disturbance" of the (convergence of the) already learnt components (and the implicitly contained parameters) [76]. *Component evolution* for expanding the predictive mappings can be achieved through the usage of concepts and methodologies from the field of evolving intelligent systems [12, 62, 77]. For instance, adding new fuzzy rules or neurons in (neuro-)fuzzy systems or neural networks is typically accomplished by checking the novelty content information of new data samples: if they indicate a new (type of) knowledge, as, e.g., verified by a high dissimilarity to already learnt structures [70], they are potential candidates for representing new operation modes/system states. Thus, the complexity and hence the non-linearity degree of the model have to be increased. However, the new knowledge could also stem from upcoming problems or failure trends in the system, which should be not integrated into the models. A proper distinction in a fully automatic way is still an unresolved problem and will be the main focus of Sect. 4.6 below.

Component pruning methods for mapping shrinkage check for (1) redundant information contained in the model components [80, 83]—e.g., consider two significantly overlapping Gaussian distributions which could be merged without loss of significant information—and (2) for parts of the model not visited for a longer time frame, thus becoming obsolete [8, 56] $\rightarrow$ pruning beneficial.

Regarding structural changes on network level (adding, pruning of arrows), this can be seen as more or less equivalent to incremental subspace learning [71]: inputs (features) can change their importance levels over time and thus may be discarded or even be reactivated in different stages of the stream learning process. The latter issue has been only loosely handled so far. There have been first attempts in the direction of feature weighing which assign the most important features higher weights than the lower important ones [85, 99]. The weights are then used during incremental model update steps to reduce the curse of dimensionality effect and thus to improve precision of the model [85], because features with low weights can be ignored during the learning process. However, the problem is that redundant features may receive similar high weights despite not contributing something additional for explaining the (causal relation to the) target. Recently, an approach in [6] partially resolves this problematic by a local feature selection per local region individually.

Input features with low weights can be seen as not represented in the network, thus the arrows to and from them can be discarded. Weights to the edges can be assigned to tell users how much influence from an input to a target is expected.

4.5.2 Self-adaptive Distribution-Based Reference Models

This form of reference models differs from causal relation networks in a way that they are established in fully unsupervised manner. Thus, there is no possibility to update linear and non-linear parameters within the context of classical recursive adaptation approaches relying on optimization functions and step-ahead prediction errors on new samples.

We now discuss incremental updating possibilities for the various reference model types discussed in Sect. 4.4.

- **For the native PCA-based approach described in Sect. 4.4.1:** the principal component space can be either updated (1) by adaptation of the loading vectors (PCA directions) one-by-one, conducting iteratively the deflation of new points over the already updated components—as done in [120]. This leads to successive step-wise (small) rotations of the feature axes; however, it may have some problems when a significant shift in the mean of the input variables takes place over time, or (2) by updating the covariance matrix Σ, which can be established in a recursive, more stable manner when including a rank-1 modification term [78]:

$$\Sigma(new) = \frac{N}{N+1}\Sigma(old) + \frac{N}{(N+1)^2}(\bar{x}(N) - \mathbf{x}_{N+1})^T(\bar{x}(N) - \mathbf{x}_{N+1}) + \Delta(N+1) \tag{4.16}$$

  where N denotes the number of samples seen so far, $\mathbf{x}_{N+1}$ the new sample and $\bar{x}(N)$ the mean over all input features up to sample N; $\Delta_{i,j}(N+1) = (\bar{x}_i(N+1) - \bar{x}_i(N))(\bar{x}_j(N+1) - \bar{x}_j(N))$, i.e. the degree of mean shifts from sample N to $N+1$ in the variables i and j multiplied with each other ($\rightarrow$ rank-1 modification). On the other hand, this requires the eigen-decomposition of the covariance matrix in each update step to obtain the updated loadings, which may be time-intensive.
- **For the partial local principal component space described in Sect. 4.4.2**, which splits regular modes already into various principal components (due to the nature of the data clouds, see Fig. 4.14), the model update concept relies on the tradeoff between adapting already available components (with the aforementioned techniques) and the evolution of new ones, based on significant novelty content of new samples. In [69], a remarkable solution is suggested, where the authors apply an adaptive fuzzy classifier which can identify new states on the fly based on a cluster evolution criterion (founded in the evolving participatory learning concept (ePL) [73]). This is directly associated with a rule evolution criterion: thus, if a new cluster is created, it automatically results in

a new rule in the fuzzy classifier. Upon the violation of the evolution criterion, a feedback from the operator is requested whenever the membership value to the closest rule is low enough, which means that the sample is significantly falling out of already available rules. Otherwise, the consequent class of the most active rule is used as mode for the new rule. The workflow of this approach is visualized in Fig. 4.18 below. This feedback comprises the type of the mode (=regular operation mode or fault class). In this sense, it works only in a *semi-supervised, half-automatic* way and not in a fully unsupervised, automatic way. A fully automatic approach is still an open problem in the evolving systems community—we will propose a solution by the usage of discrete event systems, see Sect. 4.6.

- **For the generalized data clouds described in Sect. 4.4.3**, which provides a generalized viewpoint on data distributions, the update of the sample density information to the clouds can be achieved by using the recursive density estimator concepts developed in [8, 10] and recently extended in [100]. In [29], the authors use these concepts for conducting fault and anomaly detection in on-line

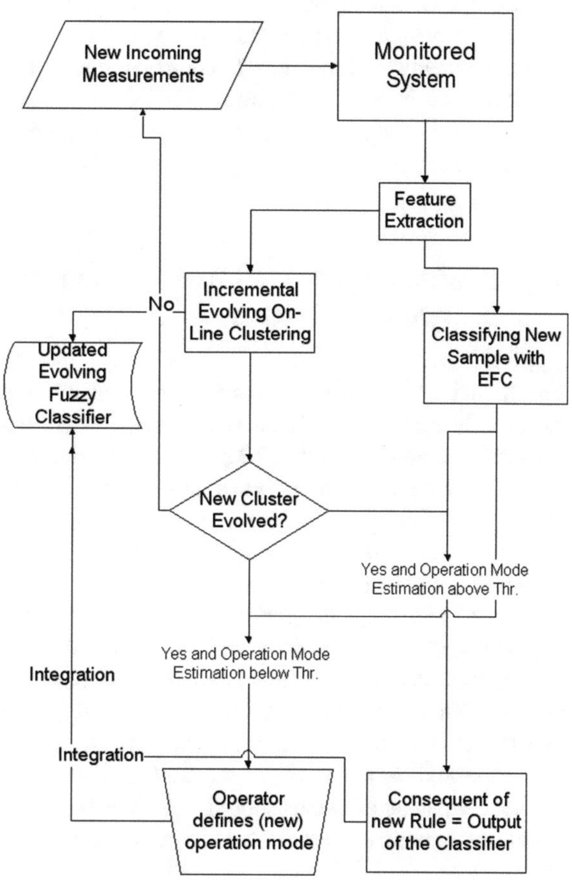

Fig. 4.18 Workflow for updating system mode classifier based on new incoming samples denoting old and new system states—according to the approach used in [69]

single-pass manner, also with the possibility to add new clouds on the fly and on demand. If significant consecutive new samples are falling outside two times the zone of influence of a cloud (and thus are treated as outliers) and if the density of the outliers (stored in the vector of past outliers) is higher than the average density (recursively calculated) of all existing clouds, then a new cloud is evolved. Each (evolved) cloud is then inspected as an operation mode, which can be a normal mode or an anomal (fault) mode. A filter stage based on the mean density over all data samples (when this falls below a threshold) is applied in order to distinguish between anomal and normal modes in advance. The evolved cloud is then always automatically associated with a new fault mode indexed by the next integer. In this sense, no operator intervention is needed to decide whether a new state is anomal or normal and whether it is a new anomal state. However, this approach requests two specific (application-based) features which are sent into the evolving cloud modeling approach for distinguishing new regular and fault modes.

- For the more complex non-linear representations, the update of their shapes and sizes becomes more complicated. In case of one-class SVMs (the most widely used approach), support vectors can be updated by the incremental techniques as suggested in past publications [72, 111]. Some embed the possibility to automatically evolve support vectors on the fly to significantly expand hulls or evolve new hulls in other parts of the feature space. But, also here the problem of an automatic appropriate distinction between new regular and fault-case modes remains open.

4.6 Distinction Between Intended and Non-intended Changes in Dynamic Systems

The model update mechanisms as discussed throughout the preliminary section are blindly using all the data stream samples recorded and available during the (dynamic) on-line process. They try to incrementally optimize parameters and evolve structures according to best principles and concepts in terms of maximizing model precision based on all these samples. Now, it may happen that changes or drifts in the system appear which point to potential upcoming problems (faults) in the system, whereas in other cases such changes are natural according to variations in the process settings or environmental influences. We call the former *non-intended changes* or *non-intended drifts* and the latter *intended changes* or *intended drifts*.

It is now an essential task to distinguish between intended and non-intended changes. This is because the former should be taken into account during model updates to assure sufficient flexibility of the model and especially to further guarantee high model performance also after the drift has occurred (see previous section), whereas the latter should not be respected and integrated into the model update as it typically would lead to a deterioration of model performance. This distinction is even more precarious in dynamic fault detection (and diagnosis) systems, as

1. the likelihood that both, dynamic changes and faults, appear unpredictably and variantly are usually much higher than in other environments and
2. an integration of fault modes would let the model forget how to detect such types of faults and thus would lead to a deterioration in fault detection performance regarding a correct detection of these types of faults which have been wrongly learnt into the models (no deviation would be visible any more, because the fault mode is part of the (wrongly) updated reference model).

Ideally, the distinction should be also performed fully autonomously and automatically without costly operator interventions. This abandons the usage of semi-supervised adaptive and evolving learning system as, e.g., shown in Fig. 4.18 requiring operator's feedback for each newly detected mode. Furthermore, the distinction should be as generically applicable as possible and not be restricted to particular applications. This abandons the usage of *expert-based pattern-like descriptions* for fault and non-fault modes, e.g., in form of *fault signatures* [5, 46] indicating the expected violation trend and/or intensity levels of residuals. Someone may consider that changes induced by faults are affected with higher intensity levels than changes induced by regular dynamics or vice versa. For instance, when inspecting Fig. 4.1 (in Sect. 4.1), someone may have the intuition to see the small drift starting around Sample #300 as a regular dynamical fluctuation (=intended change) in the system, and the much heavier drift starting after Sample #400 as a real fault case—which is true in this particular situation, and which would be even correctly handled by our residual signal analysis approach. However, this can be not generalized with sufficient accuracy, as non-intended changes may be small (e.g., 5% fault levels) but intended ones may be big in other cases.

4.6.1 Discrete Event Signals Indicating (Intended) Process Changes

The basic idea of our approach is to exploit the presence of discrete event signals in on-line processes in order to distinguish between intended and non-intended changes. A discrete event signal is a dynamic time-based signal with discrete states, the transitions of which are triggered by events. Events can not only follow physical laws, but are often induced by man-made rules or regulations from outside (e.g., an environmental change or a switch in process settings). For instance, in most conventional production systems of the today's Factories of the Future, the parameters settings or the charge types for producing particular products are changed during the production process [39, 103] from one particular setting of values to another one. A particular example of a discrete event signal from a rolling mill production process embedding changes in parameter settings is shown in Fig. 4.19, including four events leading to five states.

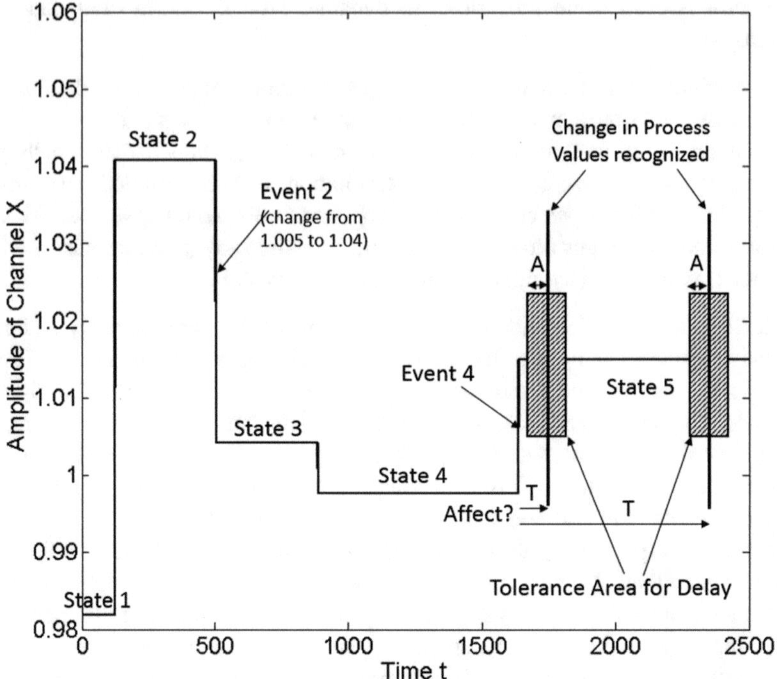

Fig. 4.19 A typical discrete event signal containing five process states induced by four events (=crisp changes) over time frame of 2500 samples (=24 h of process uptime in this case); two recognized changes in the process are indicated by vertical markers and surrounded by tolerance areas which represent the (known, elicited) delay between the happening of the events and their affect onto the process

Typically such changes affect the system in a way that relations and interdependencies modeled before are not valid or not partially valid any longer. Thus, the on-line samples recorded during or after such changes appear as significant deviations (anomalies) to the models.

4.6.2 Identification of Intended Changes by Hybridization of Discrete Events and Change Isolation

The basic idea is now

1. to check whether changing events in these discrete event signals happened before or during the time span when potential fault candidates have been elicited (by any of the methods described in Sects. 4.3 and 4.4), subject to an influencing level of delay; and

2. if so, to check whether the fault candidates (as significant violations of the reference models) occurred in channels (=influencing process variables) which can be actually affected by the "before-happening" changing events at all.

If both are true, the change is likely to be an intended one (thus the fault candidate is not confirmed as a real fault) and the model should be autonomously updated in order to expand its knowledge to new states and to avoid further false alarms. If (1) and (2) above are not the case, the change is likely to be a non-intended one and thus the likelihood of a fault is pretty high. Therefore, the operator should be alarmed, who can then give a final feedback. This leads to a reduction of false positives = false information to operators in case of intended changes happened (and no real faults).

4.6.2.1 Delay Elicitation for Process Changes

Regarding the first issue, in several cases there may be expert knowledge available knowing about the (expected) time delay between a discrete event (e.g., process setting change) and its actual effect on the process (and included variables/values). In other cases, the delay can be elicited within a design of experiment (DoE) stage [43], where various event settings (guided by the DoE) are tested and the corresponding process variables recorded over time. Then, it is a matter of establishing a predictive mapping between events and process variables in connection with a search procedure for the ideal prediction horizon in these mappings (optimizing the tradeoff between the length of the horizon and the accuracy of the mapping). The latter can then be associated with the intrinsic delay between events and process changes. The found (or known) delay L can then be matched against the time duration T which passed by between the happening of the discrete event and the actual process change detected as fault candidate by the reference model(s). A certain tolerance area A might be useful, as the detection of the changes may be also affected by some delays (depending on the accuracy of the method). If the time duration falls within the tolerance area, i.e. if

$$L \in [T - A, T + A] \tag{4.17}$$

the change is likely to be an intended one, as the duration between the discrete event and the detected process change is similar to the known delay. The level of this likelihood is verified by a change isolation component (subsequent section), which checks whether the detected process change is located in at least one process channel (contained in any of the reference models) which can be affected by a discrete event at all.

4.6.2.2 Location of Changes with Change Isolation

It is a major task to find out which channels are most likely affected by the change, or in other words where in the sensor network is the change located which leads to the violation of the reference model(s). The concept of *fault isolation* [35, 46, 107] is a promising and widely used methodology for such a task, where the violated reference models are used as basis for the search. In the context when we see each model violation as a change in the process and not necessarily as a fault, we can use this directly for *change isolation*.

In case of the residual-based approach with the usage of causal relation networks (Sect. 4.3), the change/fault isolation problematic can be ideally investigated by exploiting the multi-model nature, because various channels may appear or not appear in more than one model (as input or target) and this with different influence levels. If, for instance, a channel appears in many different *violated models* (=models where the extracted residual signal exceeds the tolerance band), it is more likely that the change happened in this channel than in another one appearing only in one violated model. However, this also may depend on the influence of this channel onto the target of the violated model(s), on the quality of the violated model(s) and on the intensity of violation(s), i.e., whether the tolerance band is slightly or significantly exceeded. According to the consideration in [107], we suggest the following isolation indicators:

- model gradients (along each input variable) in the current sample as a measure for the degree of variable influence,
- model quality, measured by the coefficient of determination R^2 over a statistical evaluation procedure (e.g., cross-validation), and
- the degree of model violation measured by the distance between the residuals and the dynamic tolerance band surrounding them (compare with Fig. 4.9).

These are accumulated over all violated models $f_1, \ldots, f_v$. Formally, the relative likelihood that the detected change appears in channel x_j based solely on a single (violated) model f_i, where channel x_j appears as input or target variable, is given by

$$ind_{ij} = |grad_{ij}| \cdot qual_i \cdot viol_i \tag{4.18}$$

with $qual_i$ the quality of the ith model measured in terms of the cross-validated R^2 obtained in the preliminary off-line training process (assuming K folds):

$$R^2(CV) = \frac{1}{K}\sum_{i=1}^{K} \frac{\sum_{k=(i-1)*(N/K)+1}^{i*(N/K+1)} (\widehat{y}(k) - \bar{y}(k))^2}{\sum_{k=(i-1)*(N/K)+1}^{i*(N/K+1)} (\widehat{y}(k) - \bar{y}(k))^2 + \sum_{k=(i-1)*(N/K)+1}^{i*(N/K+1)} (y(k) - \widehat{y}(k))^2} \tag{4.19}$$

where $\bar{y}$ denotes the average over the target values and thus represents a dummy model with a constant surface. $viol_i$ is calculated by

$$viol_i = \frac{res_i}{tolband_i} \tag{4.20}$$

i.e., it measures the proportion of the residual to the tolerance band, which is always greater than 1, as otherwise the model would not be violated. The calculation of the gradient $grad_{ij}$ depends on the model architecture/structure used for establishing the partial mappings within the causal relation networks, but usually can be elicited once the model can be represented as differentiable functional form—as is, e.g. the case for all types of linear regression models as well as neural networks, fuzzy systems, or symbolic regression formulas.

Then, the likelihood that the change affects or appears in channel x_j is given by:

$$lik_j = Agg_{\{f_i \in F\_viol(\mathbf{x}) | x_j \in f_i\}}(ind_{ij}) \tag{4.21}$$

with Agg an aggregation operator and $F\_viol(\mathbf{x})$ the set of violated models in the current sample $\mathbf{x}$, which is normalized by the maximal likelihood among all N channels involved in any of the violated models in $F\_viol(\mathbf{x})$:

$$lik_j(normed) = \frac{lik_j}{max_{k=1,\ldots,N}(lik_k)}. \tag{4.22}$$

In this sense, variables with similar but slightly lower degree than the maximum are considered as *potential isolation candidates* $iso_1, \ldots, iso_k$, as their $lik_j(normed)$ value will be close to 1; other channels with low likelihoods compared to the maximum will be assigned a value close to 0 and can be neglected.

4.6.2.3 Hybridization of Isolation Candidates with Discrete Event Signals to Identify Intended Changes

Afterwards, it can be checked whether any event $e_1, \ldots, e_l$ occurring before the detection of the process change (candidate) does have a (time-delayed) affect on any of the isolation candidates $iso_1, \ldots, iso_k$, i.e. whether

$$\exists_{j=1,\ldots,k} \left(qual_j(fpred_j(\mathbf{e}_{\text{full}}) \rightarrow iso_j) \geq thr \quad \wedge \quad \exists_{i=1,\ldots,l}(infl(e_i) \rightarrow iso_j) = sign_\alpha\right) \tag{4.23}$$

is met. The first condition checks whether there is a predictive mapping of high quality between the complete set of event signals $\mathbf{e}_{\text{full}} = (e_1, \ldots, e_l)$ and any isolated channel (with prediction horizon lying in the $[T - A, T + A]$ interval), the second one checks whether any event $e_1, \ldots, e_l$ has a significant influence on any of the isolated candidate (as could be conducted with ANOVA test [53] or gradient-based sensitivity analysis based on pre-recorded data or known due to expert knowledge).

- If (4.23) holds, the likelihood is high that the change is an intended one and the model can be updated to account for the change, to integrate the changed/expanded system behavior in its structure (according to the techniques discussed in Sect. 4.5.1).
- If (4.23) does not hold, the process change is probably not intended, thus a fault alarm is triggered to the operator.

In case of unsupervised distribution-based models, change isolation is more sophisticated high-dimensional because distribution-based models are used as reference, where all or a larger subset of channels are involved. In case of PCA or partial PCA based approaches, the loadings of the channels over the most important (highly ranked) components could be used as indicator how much important channels are for describing the distributions. These loadings may then substitute the gradients in (4.18), and the violation degrees $viol_i$ can be interpreted as the proportional distance of the sample to the surface of the cloud.

A final note goes to SCADA systems where events may appear with a very high frequency and thus the time fraction for model maintenance may become dominant. On the other hand, events only concern the maintenance whenever they are significant, i.e. are able to trigger a process change (=potential fault candidate) detected as such by one of the methods described above (thus, others are automatically filtered). Second, maintenance is done by updating the models in a single-pass, sample-wise manner by using the incremental learning and evolving methods described throughout Sects. 4.3 and 4.4: from our experience, these are all very fast methods operating in real-time even in high-dimensional cases, much faster than re-calibration cycles based on past samples and sliding windows, such that no significant delay for model updating can be expected. Similar considerations hold for the change isolation component, as the values in (4.18) are fast to compute (gradients can be approximately calculated by difference quotients).

Acknowledgements The author acknowledges the Austrian research funding association (FFG) within the scope of the "IKT of the future" programme, project "Generating process feedback from heterogeneous data sources in quality control (mvControl)" (contract # 849962).

References

1. M. Affenzeller, S. Winkler, S. Wagner, A. Beham, *Genetic Algorithms and Genetic Programming: Modern Concepts and Practical Applications* (Chapman & Hall, Boca Raton, FL, 2009)
2. S. Agarwal, D. Starobinski, A. Trachtenberg, On the scalability of data synchronization protocols for PDAs and mobile devices. IEEE Netw. **16**(4), 22–28 (2002)
3. J. Aguilar-Martin, Qualitative control, diagnostic and supervision of complex processes. Math. Comput. Simul. **36**(2), 115–127 (1994)

4. D. Albesano, R. Gemello, P. Laface, F. Mana, S. Scanzio, Adaptation of artificial neural networks avoiding catastrophic forgetting, in *International Joint Conference on Neural Networks 2006* (2006), pp. 1554–1561
5. C. Alcala, S. Qin, Reconstruction-based contribution for process monitoring. Automatica **45**(7), 1593–1600 (2009)
6. S. Alizadeh, A. Kalhor, H. Jamalabadi, B. Araabi, M. Ahmadabadi, Online local input selection through evolving heterogeneous fuzzy inference system. IEEE Trans. Fuzzy Syst. **24**(6), 1364–1377 (2016)
7. M. Andersson, A comparison of nine pls1 algorithms. J. Chemometr. **23**, 518–529 (2009)
8. P. Angelov, Evolving Takagi-Sugeno fuzzy systems from streaming data, eTS+, in *Evolving Intelligent Systems: Methodology and Applications*, ed. by P. Angelov, D. Filev, N. Kasabov (Wiley, New York, 2010), pp. 21–50
9. P. Angelov, R. Yager, A new type of simplified fuzzy rule-based system. Int. J. Gen. Syst. **41**(2), 163–185 (2012)
10. P. Angelov, X. Zhou, Evolving fuzzy-rule-based classifiers from data streams. IEEE Trans. Fuzzy Syst. **16**(6), 1462–1475 (2008)
11. P. Angelov, V. Giglio, C. Guardiola, E. Lughofer, J. Luján, An approach to model-based fault detection in industrial measurement systems with application to engine test benches. Meas. Sci. Technol. **17**(7), 1809–1818 (2006)
12. P. Angelov, D. Filev, N. Kasabov, *Evolving Intelligent Systems — Methodology and Applications* (Wiley, New York, 2010)
13. K. Aström, B. Wittenmark, *Adaptive Control*, 2nd edn. (Addison-Wesley Longman, Boston, MA, 1994)
14. B. Babadi, N. Kalouptsidis, V. Tarokh, SPARLS: the sparse rls algorithm. IEEE Trans. Signal Process. **58**(8), 4013–4025 (2010)
15. D. Bartholomew, F. Steele, J. Galbraith, I. Moustaki, *Analysis of Multivariate Social Science Data*. Statistics in the Social and Behavioral Sciences Series, 2nd edn. (Taylor & Francis, London, 1993)
16. S. Bay, K. Saito, N. Ueda, P. Langley, A framework for discovering anomalous regimes in multivariate time-series data with local models, in *Symposium on Machine Learning for Anomaly Detection*, Stanford, 2004
17. A. Bifet, G. Holmes, R. Kirkby, B. Pfahringer, MOA: massive online analysis. J. Mach. Learn. Res. **11**, 1601–1604 (2010)
18. P. Boltryk, C.J. Harris, N.M. White, Intelligent sensors - a generic software approach. J. Phys. Conf. Ser. **15**, 155–160 (2005)
19. L. Breiman, J. Friedman, C. Stone, R. Olshen, *Classification and Regression Trees* (Chapman & Hall, Boca Raton, FL, 1993)
20. E. Brusa, L. Lemma, D. Benasciutti, Vibration analysis of a sendzimir cold rolling mill and bearing fault detection. Proc. Inst. Mech. Eng. C J. Mech. Eng. Sci. **224**(C8), 1645–1654 (2010)
21. P. Caleb-Solly, J. Smith, Adaptive surface inspection via interactive evolution. Image Vis. Comput. **25**(7), 1058–1072 (2007)
22. V. Carvalho, W. Cohen, Single-pass online learning: performance, voting schemes and online feature selection, in *Proceedings of the 12th ACM SIGKDD International Conference on Knowledge Discovery and Data Mining* (ACM Press, New York, 2006), pp. 548–553
23. C. Cernuda, E. Lughofer, P. Hintenaus, W. Märzinger, T. Reischer, M. Pawlicek, J. Kasberger, Hybrid adaptive calibration methods and ensemble strategy for prediction of cloud point in melamine resin production. Chemom. Intell. Lab. Syst. **126**, 60–75 (2013)
24. C. Cernuda, E. Lughofer, G. Mayr, T. Röder, P. Hintenaus, W. Märzinger, J. Kasberger, Incremental and decremental active learning for optimized self-adaptive calibration in viscose production. Chemom. Intell. Lab. Syst. **138**, 14–29 (2014)
25. V. Chandola, A. Banerjee, V. Kumar, Anomaly detection: a survey. ACM Comput. Surv. **41**(3), 1–58 (2009)

26. H. Chen, P. Tino, X. Yao, A. Rodan, Learning in the model space for fault diagnosis. IEEE Trans. Neural Netw. Learn. Syst. **25**(1), 124–136 (2014)
27. G. Claeskens, N. Hjort, *Model Selection and Model Averaging* (Cambridge University Press, Cambridge, 2008)
28. L. Cohen, G. Avrahami-Bakish, M. Last, A. Kandel, O. Kipersztok, Real-time data mining of non-stationary data streams from sensor networks. Inf. Fusion **9**(3), 344–353 (2008)
29. B. Costa, P. Angelov, L. Guedes, Fully unsupervised fault detection and identification based on recursive density estimation and self-evolving cloud-based classifier. Neurocomputing **150**(A), 289–303 (2015)
30. D. Cox, D. Hinkley, *Theoretical Statistics* (Chapman & Hall, London, 1974)
31. J.C. da Silva, A. Saxena, E. Balaban, K. Goebel, A knowledge-based system approach for sensor fault modeling, detection and mitigation. Expert Syst. Appl. **39**, 10977–10989 (2012)
32. S.J. Delany, P. Cunningham, A. Tsymbal, L. Coyle, Curvilinear component analysis: a self-organizing neural network for nonlinear mapping of datasets. IEEE Trans. Neural Netw. **8**(1), 148–154 (1997)
33. S. Ding, *Model-Based Fault Diagnosis Techniques: Design Schemes, Algorithms, and Tools* (Springer, Berlin, 2008)
34. D. Dou, S. Zhou, Comparison of four direct classification methods for intelligent fault diagnosis of rotating machinery. Appl. Soft Comput. **46**, 459–468 (2016)
35. H. Efendic, A. Schrempf, L.D. Re, Data based fault isolation in complex measurement systems using models on demand, in *Proceedings of the IFAC-Safeprocess 2003* (IFAC, Washington, DC, 2003), pp. 1149–1154
36. B. Efron, R. Tibshirani, Improvements on cross-validation: the .632+ bootstrap method. J. Am. Stat. Assoc. **92**(438), 548–560 (1997)
37. C. Eitzinger, W. Heidl, E. Lughofer, S. Raiser, J. Smith, M. Tahir, D. Sannen, H. van Brussel, Assessment of the influence of adaptive components in trainable surface inspection systems. Mach. Vis. Appl. **21**(5), 613–626 (2010)
38. D. Evans, A. Jones, A proof of the gamma test. Roy. Soc. **458**, 2759–2799 (2002)
39. X. Fang, J. Du, Z. Wei, P. He, H. Bai, X. Wang, B. Lu, An investigation on effects of process parameters in fused-coating based metal additive manufacturing. J. Manuf. Process. **28**(2), 383–389 (2017)
40. D. Fernández-Francos, D. Martínez-Rego, O. Fontenla-Romero, A. Alonso-Betanzos, Automatic bearing fault diagnosis based on one-class v-svm. Comput. Ind. Eng. **64**(1), 357–365 (2013)
41. L. Fortuna, S. Graziani, A. Rizzo, M. Xibilia, *Soft Sensor for Monitoring and Control of Industrial Processes* (Springer, London, 2007)
42. H. Fourati, *Multisensor Data Fusion: From Algorithms and Architectural Design to Applications* (CRC Press, Taylor & Francis Group, Boca Raton, FL, 2015)
43. G. Franceschini, S. Macchietto, Model-based design of experiments for parameter precision: state of the art. Chem. Eng. Sci. **63**(19), 4846–4872 (2008)
44. J. Gama, *Knowledge Discovery from Data Streams* (Chapman & Hall/CRC, Boca Raton, FL, 2010)
45. V. Giurgiutiu, *Structural Health Monitoring: Fundamentals and Applications: With Piezoelectric Wafer Active Sensors* (Academic, San Diego, CA, 2007)
46. D. Gorinevsky, Bayesian fault isolation in multivariate statistical process monitoring, in *Proceedings of the American Control Conference*, San Francisco, CA (2011), pp. 1963–1968
47. D. Gustafson, W. Kessel, Fuzzy clustering with a fuzzy covariance matrix, in *Proceedings of the IEEE CDC Conference 1979*, San Diego, CA (1979), pp. 761–766
48. I. Guyon, A. Elisseeff, An introduction to variable and feature selection. J. Mach. Learn. Res. **3**, 1157–1182 (2003)
49. F. Harrel, *Regression Modeling Strategies* (Springer, New York, 2001)
50. L. Hartert, M. Sayed-Mouchaweh, Dynamic supervised classification method for online monitoring in non-stationary environments. Neurocomputing **126**, 118–131 (2014)

51. T. Hastie, R. Tibshirani, J. Friedman, *The Elements of Statistical Learning: Data Mining, Inference and Prediction*, 2nd edn. (Springer, New York, 2009)
52. W. Heidl, S. Thumfart, E. Lughofer, C. Eitzinger, E. Klement, Machine learning based analysis of gender differences in visual inspection decision making. Inf. Sci. **224**, 62–76 (2013)
53. T. Hill, P. Lewicki, *Statistics: Methods and Applications* (StatSoft, Tulsa, 2007)
54. G. Hinton, S. Roweis, Stochastic neighbor embedding. Adv. Neural Inf. Process. Syst. **15**, 833–840 (2003)
55. M. Hisada, S. Ozawa, K. Zhang, N. Kasabov, Incremental linear discriminant analysis for evolving feature spaces in multitask pattern recognition problems. Evol. Syst. **1**(1), 17–27 (2010)
56. G. Huang, P. Saratchandran, N. Sundararajan, An efficient sequential learning algorithm for growing and pruning RBF (GAP-RBF) networks. IEEE Trans. Syst. Man Cybern. B Cybern. **34**(6), 2284–2292 (2004)
57. R. Isermann, *Fault Diagnosis Systems: An Introduction from Fault Detection to Fault Tolerance* (Springer, Berlin, 2009)
58. R. Isermann, P. Ballé, Trends in the application of model-based fault detection and diagnosis of technical processes. Control. Eng. Pract. **5**(5), 709–719 (1997)
59. H. Jiang, C. Kwong, W. Ip, T. Wong, Modeling customer satisfaction for new product development using a PSO-based ANFIS approach. Appl. Soft Comput. **12**(2), 726–734 (2013)
60. I. Jolliffe, *Principal Component Analysis* (Springer, Berlin, 2002)
61. R. Kalman, A new approach to linear filtering and prediction problems. Trans. ASME J. Basic Eng. **82**, 35–45 (1960)
62. N. Kasabov, *Evolving Connectionist Systems: The Knowledge Engineering Approach*, 2nd edn. (Springer, London, 2007)
63. B. Khaleghi, A. Khamis, F.O. Karray, S.N. Razavi, Multisensor data fusion: a review of the state-of-the-art. Inf. Fusion **14**(1), 28–44 (2013)
64. I. Khamassi, M. Sayed-Mouchaweh, M. Hammami, K. Ghedira, Discussion and review on evolving data streams and concept drift adapting. Evol. Syst. (2017, on-line and in press). https://10.1007/s12530-016-9168-2
65. J. Korbicz, J. Koscielny, Z. Kowalczuk, W. Cholewa, *Fault Diagnosis - Models, Artificial Intelligence and Applications* (Springer, Berlin, 2004)
66. P. Laskov, C. Gehl, S. Krüger, K. Müller, Incremental support vector learning: analysis, implementation and applications. J. Mach. Learn. Res. **7**, 1909–1936 (2006)
67. D. Leite, R. Palhares, C.S. Campos, F. Gomide, Evolving granular fuzzy model-based control of nonlinear dynamic systems. IEEE Trans. Fuzzy Syst. **23**(4), 923–938 (2015)
68. A. Lemos, Adaptive fault detection and diagnosis using evolving intelligent systems, in *Proceedings of the IEEE Evolving and Adaptive Intelligent Systems Conference (EAIS) 2016*, Natal (2016)
69. A. Lemos, W. Caminhas, F. Gomide, Adaptive fault detection and diagnosis using an evolving fuzzy classifier. Inf. Sci. **220**, 64–85 (2013)
70. G. Leng, X.J. Zeng, J. Keane, An improved approach of self-organising fuzzy neural network based on similarity measures. Evol. Syst. **3**(1), 19–30 (2012)
71. Y. Li, On incremental and robust subspace learning. Pattern Recogn. **37**(7), 1509–1518 (2004)
72. Z. Liang, Y. Li, Incremental support vector machine learning in the primal and applications. Neurocomputing **72**(10–12), 2249–2258 (2009)
73. E. Lima, M. Hell, R. Ballini, F. Gomide, Evolving fuzzy modeling using participatory learning, in *Evolving Intelligent Systems: Methodology and Applications*, ed. by P. Angelov, D. Filev, N. Kasabov (Wiley, New York, 2010), pp. 67–86
74. D. Liu, Y. Zhang, Z. Yu, M. Zeng, Incremental supervised locally linear embedding for machinery fault diagnosis. Eng. Appl. Artif. Intell. **50**(C), 60–70 (2016)
75. L. Ljung, *System Identification: Theory for the User* (Prentice Hall PTR, Prentice Hall, Upper Saddle River, NJ, 1999)

76. E. Lughofer, FLEXFIS: a robust incremental learning approach for evolving TS fuzzy models. IEEE Trans. Fuzzy Syst. **16**(6), 1393–1410 (2008)
77. E. Lughofer, *Evolving Fuzzy Systems — Methodologies, Advanced Concepts and Applications* (Springer, Berlin, 2011)
78. E. Lughofer, On-line incremental feature weighting in evolving fuzzy classifiers. Fuzzy Sets Syst. **163**(1), 1–23 (2011)
79. E. Lughofer, eVQ-AM: an extended dynamic version of evolving vector quantization, in *Proceedings of the 2013 IEEE Conference on Evolving and Adaptive Intelligent Systems (EAIS)*, Singapore (2013), pp. 40–47
80. E. Lughofer, On-line assurance of interpretability criteria in evolving fuzzy systems — achievements, new concepts and open issues. Inf. Sci. **251**, 22–46 (2013)
81. E. Lughofer, C. Guardiola, On-line fault detection with data-driven evolving fuzzy models. J. Control Intell. Syst. **36**(4), 307–317 (2008)
82. E. Lughofer, S. Kindermann, SparseFIS: data-driven learning of fuzzy systems with sparsity constraints. IEEE Trans. Fuzzy Syst. **18**(2), 396–411 (2010)
83. E. Lughofer, J.L. Bouchot, A. Shaker, On-line elimination of local redundancies in evolving fuzzy systems. Evol. Syst. **2**(3), 165–187 (2011)
84. E. Lughofer, C. Eitzinger, C. Guardiola, On-line quality control with flexible evolving fuzzy systems, in *Learning in Non-stationary Environments: Methods and Applications*, ed. by M. Sayed-Mouchaweh, E. Lughofer (Springer, New York, 2012), pp. 375–406
85. E. Lughofer, C. Cernuda, S. Kindermann, M. Pratama, Generalized smart evolving fuzzy systems. Evol. Syst. **6**(4), 269–292 (2015)
86. E. Lughofer, S. Kindermann, M. Pratama, J. Rubio, Top-down sparse fuzzy regression modeling from data with improved coverage. Int. J. Fuzzy Syst. **19**(5), 1645–1658 (2017)
87. E. Lughofer, R. Richter, U. Neissl, W. Heidl, C. Eitzinger, T. Radauer, Explaining classifier decisions linguistically for stimulating and improving operators labeling behavior. Inf. Sci. **420**, 16–36 (2017)
88. S. Mahadevan, S. Shah, Fault detection and diagnosis in process data using one-class support vector machines. J. Process Control **19**(10), 1627–1639 (2009)
89. L. Manevitz, M. Yousef, One-class svms for document classification. J. Mach. Learn. Res. **2**, 139–154 (2001)
90. T. McConaghy, Fast scalable, deterministic symbolic regression technology, in *Genetic Programming Theory and Practice IX*, ed. by R. Riolo et al. Genetic and Evolutionary Computation (Springer Science+Business Media, Heidelberg, 2011), pp. 235–260
91. P. Mcnicholas, T. Murphy, Parsimonious gaussian mixture models. Stat. Comput. **18**(3), 285–296 (2008)
92. L. Mendonça, J. Sousa, J.S. da Costa, An architecture for fault detection and isolation based on fuzzy methods. Expert Syst. Appl. **36**(2), 1092–1104 (2009)
93. T. Nakamura, A. Lemos, A batch-incremental process fault detection and diagnosis using mixtures of probabilistic PCA, in *Proceedings of the Evolving and Adaptive Intelligent Systems (EAIS) Conference 2014* (IEEE Press, Linz, 2014)
94. O. Nelles, *Nonlinear System Identification* (Springer, Berlin, 2001)
95. L. Ngia, J. Sjöberg, Efficient training of neural nets for nonlinear adaptive filtering using a recursive Levenberg-Marquardt algorithm. IEEE Trans. Signal Process. **48**(7), 1915–1926 (2000)
96. P. Odgaard, B. Lin, S. Jorgensen, Observer and data-driven-model-based fault detection in power plant coal mills. IEEE Trans. Energy Convers. **23**(2), 659–668 (2008)
97. W. Penny, S. Roberts, Error bars for linear and nonlinear neural network regression models (1998). http:\citeseer.nj.nec.com/penny98error.html
98. K. Pichler, E. Lughofer, M. Pichler, T. Buchegger, E. Klement, M. Huschenbett, Fault detection in reciprocating compressor valves under varying load conditions. Mech. Syst. Signal Process. **70–71**, 104–119 (2016)
99. M. Pratama, S. Anavatti, E. Lughofer, C. Lim, An incremental meta-cognitive-based scaffolding fuzzy neural network. Neurocomputing **171**, 89–105 (2016)

100. M. Pratama, J. Lu, E. Lughofer, G. Zhang, M. Er, Incremental learning of concept drift using evolving type-2 recurrent fuzzy neural network. IEEE Trans. Fuzzy Syst. **25**(5), 1175–1192 (2017)
101. M. Sayed-Mouchaweh, E. Lughofer, *Learning in Non-stationary Environments: Methods and Applications* (Springer, New York, 2012)
102. M. Sayed-Mouchaweh, E. Lughofer, Decentralized fault diagnosis approach without a global model for fault diagnosis of discrete event systems. Int. J. Control. **88**(11), 2228–2241 (2015)
103. M. Schrenk, S. Krenn, M. Ripoll, A. Nevosad, S. Paar, R. Grundtner, G. Rohm, F. Franek, Statistical analysis on the impact of process parameters on tool damage during press hardening. J. Manuf. Process. **23**, 222–230 (2016)
104. A. Seifi, H. Riahi-Madvar, Input variable selection in expert systems based on hybrid gamma test-least square support vector machine, ANFIS and ANN models, in *Advances in Expert Systems* (Springer, Berlin, 2012)
105. F. Serdio, E. Lughofer, K. Pichler, T. Buchegger, H. Efendic, Residual-based fault detection using soft computing techniques for condition monitoring at rolling mills. Inf. Sci. **259**, 304–320 (2014)
106. F. Serdio, E. Lughofer, K. Pichler, M. Pichler, T. Buchegger, H. Efendic, Fault detection in multi-sensor networks based on multivariate time-series models and orthogonal transformations. Inf. Fusion **20**, 272–291 (2014)
107. F. Serdio, E. Lughofer, K. Pichler, M. Pichler, T. Buchegger, H. Efendic, Fuzzy fault isolation using gradient information and quality criteria from system identification models. Inf. Sci. **316**, 18–39 (2015)
108. F. Serdio, E. Lughofer, A.C. Zavoianu, K. Pichler, M. Pichler, T. Buchegger, H. Efendic, Improved fault detection employing hybrid memetic fuzzy modeling and adaptive filters. Appl. Soft Comput. **51**, 60–82 (2017)
109. M. Shabanian, M. Montazeri, A neuro-fuzzy online fault detection and diagnosis algorithm for nonlinear and dynamic systems. Int. J. Control. Autom. Syst. **9**(4), 665–670 (2011)
110. A. Shaker, E. Lughofer, Self-adaptive and local strategies for a smooth treatment of drifts in data streams. Evol. Syst. **5**(4), 239–257 (2014)
111. A. Shilton, M. Palaniswami, D. Ralph, A. Tsoi, Incremental training of support vector machines. IEEE Trans. Neural Netw. **16**(1), 114–131 (2005)
112. S. Simani, C. Fantuzzi, R. Patton, *Model-Based Fault Diagnosis in Dynamic Systems Using Identification Techniques* (Springer, Berlin, 2002)
113. C. So, S. Ng, S. Leung, Gradient based variable forgetting factor RLS algorithm. Signal Process. **83**(6), 1163–1175 (2003)
114. M. Stone, Cross-validatory choice and assessment of statistical predictions. J. R. Stat. Soc. **36**(1), 111–147 (1974)
115. M. Tamura, S. Tsujita, A study on the number of principal components and sensitivity of fault detection using PCA. Comput. Chem. Eng. **31**(9), 1035–1046 (2007)
116. M. Tipping, C. Bishop, Probabilistic principal component analysis. J. R. Stat. Soc. Ser. B Stat. Methodol. **61**(3), 611–622 (1999)
117. L. Wang, R. Gao, *Condition Monitoring and Control for Intelligent Manufacturing* (Springer, London, 2006)
118. W. Wang, J. Vrbanek, An evolving fuzzy predictor for industrial applications. IEEE Trans. Fuzzy Syst. **16**(6), 1439–1449 (2008)
119. J. Wang, P. Zhao, S. Hoi, R. Jin, Online feature selection and its applications. IEEE Trans. Knowl. Data Eng. **26**(3), 698–710 (2004)
120. J. Weng, Y. Zhang, W.S. Hwang, Candid covariance-free incremental principal component analysis. IEEE Trans. Pattern Anal. Mach. Intell. **25**(8), 1034–1040 (2003)
121. Y. Xu, K. Wong, C. Leung, Generalized recursive least square to the training of neural network. IEEE Trans. Neural Netw. **17**(1), 19–34 (2006)
122. M. Yang, V. Makis, ARX model-based gearbox fault detection and localization under varying load conditions. J. Sound Vib. **329**(24), 5209–5221 (2010)

123. X.Q. Zeng, G.Z. Li, Incremental partial least squares analysis of big streaming data. Pattern Recogn. **47**, 3726–3735 (2014)
124. Y.Q. Zhang, Constructive granular systems with universal approximation and fast knowledge discovery. IEEE Trans. Fuzzy Syst. **13**(1), 48–57 (2005)
125. Y.Q. Zhang, Combining uncertainty sampling methods for supporting the generation of meta-examples. Inf. Sci. **196**(1), 1–14 (2012)
126. E. Zhong, W. Fan, Q. Yang, O. Verscheure, J. Ren, Cross validation framework to choose amongst models and datasets for transfer learning, in *Proceedings of the Joint European Conference on Machine Learning and Knowledge Discovery in Databases*, Banff, AB (2010), pp. 547–562
127. P. Zhu, B. Chen, J. Principe, A novel extended kernel recursive least squares algorithm. Neural Netw. **32**, 349–357 (2012)

Chapter 5
Critical States Distance Filter Based Approach for Detection and Blockage of Cyberattacks in Industrial Control Systems

Franck Sicard, Éric Zamai, and Jean-Marie Flaus

5.1 Introduction

In this section, components and architecture of ICS will be described as well as the main purpose of these systems. Then, vulnerabilities of ICS and their causes will be explained in detail. Thereafter, a history of main successful attacks against control-command systems will be drawn. Finally, details on ICS's specificities and problematics of cybersecurity in control-command systems will be given.

5.1.1 Industrial Control Systems (ICS)

Industrial Control Systems (ICS) are a combination of cyber and physical layers that act together to achieve an objective in industrial environments. Nowadays, they are integrated in many sectors with critical infrastructures such as: energy production and distribution (electricity, water, oil and gas, etc.), manufacturing systems, transportation systems, health services, or defense [1]. Different typologies and materiel architectures can be found for describing ICS; however, CIM (Computer-Integrated Manufacturing) architecture formalizes ICS in several hierarchical layers [2] as represented in Fig. 5.1. Each layer has different information according to processing capacity and decision-making power of the component. In this paper, the term ICS architecture can be replaced by SCADA architecture; more details are presented in Sect. 5.1.1.3.

F. Sicard (✉) · É. Zamai · J.-M. Flaus
Univ. Grenoble Alpes, CNRS, Grenoble INP, G-SCOP, 38000 Grenoble, France
e-mail: franck.sicard@grenoble-inp.fr

M. Sayed-Mouchaweh (ed.), *Diagnosability, Security and Safety of Hybrid Dynamic and Cyber-Physical Systems*, https://doi.org/10.1007/978-3-319-74962-4_5

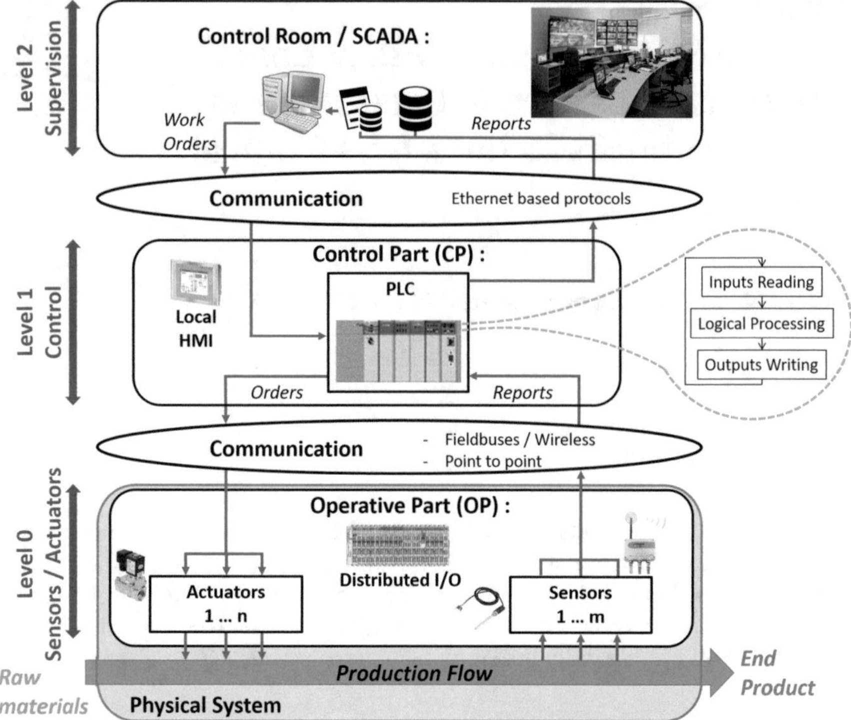

Fig. 5.1 Illustration of an ICS architecture from a functional point of view with CIM real time levels

5.1.1.1 Level 0: Sensors and Actuators

The purpose of an ICS is to transform raw materials into finished product by interacting on the production flow. In other words, the control system has to bring the process from an initial to final state (production flow) by acting on it to ensure productivity and reliability.

Sensors and actuators link cyber and physical layers. Sensors convert physical measurements into electrical signal transmitted to the upper layer (level 1). Actuators convert electrical signals from numerical layer into mechanical motion in order to act on the production flow. These components constitute the Operative Part (OP) that is equivalent to level 0 in CIM architecture. The OP and the production flow are called the physical system.

5.1.1.2 Level 1: Control

This layer has to acquire data from the field devices (level 0), control actuators (level 0), and communicate with operators (local HMI) and control room (level 2). The main device is the Programmable Logic Controller (PLC) that controls the process in real time. For this purpose, PLC repeats a scan cycle composed of three steps: (1) PLC acquires sensors measurements of the physical system, (2) logic program embedded in PLC (control law) runs and determines actions that have to be applied on the system then (3) orders are sent to actuators.

Level 1 is also called Control Part (CP).

5.1.1.3 Level 2: Supervision (SCADA)

Level 2 has several objectives in an ICS. It gathers data from lower layers in order to show an image of the process. Through HMI and SCADA, operators and managers are able to oversee the production flow. Work orders are transmitted to level 1 to adjust control law or to avoid damages on the process. Finally, level 2 communicates with the upper levels. This layer is called the control room. The term "SCADA" is also found in the literature, it can design the ICS architecture: SCADA is thus a synonym of ICS and refers to SCADA architecture; or level 2 of CIM architecture and it refers to level 2 (supervision). If that is the case, SCADA represents only the supervision part of the ICS.

5.1.1.4 Communication Networks

Communication networks connect the different layers. Originally, ICS used analog or discrete inputs/outputs (I/O) or proprietary protocols (fieldbuses) to communicate but in recent architectures, devices draw on TCP/IP protocols mainly. Increasing the volume and speed of data transported is the main benefit. However, by introducing technologies inherited from Information Technology (IT), new vulnerabilities appear in ICS (see Sect. 5.1.2). Moreover, wireless protocols, such as WirelessHART or ISA100, are gaining more and more attention from both academic and industrial point of view. In the meantime, these protocols make the cybersecurity issues even more critical.

Between level 1 and level 0, communication [2, 3] has to transport few volumes of data (Bytes/Bits) but with a shortened response time (10 ms/ms). Point-to-point connection or fieldbuses are used to communicate. These buses are mainly proprietary protocols, such as Profibus (Siemens) or Fipway[1] (Schneider Electric), specific application protocols like AS-i, CAN, or open protocols, as Modbus. Wireless protocols are increasingly being used to communicate in low layers in cyberphysical systems [4], mainly for transmitting sensors values (SmartGrid, Industry 4.0, etc.).

On the contrary, levels 2 and 1 have to communicate with a large volume (Kb/Bytes) of data with a longer response time (min-s/ms). These protocols

[1]Fipway is no longer supported by Schneider Electric that focuses on Ethernet based protocols.

[2, 3], such as Profinet (Siemens), are based on Ethernet TCP/IP protocol. Nowadays, protocols allowing connection of all the architecture are developed to unify industrial communication protocols. This is the purpose of EtherNet/IP [5] that is based on Ethernet TCP/IP protocol respecting recent norms as IEEE 802.3 or EtherCAT [6] that is an Ethernet fieldbus facilitating connection with upper level of CIM architecture and normalized by IEC and ISO.

5.1.1.5 Other Levels

Our study only focuses on level 0, 1, and 2 of the CIM architecture. Upper levels (3 and 4) referring to scheduling and global management of production flow will not be discussed.

5.1.2 Vulnerabilities and Attacks against ICS

Since the beginning of the century, ICS are targeted by hackers who exploit vulnerabilities on components or the architecture in order to perform cyberattacks [7, 8]. The interest of hackers concerning industrial systems comes from the fact that they are designed to solve production issues (productivity and safety) without taking into account security issues. Thus, the use of IT technologies introduces new vulnerabilities. Moreover, cyberattacks can inflict important damages to the process (production shutdown, repair time, recovery time, etc.), the environment (negative impact on human, heath, ecology, social, etc.), and company (financial loss, negative image, etc.) [1, 7].

An exhaustive list of cyberattacks can be found in McLaughlin [8], Khorrami [9], and RISI database [10]. However, we will focus on three major attacks:

- Maroochy Shire sewage spill (Australia, 2000) [11]: it is the first attack on ICS in the sense that intrusion was performed and processes were impacted. A former employee used his access to SCADA equipment and poured 800,000 L of raw sewage causing huge ecological damages.
- Stuxnet (Iran, 2010) [12]: Stuxnet is the reference of what a cyberattack can be. This computer worm was introduced in the industrial network via a USB key. It targeted PLC Siemens S7 via the infection of a computer where Siemens software Step7 (programming software) was installed. Thus, Stuxnet sent false orders to wear centrifuges out prematurely. Meanwhile, data sensors were spoofed to prevent operators from detecting abnormal operations.
- Steel mill plant attack and Power grid attack (Germany, 2014 and Ukraine, 2015) [13, 14]: these attacks are a milestone because attackers used an office network to get into an industrial network and damage the process.

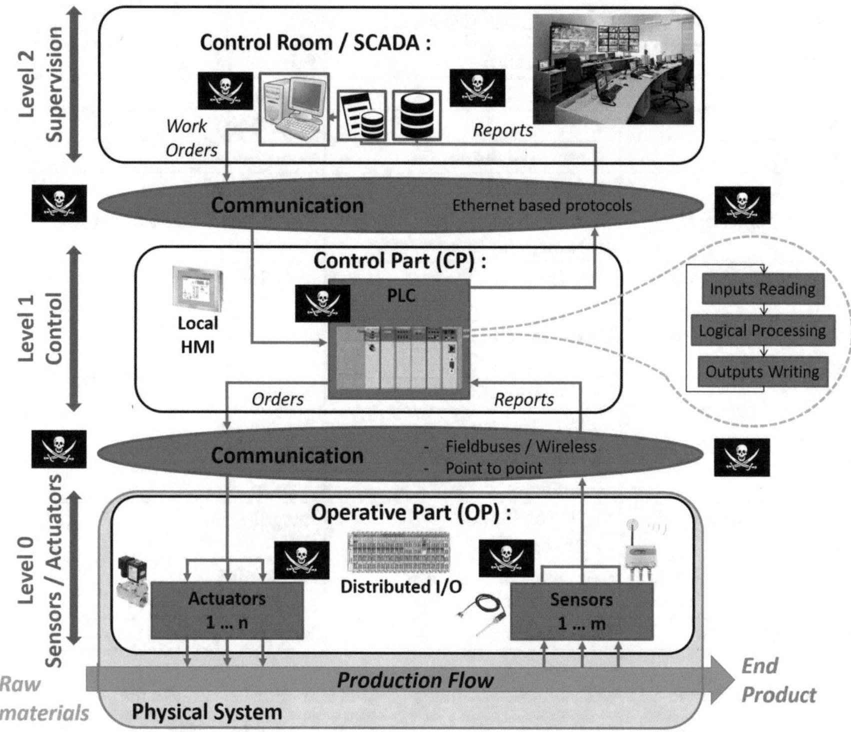

Fig. 5.2 Main vulnerabilities and potential attack surfaces on ICS

An updated list of attacks and vulnerabilities is available on ICS-CERT [15]. Moreover, aside from classical IT attacks as DOS (Denial of Service) or DDOS (Distributed Denial of Service), Man-in-the-middle (MITM) or replay attack that affect level 2, other attacks inherent to levels 1 and 0 can be performed on an ICS. Thus, *random attacks* send information without taking into account the process states or process knowledge; *sequential attacks* [16, 17] break the sequential logic of the control law and *false data injection attacks* [18] where data is intercept and spoofed. To complete this range of attacks, we introduce attacks on PLC developed in [8, 9] as modification of control law, alteration of configuration, or attacks on the firmware.

Attacks can be performed because ICS have vulnerabilities that can be exploited. Indeed, components, architecture, and environment of a control-command system can be used to perform an attack as presented in [8, 19]. Figure 5.2 and Table 5.1 summarize vulnerabilities and attacks presented in this section.

Table 5.1 Some vulnerabilities of an ICS classified by layers

Location	Vulnerabilities	Effects
SCADA	Code injection	Integrity
SCADA	DOS	Availability
SCADA	Information disclosure	Confidentiality
SCADA	Buffer overflow	Availability
Control room/HMI	Various attacks (virus, worm, malware, ransomware, etc.)	Integrity/confidentiality/availability
Communication levels 2–1	Data spoofing	Integrity
Communication levels 2–1	Communication removal	Availability
PLC	Firmware modification	Integrity
PLC	Buffer overflow	Availability
PLC	Configuration modification	Integrity
PLC	Alteration of control law	Availability
Communication levels 1–0	Data spoofing for orders/reports	Integrity
Communication levels 1–0	Communication removal	Availability
Smart equipment	Data spoofing	Integrity

Table 5.2 Differences between industrial control systems and information systems

Information technology (IT)	Operational technology (OT)
Controls only digital systems	Controls cyber physical systems
No real-time constraints	Strong real-time constraints
Priorities: confidentiality, integrity, availability	Priorities: availability, integrity, confidentiality
Protocols and communication technologies standardized	Heterogeneous stacking of protocols and technologies
Large material resource (memory)	Limited material resource
Running can be suspended, frequent updates	Continuous production (24/24, 7/7), infrequent updates

5.1.3 Problematic

Part of the problem lies in the fact that IT technologies have been deployed in ICS without taking into account security aspects. However, ICS have specificities compared to Information System (IS) making it difficult or impossible to use "classical" solutions. Table 5.2 presents some differences between ICS and classical IS.

Moreover, solutions applied in traditional IS cannot be used in ICS. Indeed, cryptographic solutions raise the question of compliance with real-time constraints and management of encryption keys. Antivirus uses processing time and needs

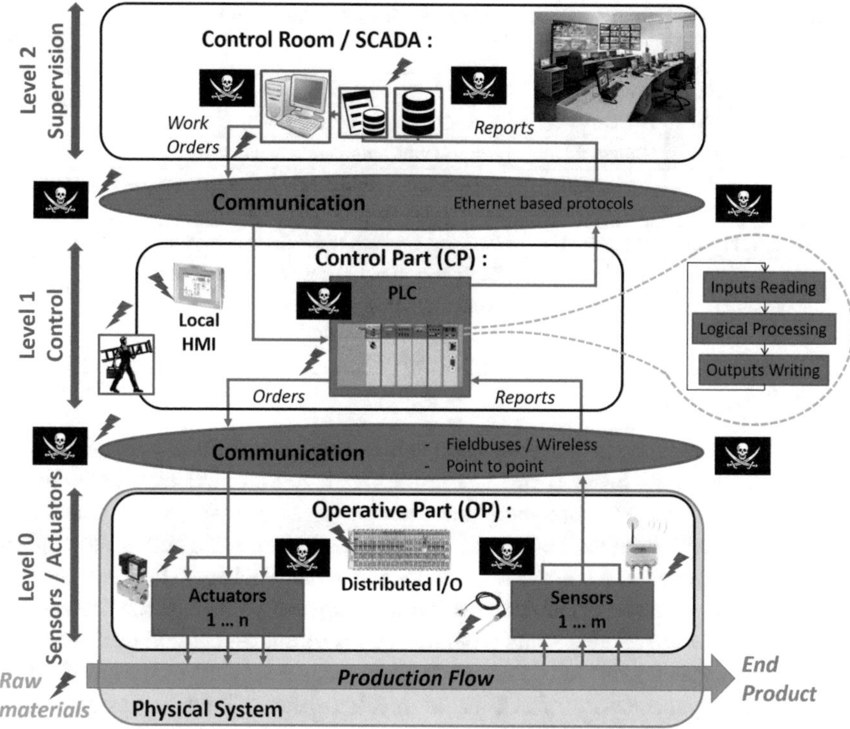

Fig. 5.3 Superposition of attacks and failures in ICS architectures

updated viral databases. Employee training, regular backup and enhanced access control need time to be applied and to become efficient. Finally, firewalls and IDS are key components for ICS security, operating at all levels. However, bandwidth, integration in the architecture, response time, lack of knowledge of industrial protocols, and limited availability of resources in ICS complicate their integration.

Our study focuses on levels 1 and 0 of ICS architecture in order to bring automation-specific knowledge. Thus, the main objective of our work is the detection of cyberattacks between OP and CP. Indeed, as ICS architecture is vulnerable, blocking orders between PLC and actuators is the last occasion to stop any illicit action. Anticipation of deviations in order to avoid blockage moves in the same vein. However, the main problematic of our work is how we can distinguish an attack from a failure in a fallible environment. Indeed, an attack, as a failure, has a service lost effect but has also a notion of intentionality. Nguyen [20] identifies four sources of failure: human factor, equipment, recipes, and product. The scientific problem of our work is the distinction of attacks in an environment subject to hazard as presented in Fig. 5.3 and Table 5.3. A technological problem appears regarding the location of the detection and its integration in the ICS architecture.

Table 5.3 Sources, causes, and location of failures in an ICS based on Nguyen [20]

Location	Source	Causes
Levels 2, 1, and 0	Human factor	Lack of overall vision
Levels 2, 1, and 0	Human factor	Fallible in tasks voluntarily or not
Levels 2, 1, and 0	Human factor	Irreplaceable/adaptation capacities
Levels 2, 1, and 0	Equipment	Breakage of components related to incorrect assembly, bad manufacture or incorrect use
Levels 2, 1, and 0	Equipment	Fast and uncheck technological evolutions
Level 2	Work order	Tests performed in unstressed environments
Level 0	Product	Non-conforming raw materials • Damaged equipment • Premature wear

In this chapter, we propose an approach to protect ICS from cyberattacks by using filters between levels 0 and 1. These filters based on models of both control and operative parts are analyzing information exchanged to detect anomalies. Mechanisms exploiting the notion of distance between states are able to stop orders to secure the system and anticipate deviations to avoid blockages. In this approach, after detecting an anomaly, discrimination is performed with trajectory concept to distinguish between an attack and a failure. Limits of this aspect will be discussed.

This chapter is organized as follows. In Sect. 5.2, a brief state of the art of works dealing with cybersecurity of ICS issues has been done. Then, the principles of the proposed approach are detailed. Methodology to obtain filters, the principle of functioning, and the notion of distance and its improvements are explained. Section 5.4 is a study case, after applying our approach on an example, advantages and drawbacks are discussed, in particular about the discrimination of detected anomalies. We finish the chapter with a conclusion and perspectives on future works.

5.2 State of the Art in Cybersecurity

As explained in Sect. 5.1.3, our approach to secure ICS from cyberattacks is based on knowledge of automation between level 1 and 0. Thus, approaches in the field of systems monitoring, especially for detection issues, can provide solution to secure ICS. This section presents approaches in safety and security fields that can be efficient in order to protect ICS from cyberattacks.

5.2.1 From a Security Point of View

In the field of security research, Intrusion Detection System (IDS) is generally proposed to detect attacks against computer systems and networks. Denning [21]

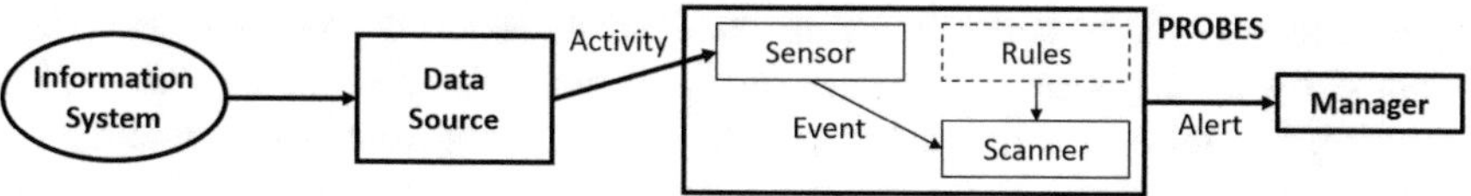

Fig. 5.4 Illustration of main functions and running of an IDS probe

in 1987 provides an IDS framework that is still used and efficient in IS. IDS is a posteriori security measure and detection occurs after an intrusion. An IDS automatically identifies violation of a system security policy based on confidentiality, availability, and integrity. To do so, IDS is based on data acquired by probes from the environment to protect. Thus, location of probes in the network is essential to ensure good detection. Fig. 5.4 presents main functions of an IDS probe. As formalized by Mitchell and Chen [22], different aspects of IDS must be considered toward the choice of detection method and data source. Some IDSs detect deviations from behavior model (behavioral approach) and others rely on abnormal behavior knowledge database (signature approach). IDSs that monitor network are called *Network Intrusion Detection System* (NIDS) and those using host data are called *Host Intrusion Detecting System* (HIDS).

Studies based on NIDS take an interest in specification on message structure used in industrial protocols. Thus, Cheung et al. [23] propose to specify function codes, exception codes, and protocol identifiers. Goldenberg and Wool [24] model exchanges between PLC and control room without taking into account semantic. Barbosa [25] studies frequency and number of exchanged packets in the network based on the hypothesis that an intrusion degrades exchange periodicity. Finally, in [26], Barbosa presents an approach based on flow whitelisting. Normal behavior of the network needs to be identified on the client address, server address, and communication port and communication protocol.

Some works focus on components with processing capacities inside ICS architecture. HIDS is investigated less than NIDS for ICS because of the specificities of components and availability constraints of the resources. Zimmer et al. [27] propose an approach based on execution time analysis for tasks in components with real time constraints. Bellettini and Rrushi [28] study a mechanism for detecting incorrect access to memory. A state-machine is defined to follow the evolution of memory. Finally, McLaughlin [29] develops several approaches for verifying the code integrity or orders before they are sent to the system.

Signature approach is based on recognition of specific behavior as used in Pan et al. [30]. The main issue of this approach is the non-detection of *zero-day* attacks. By definition, these attacks are unknown when these are performed. It involves to regularly update the database to be protected. Behavioral approach, or model-based approach, seems to be more adapted because ICS control physical process. Specifications of the system are studied to define rules to insure safety, reliability, and security of the system. These specifications can be expressed with mathematical equations (quantitative method) or with models (qualitative method). Although, quantitative methods allow knowing state variables of the process. However, complexity of the system, non-linearities, or the lack of available data makes mathematical modeling difficult even impossible. Qualitative methods describe the system through process modeling (what we can do) and control modeling (what we want to do). An interesting use of process knowledge can be found in [31, 32]. After identifying safety area for the process by setting ranges of running for each variable, authors use the concept of distance from critical state to detect attacks. Evaluating the distance defines the increasing closeness between safe state and critical area. Four fundamental hypothesis are needed: (1) the set of critical states has to be known and relatively small for sub-systems, (2) hackers must interfere with system state in order to damage the process, (3) monitoring evolution of critical states allows detecting attacks using legal orders, (4) by monitoring system state, failure or cyberattacks can be distinguished. More details about distance concept are given in Sect. 5.3.1.4.

5.2.2 *From a Safety Point of View*

As explained in the introduction, our problematic joins the already established and classic problem of detection where different approaches are known to detect failures in a system. In Zamaï [33], three approaches attract our attention:

- Reference model [34]: process model is the reference for the control part and defines normal behavior of the system. Before each transmission from CP to OP, the order is tested on the reference model in order to check whether all the conditions are met for sending this request. Failures in OP can be detected by comparing report received and the state in reference model. This method guarantees an exact representation of current state of the system.
- Emulator approach [35]: it compares data from the operative part (process) and emulator (model). Any inconsistency shows a failure in the process. This approach does not detect any errors on the command.
- Filter approach [36]: it is based on two validation blocks called filters. The filter between CP and OP is control filter. The block ensures the consistency of orders emitted with respect to the predicted orders and the current state of the system (detection of command failure). The second block is called report filter and it is located between OP and CP. The filter confirms achievement of services requested to the process (detection of process failure).

Filter approach, represented in Fig. 5.5, seems to be an interesting solution with several advantages as evaluating the accuracy of information exchanged before execution by actuators for control filter (control failures) or taking into account by CP for report filter (process failure). Moreover, filters allow implementing models and thus bringing process knowledge in detection mechanism. Finally, integration of filters in the actual ICS architecture is rather non-intrusive. However, modifications

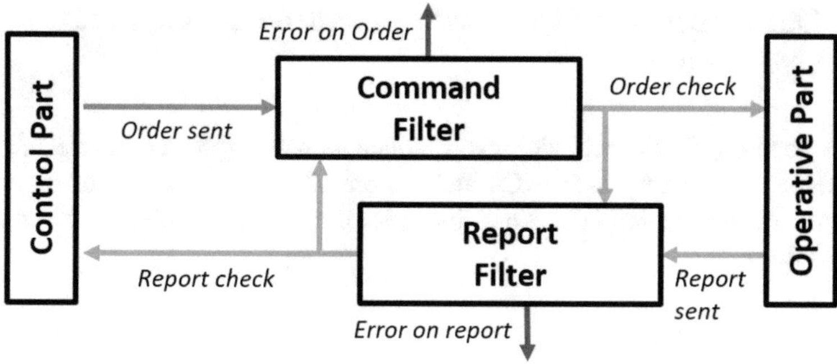

Fig. 5.5 Illustration of the filter approach defined by D. Cruette

on the location of the filters have to be made. Indeed, filters cannot be located inside PLC, as in failure detection, because PLC is vulnerable to cyberattacks (see Sect. 5.1.2).

5.2.3 Safety and Security: Mutual Reinforcement

As explained in [7], safety and security are distinct but related concepts. Safety studies accidental risks (failure, accident, etc.) affecting environment of the system, goods and people [37]. Security deals with hazard in relation to malicious acts and cybersecurity focuses on malicious acts by numerical vector. Four varieties of interdependence are listed in [7]:

- Conditional dependence: respect of safety requirement determines the level of safety and vice versa,
- Mutual reinforcement: security measures contribute to safety of the system and vice versa,
- Antagonism: security or safety requirements or measures lead to conflicting situations,
- Independent: no interactions.

In our study, we will use benefits of IDS combining with filter approach in a mutual reinforcement. Indeed, behavioral and host based IDS provides good solution with the notion of distance to critical state. Indeed, *zero*-day attacks are detected and models of the system can be used. Thus, risk assessment, required for safety, will identify these critical states. Moreover, as probes in a network, filters will be located before actuators (control filter) and PLC (report filter) in order to minimize attack surface and secure data transiting. Finally, filter approach has proved its efficacy in safety field for detecting failures and allows checking, and even blocking, orders and reports in the ICS. Our approach leads to a solution for protecting ICS against cyberattacks on levels 1–0 unlike other approaches in the literature or industrial solutions that focus on the protection of levels 2–1.

5.3 Our Approach: Filters with Distance Concept for ICS Cybersecurity

In this section, filters methodology of conception is explained as well as their integration and running in the ICS architecture. Finally, the notion of distance is explained in order to anticipate deviations and, if necessary, to distinguish an attack from a failure.

5.3.1 Methodology

This methodology is based on three steps: (1) the two first are off-line: critical states are identified, (2) models of the system are built and sequences leading to prohibited states are explored; (3) the third is online: this is the detection mechanism based on the notion of distance. Step zero illustrates the integration of filters in the ICS architecture. The main objective is to obtain filters on the control and on the report based on models that analyze information exchanged by the system. If so, information may be blocked by one of these blocks to prevent damages. After detection of an anomaly, filters will try to determine whether it is a failure or an attack.

5.3.1.1 Step 0: Integration in the ICS Architecture

As presented in Sect. 5.2, filters are based on IDS and filter approach. In the first approach, probes are deployed in the network in order to analyze exchanged data with system policy. Thus, no limitation on the number of probes exists in IS. In ICS, filters need to be easily implemented in the architecture with a minimum of probes in order to preserve material architecture. Filter approach completes this objective with two verification blocks: control filter and report filter. However, filters are implemented directly in the PLC initially. As seen in Sect. 5.1, PLC is vulnerable to cyberattacks in several ways. Filters may be corrupted and an attack could be performed on the system.

To be efficient, filters have to be located outside the PLC, or any electronic devices connected with the industrial network, and as close as possible to components that have to be protected. Connection with actuators is made by digital link or industrial protocol but isolated from the one used in the ICS. This disposition allows reducing the attack surface for a hacker. Actuators are sort of separated from the ICS and the control filter is like a power switch. For example, if a PLC communicates by digital outputs and Modbus protocol with actuators, control filter is placed as close as possible to devices (last distributed I/O module). This validation block communicates with actuators by digital outputs and another Modbus channel. So that, actuators are protected by the control filter.

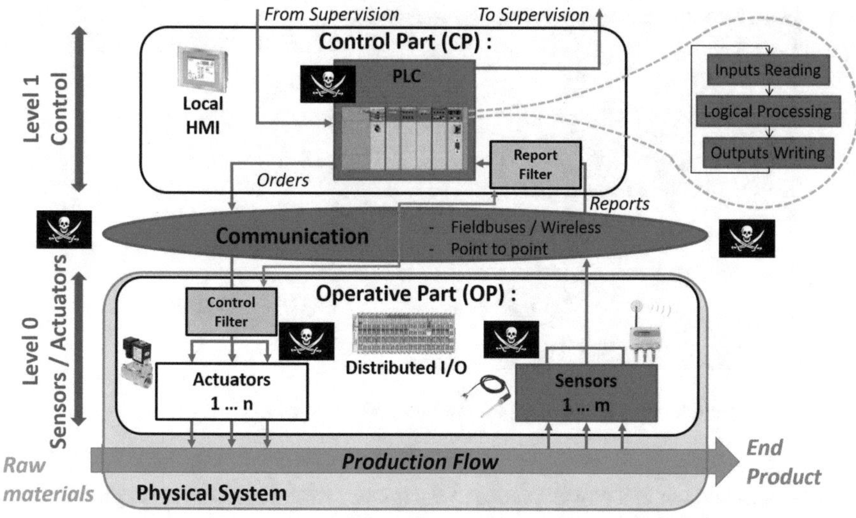

Fig. 5.6 Implementation of filters in ICS architecture in order to reduce attack surface for hackers

Report filter acquires data from sensors in the same way as control filter and transmits them to the PLC. Attackers may corrupt this communication as well as data inside PLC before logical processing. However, the main purpose of this block is to be a trust anchor for control filter by transmitting current state of the physical system and distance with predicted reports. That is also why only control filter can block information in our approach.

Finally, communication between the two filters is on a different network isolated from industrial network. Attackers cannot affect confidentiality, availability or integrity of data exchanged by the control or the report filter. Moreover, filters are considered as invulnerable to cyberattacks, for example by restricting memory access or conditions for uploading models. Algorithms inside filters and communication between them cannot be attacked which thus guarantee a reliable security solution for ICS. Illustration of filters implementation is available in Fig. 5.6.

The developed approach proposes simple, reliable, and secure implementation in the ICS architecture between levels 1 and 0. However, detection mechanism is now based on algorithms presented in the following sections.

5.3.1.2 Step 1: Risk Assessment

This first step is based on work done in risk assessment. This analysis is made on a critical part of the process to identify failure modes and to determine causes leading to these critical states. Fourastier and Pietre-Cambacedes [7] define this analysis by the verification of the ability of an entity to satisfy one or more requirement under given conditions.

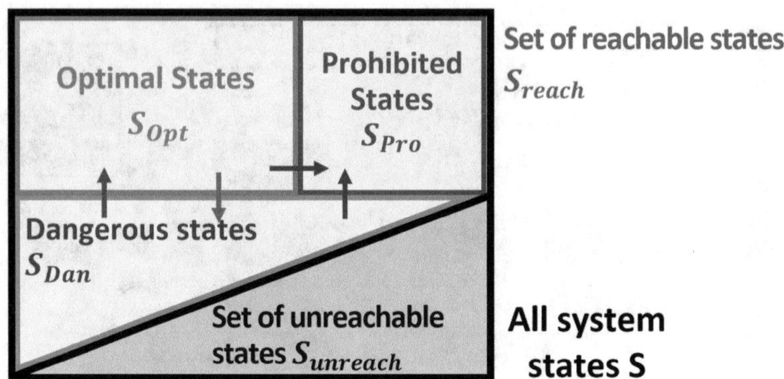

Fig. 5.7 Illustration of the different set of states possible in an ICS

Risk assessment is a powerful tool to design our filters and prevents cyberattacks. Indeed, when hackers launch an attack on an ICS, they always try to degrade the process (final product, production equipment, configurations, etc.) by bringing the system into critical states. Moreover, contrary to failure, an attack will always try to deteriorate the service (production flow) of the system.

ICS can be defined by a unique type of state as presented by Mitchell and Chen [22] and in Fig. 5.7. Risk assessment identifies critical states of the system and the necessary parameters for modeling the system. Indeed, this approach is designed to protect the most sensitive parts of ICS and not the entire infrastructure. For our study, EBIOS norm [38], recommended by the ANSSI, is used.

In this chapter, lowercases denote states and capital letters refer to sets of state. Thus, a state s_i that is in the set of system states S is composed by sensors and actuators values and step states of the system. This state s_i represents the current state of the system. An action a_j that is in the set of possible system actions A is PLC output and corresponds to orders that can be applied on production flow by actuators. A given state s_i can be reachable or not depending on effects of actions. This study will only focus on the set of reachable states S_{Reach}. Such states can be in only one subset at the same time:

- Optimal State s_{Opt} with $S_{\mathrm{Opt}} \subset S_{\mathrm{Reach}}$ respects the running of the process and the constraint imposed by the control law in the PLC. PLC is programmed to keep the system in this subset of safe state.
- Dangerous State s_{Dan} with $S_{\mathrm{Dan}} \subset S_{\mathrm{Reach}}$. In this subset, constraints imposed by control law are violated without inflicting critical damages on the process.
- Prohibit State s_{Pro} with $S_{\mathrm{Pro}} \subset S_{\mathrm{Reach}}$. Constraints of control law and integrity of the process are transgressed causing significant degradation on the system and its environment. This subset has to be avoided and the system stopped before being in this subset.

To conclude, A_{Opt} represents the set of orders respecting the control law and S_{Pre} and A_{Pre} respectively the set of states and orders predicted by filters.

5.3.1.3 Step 2: System States Exploration

Previous step identifies the fact that an ICS is in a unique subset of states at each time. Risk assessment makes the highlight on subset of critical states. These states damage the system (production flow, equipment, etc.) and its environment. Once causes and consequences are identified, we describe a pair {state s_i; action a_j} such that if action $a_j \in A$ is applied on system state $s_i \in S_{\text{Reach}} \backslash \{S_{\text{Pro}}\}$ the resulting state s_{i+1} belongs to prohibited states subset S_{Pro}. This pair is called context.

It should be noted that knowing action effects of the system allows detecting context leading to critical state. So that, the objective of this second step is to explore all the different states of the system, categorize them in the different subsets, and list all the contexts. The a priori knowledge of our system allows to identify sequences leading to one or several prohibited states and Sect. 5.3.1.4 will illustrate how distance between the current state and the forbidden state is calculated. All the necessary parameters for modeling the physical system stem from the risk assessment step. Indeed, detection mechanisms are based on this step where critical states that have to be detected are deduced as well as parameters representing the system.

To do so, control model and process model of the system are required. The first model organizes and schedule actions that have to be done on the system in order to reach the final state of the process in the fastest and safest way. It represents what we want to achieve on the system. We choose to describe control model with Petri nets. This representation allows expressing constraints easily, especially sequence properties (task scheduling), parallelism (execution of several tasks at the same time), mutual exclusion (execution of an activity prevents execution of others), and synchronization (waiting the end of one, or several, specific activity before executing others). Petri net is a place/transition net. Graphical and mathematical writings allow expressing pre and post conditions explicitly as well as the marking to fire a transition. For control model, places represent actions that are sent to the operational part (OP) and transitions are reports transmitted by sensors. Process model details all reachable states based on the description of actions and their effects on the system. It represents what is possible to do on the process. In the process model, a finite number of actions can be executed by the control part on the operative part so that the number of states is also limited. Modeling by using automaton is adapted to represent the process. In this work, only deterministic finite automaton M is used to illustrate detection mechanisms of cyberattacks. However, for more complex systems, hybrid automata can also be used as in [39] or [40]. Such automaton is a formal model for a dynamic system with discrete and continuous component. The following quintuplet defines the automaton M:

$$M = \{S_n, A_m, \delta, S_0, S_{\text{final}}\} \tag{5.1}$$

- S_n is the finite set of possible states for the system, in other words S_{Reach} where n denotes total number of states s_i. Vector s_i representing the state of the system can be in the subset of optimal states S_{Opt}, dangerous states S_{Dan} or prohibited states S_{Pro},

- A_m is the finite set of orders that can be performed by the system. That is the set A defined in previous part with m representing number of actions,
- δ is the transition function that represents relation between states and orders. These relationships will change the automaton by simulating the evolution of the system. To do that, each action has to be modeled by its effect on the process (see paragraph below),
- $S_0 \subset S_{\text{Reach}}$ is the set of initial states of the system,
- S_{final} is the set of final states. When one of these final states is reached, evolution in the automaton is stopped.

To obtain all reachable states of the system, we first need to define the transition function that models effects, conditions and associated constraints of each actuator on the production flow. To do so, works of Henry [41] are used. Behavior of each operation of the process is listed. Therefore, we define a transition function δ which associates a state s' to a pair {state s; action a}. We denote S_i the set of states before applying function δ and S_{i+1} the set of states obtained after function δ such that:

$$\forall s \in S_i \subset S_n, \forall a \in A_m \quad \text{such that} \quad s' = \delta(s, a) \in S_{i+1} \subset S_n \tag{5.2}$$

Based on Eq. (5.2), we generalize the transition function δ to the set of states. Thus, the set of child states S_{i+1} is obtained by applying the set of action $\mathrm{A_i}$ to a set of states $\mathrm{S_i}$ such that:

$$S_{i+1} = \{s' = \delta(s, a) \mid s \in S_i \subset S_n, a \in A_m\} \tag{5.3}$$

Algorithm computes image of S_i through transition function δ. So that, the automaton M represents evolution of the process. M starts with the initial state s_0 ($S_i = s_0$) and compute evolutions S_{i+1} by applying each possible orders $a_i \in A_m$. Then, set S_{i+1} is reduced if some state s' has already been explored during previous iteration, in other words $s' \in \cup S_i$. These states are not considered because they have been already met which means that this path has already been studied: similar actions lead to similar states. Thus, for each state of $\mathrm{S_i}$, images through δ are computed. Algorithm ends when final states are found. States of S_{i+1} are composed of:

- Initial state S_0: $S_{i+1} \cap S_0 \neq \{\emptyset\} \leftrightarrow \exists s' \in S_{i+1}$ such that $s' \in S_0$,
- Prohibited states $s' \in S_{\text{Pro}} \subset S_{\text{Reach}}$ because when these states are reached no further evolution of the process is possible (production flow is severely impacted): $S_{i+1} \cap S_{\text{Pro}} \neq \{\emptyset\} \leftrightarrow \exists s' \in S_{i+1}$ such that $s' \in S_{\text{Pro}}$. Let F be the set of intersection between S_{i+1} and S_{Pro},
- Loops ω: one or several states s' have already been met, which means that this path has been studied: $S_{i+1} \subset \cup S_i \leftrightarrow \exists s' \in S_{i+1}$ s.t. $s' \in \cup S_i$. These states are called previous states s_{prev} of the system and composed the set S_{prev}. Let P be the set of intersection of S_{i+1} and $\cup S_i$.

This algorithm explores all possible combinations for the system by applying every possible sequence on the process model. Combinatory explosion is prevented by deleting states already computed and the choice of parameters for modeling the system. Thus, a unique branch is explored at each iteration.

Contexts leading to prohibited states s_{Pro} are listed in the set S_{Pro}. These contexts can be easily identified because state $s' \in S_{i+1}$ as defined in Eq. (5.2) also belongs to prohibited state subset $s' \in S_{\mathrm{Pro}}$. The concept of finding contexts that lead to critical states in order to compute a notion of distance is closely similar with the theory of prognosability developed in Genc and Lafortune [42] or Chen and Kumar [43]. A prognoser that predicts a fault at least m-steps before the occurrence is built. These works provide significant results for failure prognosis but they do not take into account the context of cyberattacks explained in previous sections. Indeed, the normal and abnormal behaviors of the system must be defined in the approaches, that is not the case in our approach. In our approach, we just have to specify normal behaviors. Despite the similarities, our work focuses more on the distance from current state to the nearest critical state (on line prediction of system trajectory) rather than identification of contexts leading to prohibited states (safeguard in case of attack that cannot or has not been predicted). Optimal states s_{Opt} are obtained by applying control law orders A_{Opt} on the process model. Set of dangerous states S_{Dan} regroups all states that are not classified in previous subsets S_{Pro} and S_{Opt}.

At the end of this step, control model and process model are designed and sequences leading to prohibited states have been identified. Following step makes use of these results to build detection mechanisms.

5.3.1.4 Step 3: Detection Mechanisms

The first step identifies context leading to critical states of the system and the second step allows us to write an algorithm for classifying all reachable states and finding sequences leading to prohibited states. During this third stage, detection mechanisms are implemented into filters in order to block "wrong" orders and as far as possible anticipate deviations. These two objectives are linked to the detection part of our filter. Distinction between failures and cyberattacks is performed after detection of an anomaly.

"Static" Rules: Immediate Blocking

The purpose of this part of the filter is to identify contexts leading to critical states. To do so, current context, which is the current pair order and state, is compared with contexts contained in Matlab structure Forbidden_States.mat obtained by automaton M during step 2. These contexts tally to the states preceding prohibited states and associated to an action. If the current context matches with one of the contexts contained in this structure then order is blocked. This mechanism is implemented only in the control filter. Indeed, this block is the last bulwark in the ICS architecture to prevent an order to be executed. Report filter could also stop reports but there is no immediate danger by sending wrong report to CP. Indeed, if this report results in a wrong order leading to critical state, it will be blocked by control filter. So that, no such detection mechanism is implanted in report filter. Rules R are a set of contexts leading to prohibited states such that:

$$R = \text{true} \leftrightarrow \exists s' \in S_{i+1} \subset S_n, \exists s \in S_i \subset S_n, \exists a \in A_m \quad \text{such that} \quad s' = \delta\,(s, a) \in S_{\text{Pro}} \tag{5.4}$$

Rules are a static security measure that protects the system against immediate danger by blocking wrong order. However, step two provides models of control and process that is used in the next section to anticipate deviations. Indeed, with this paragraph, we anticipate by using the algorithm with a short prediction horizon.

"Dynamic" Detection: Anticipation of Possible Deviations

The main objective of this part of the filter is to identify deviations and send alarms to operator. To do so, prediction horizon of our approach is increased by improving algorithm proposed in step 2. Filters do not look at the previous states before forbidden states but go up in the arborescence to detect possible deviation earlier. Therefore, some actions can be defined as legitimate or not by knowing the current state of the system. In this sense, notion of distance from critical states is defined and informs about proximity of current state with prohibited states. Carcano [31] has announced an interesting concept of distance but that cannot be used for discrete systems; we will explain why after having defined it. Distance is based on the comparison of the current state s and the prohibited states represented by the critical formula Φ. Φ is defined as the set of constraints c_i that the system has to respect. Distance d is defined as the minimum $d_{\min}$ between two types of distance computation (d_1 and d_v) described below such that:

$$d : \Re^n \times \Re^n \to \Re^+, s \in \Re^n$$

$$d\,(s, \Phi) = \min_j d_{\min}\left(s, C_j\right) \text{ with } d_{\min} = d_1 = \sum_i^n |s_i - c_i| \;\; or$$

$$d_{\min} = d_v = \#\,\{i | s_i \neq c_i\} \tag{5.5}$$

d_1 computes the gap between two states components by components and d_v counts difference between two vectors based on the number of different components. Distance to critical state is the minimal distance d_1 or d_v between current state and critical formula. This notion is interesting because distance between current state and prohibited state can be computed that provides information to operators. Moreover, distance to optimal states can also be computed in order to know if the process follows optimal trajectory. To conclude, based on the models of second step, deviations are detected faster than with static rules. However, this notion of distance gives a relevant information only for continuous values. Indeed, for discrete systems with all-or-nothing sensors, distance information is useless. If actions applied on the system deviate from control law, then the system directly goes into dangerous or forbidden state. No information is provided by such a distance. For example, a system can be at a distance of 10 from a prohibited state and sending an order may lead to a forbidden state. To develop detection mechanisms for systems modeled with hybrid automata, this notion of distance has to be extended to discrete values.

Based on the models developed in Sect. 5.3.1.3, notion of distance is enriched by the number of actions to apply before being in a prohibited state. So that, algorithm is improved to compute the shortest possible way from current state to prohibited states, in other words the minimal number of actions that can be applied on the process before a prohibited state. Adaptation of second step algorithm, presented in Algorithm 5.1, returns the number of stages before a prohibited state, sequence of orders that leads to this state and states path. So that, drawing from (2), we denote a set transition function Δ such that:

$$\text{For } a \in A_m, \forall S' \subset S_n, \forall S \subset S_n, S' = \Delta(S, a) \tag{5.6}$$

Thus, distance notion can be defined as minimum number of iterations to reach a prohibited state from a state $s \in S_i \subset S_n$:

$$D(s|S_{\text{Pro}}) = \min_n \ \Delta^n(S_i, a) \in S_{\text{Pro}} \forall s \in\in S_i \subset S_n, \forall a \in A_m \tag{5.7}$$

This solution provides information about critical states as the same way as prediction of checkmate during chess game.

Algorithm 5.1 Algorithm to find the nearest prohibited state from state $s \in S_i \subset S_n$ **(shortest possible way)**

```
Function Find nearest Prohibited State
Iteration = 0
WHILE (Stop ≠ 0)
        Iteration = Iteration +1
        S' = Δ(Si, A)
        IF (S'∩SPrev ≠ {Ø})THEN
                S' = S'\P
                SPrev = SPrev ∩ S'
        ELSE
                SPrev = SPrev ∩ S'
        ENDIF
        IF (S'∩SPro ≠ {Ø}) THEN
                S' = S'\F
                Stop = 0
        ENDIF
Si = S'
DONE
End
```

At the end of this algorithm, the way and the distance $D(s|S_{\mathrm{Pro}})$ from the starting state s to the nearest prohibited state s_{Pro} is computed.

To conclude, detection of anomaly is efficient and filters protect the system against them: blockage of immediate danger and anticipation of deviations. When data is received to be checked, filters proceed as follows:

- Estimation of the context (blockage of immediate danger),
- Reconstruction of current state, computation of shortest way to prohibited states for discrete values and distance in the sense of Carcano for continuous components. So that, filters compute actions that have not to be performed by the system and distance to forbidden states,
- Check between actual data and result of previous computation (sending an alert to supervision if needed).

Two types of attacks have to be distinguished: brutal and sequential attacks. Contrary to the last type of attacks, in brutal attacks, hackers want to bring the system into a prohibited state as soon as possible. Rules, based on the contexts and Eq. (5.4), protect the system against brutal attacks. By definition, these attacks are not easily predictable. The concept of distance is more appropriate for sequential attacks to detect deviations from the control law (see Algorithm 5.1).

In case of detection, previous sequences of control are checked to determine if the cause of this detection is a failure or an attack. Main assumption for discriminating is to assume that an attacker will always seek to damage the system immediately or later. So that, distance will always tend to toward 0. On the opposite, a failure is a non-intentional event that punctually will tend the distance toward 0. Thus, discrimination is based on models developed during step 2. Indeed, the normal behavior of a system is characterized, in Fig. 5.8, by a trajectory always equal to zero for distance between sent orders and expected orders (top left figure). Equally, distance that measures gap between predicted states and optimal ones is equal to zero (middle left figure). Finally, trajectory of shortest path to a forbidden state

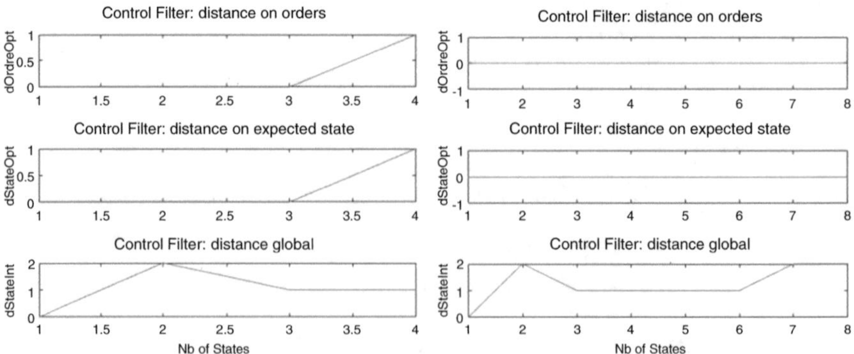

Fig. 5.8 Trajectories computed by control filter for a system with a normal behavior (right) and a system under attack (left)

evolves with states modifications (bottom left figure). During an attack, if filters detect that sent order is different from optimal order, the distance is different from 0 (top right figure). Thus, optimal state expected is different from current state and distance increases (middle right figure) as well as the distance of shortest path to forbidden state is decreasing (bottom right figure). The notion of distance combined with control law model allows detecting some attacks on ICS; however, if an attacker mimics failure or launches his attack at a specific moment, our approach will block the attack but will not distinguish with failure. This aspect is detailed in Sect. 5.4 "Application on an example".

Concept of trajectory, which is the evolution of distance across time, has to be developed in future works. Correlation algorithms with previous states and control model will also be developed.

5.3.2 About Filters for ICS Cybersecurity

As presented in Sect. 5.3.1.1 filters are divided into two validation blocks: a control filter that analyzes orders and report filter which correlates data sent by sensors with information of the other filter. In this section, details about filters and their implementation are provided.

Control filter receives an order to evaluate from PLC and information from report filter (list of predicted orders and last validated report). Filter analyzes the order with rules and trajectory detection mechanisms. Finally, order can be:

- Validated. Thus, order is sent to actuators and data is transmitted to report filter as the validated order and the list of expected report,
- Validated with an alarm. The same previous procedure is done but with an alarm sent to operators,
- Stopped. Thus, the filter blocks order, it is not transmitted to the operative part and an alert is sent to report filter and control room.

Report filter received data from sensors of operational part and information of control filter (list of expected report and last order validated). Algorithm computes distance to prohibited states. Then, report is sent to control part and data are transmitted to control filter (list of orders expected and last report validated).

Integration of these filters is presented in Sect. 5.3.1.1, control filter is located just before actuators and report filter just before PLC, as trust anchor for the other filter. The communication between both filters is considered as secured (on a different network than industrial network for example). Programming of filters will reduce vulnerabilities with interpreter inside validation blocks to secure the code with models.

Illustration of control and report filters is available in Fig. 5.9.

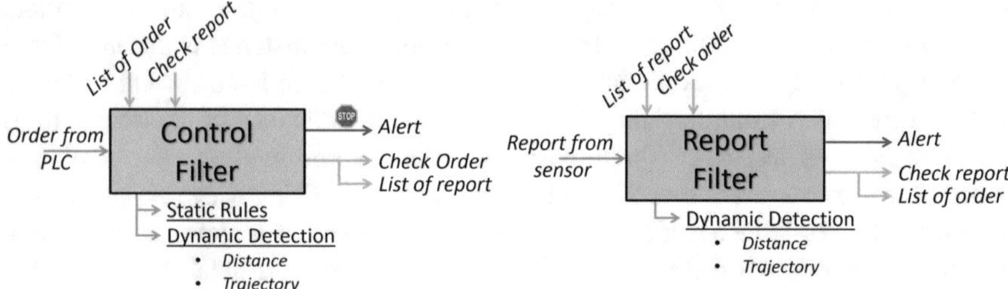

Fig. 5.9 Representation of inputs/outputs and processing for control and report filters

5.4 Application on an Example

To illustrate this work, a well-known example of the literature is used as provided in [17]. The system, illustrated in Fig. 5.10, is composed of three tanks. Two tanks T_1 and T_2 of infinite capacity contain respectively product A and product B. Each tank discharges its product into a melting tank T_3 in order to produce a product C. The filling stage is done by opening valve V_1 and V_2. Level sensors show the height of the product in the tank. Three sensors are used in this system: H_0 indicates the draining of tank T_3, H_1 the quantity of product A, and H_2 the quantity of product B. Control law imposes a filling by T_1 then T_2. When sensor H_2 is activated, valve V_3 opens to drain T_3. A new cycle of production starts.

The step of risk assessment identifies only one forbidden state for this system that is reached when level in tank T_3 exceeds sensor H_2. Moreover, necessary parameters to describe the process have been identified. So that, the system is defined by vectors and sets as follow:

- State vector s represents correct behavior of the system, faults are not considered in this model. Thus, the vector s is composed by sensors and actuators values as $s =$ [Sensors values; Actuators states] $= (H_i, V_1, V_2, V_3)$ with $H_i \in \{0 \ldots 2\}$, $V_1 \in \{0, 1\}$, $V_2 \in \{0, 1\}$, $V_3 \in \{0, 1\}$ where H_i indicates which sensor is activated in the state vector (0 no sensor is activated, 1 sensor H_1 is activated, and 2 sensor H_2 is activated). The set of states S includes at most 25 states (24 possible states and the prohibited state),
- Order vector represents actions that can be executed on the system. $a = (a_1, a_2, a_3)$ with $a_i \in \{0$ closes valve V_i, 1 opens valve $V_i\}$ $i \in \{1 \ldots 3\}$,
- Reachable state set S_{Reach} is composed of 13 states which contains 6 optimal states $S_{\text{Opt}} = \{[0\ 0\ 0\ 0], [1\ 1\ 0\ 0], [1\ 0\ 0\ 0], [2\ 0\ 1\ 0], [2\ 0\ 0\ 0], [0\ 0\ 0\ 1]\}$, 1 forbidden state $S_{\text{Pro}} = s_{\text{Overflow}}$ and 6 dangerous states $S_{\text{Dan}} = S_{\text{Reach}} \backslash (S_{\text{Opt}} \cap S_{\text{Pro}}) = \{[1\ 0\ 1\ 0], [2\ 1\ 1\ 0], [0\ 1\ 0\ 1], [0\ 0\ 1\ 1], [0\ 1\ 1\ 1], [2\ 1\ 0\ 0]\}$,
- unreachable state set $S_{\text{Unreach}} = S \backslash S_{\text{Reach}}$ contains 12 states,

In this example, we assume that the execution of an order on the system immediately fills/drains the tank and thus the direct activation of the upper/lower level sensor. By removing this simplifying assumption, which facilitates the illustration

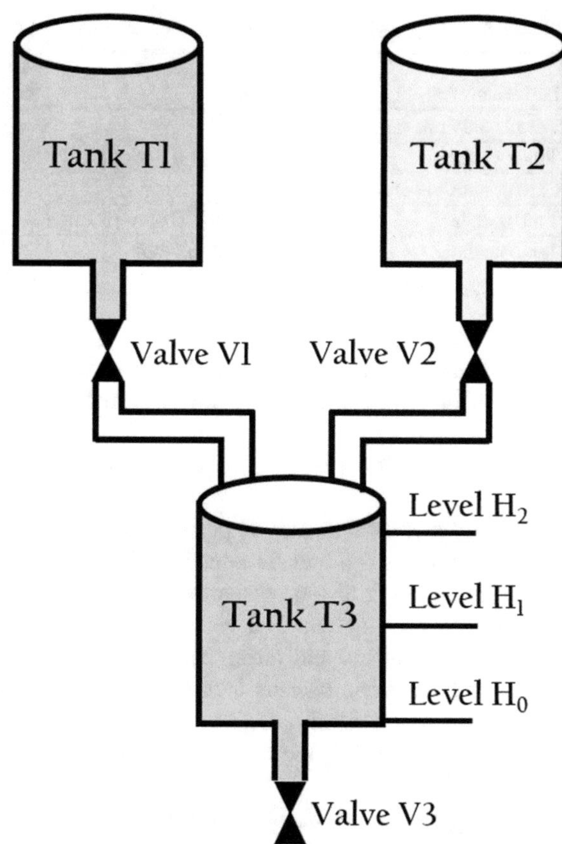

Fig. 5.10 Illustration of application example: 2 tanks discharging products in tank T_3

of the presented detection algorithms, the exploration step finds: 24 out of 25 possible states are reachable states with 9 optimal states, 14 dangerous states, and 1 prohibited state.

Prohibited and dangerous states are different in the sense that we will try to avoid the first one (for example, an overflow characterized by $\delta([2\ 0\ 0\ 0], [1\ 0\ 0]))$ while the others are a transition between optimal and forbidden states (these states are neither critical nor optimal). For example, the filing state [2 1 1 0] of tank T_3 by valves V_1 and V_2 while T_3 is empty. The control model is directly designed from the control law of the system, described from explanations at the beginning of this section. The process model is obtained by the automaton M and is used to obtain contexts and sequences leading to prohibited states regrouped in Table 5.4:

$$M = \{S_{25}, A_8, \delta, s_0, S_{\text{final}}\} \tag{5.8}$$

with S_{25} the set of possible states for the system, A_8 the set of orders that can be applied (all possible combinations), s_0 the initial state ($s_0 = [0\ 0\ 0\ 0]$), S_{final} the set

Table 5.4 Risk assessment of the system

Forbidden states S_{Pro}	Order $a \in A_8$	State $s \in S_{25}$ of the process before the action $a \in A_8$
Case 1: only one order		
Tank overflow	Open V_1 or V_2: [1 0 0] or [0 1 0]	Sensor H_2 activated: $s = [2\ x\ x\ x]$
Case 2: multiple orders		
Tank overflow	Open V_1 or V_2: [1 0 0] or [0 1 0]	Sensor H_2 activated: $s = [2\ x\ x\ x]$
Tank overflow	Open V_1 and V_2: [1 1 0]	Sensor H_2 or H_1 activated: $s = [2\ x\ x\ x]$ or $s = [1\ x\ x\ x]$

of final states composed by the initial state s_0, prohibited state s_{Overflow} or a loop and δ the transition function. For example, a final state can be characterized by:

$$\forall s \in S_{25} \backslash S_{\text{Pro}}, a_3 \in A_8 \quad s.t. \quad s' = \delta(s, a_3) = s_0 \tag{5.9}$$

The exploration state step is performed: only 13 states are reachable in the set of possible states S_{25} and 69 contexts, that lead to critical states, are identified. Then, during the third step, distances to critical states are computed and detection mechanisms are implemented in filters depending on the location in the ICS architecture (contexts and distance). A "Man in the middle" attack is simulated on our system. When hackers launch an attack, all the orders are intercepted and replaced by a predefined order. In this example, order $a \in A_8$ is replaced by order $a_{\text{attack}} \in A_8$ and replays the first action of the control law $a_1 = [1\ 0\ 0]$ that opens valve V_1.

During the first cycle, filters detect no deviation because attack corresponds to optimal order. The system goes from an optimal state (initial state) to another optimal state (filling of tank). Everything is good until sensor H_1 is activated. Then, PLC sends order $a_2 = [0\ 1\ 0]$ for opening valve V_2 but this action is replaced by a_{attack}. Order is sent because context does not lead to prohibited state. Moreover, from a distance point of view, control filter does not detect an issue because filling state is reached. As algorithm takes into account sequences of actions an alert is raised to warn operators that a bad order has been sent to the process. When sensor H_2 is activated, order $a_3 = [0\ 0\ 1]$ is sent to drain tank T_3. When control filter received this order, it is blocked because one of the security rules has been violated and application of a_{attack} leads to prohibited state. Discrimination is also possible because the control law was not respected several times during this attack. So that, distance between received and expected orders is not equal to zero.

Now, anticipation of attack is studied. Let us consider an attack sequence that respects the control law but an order of filling is sent instead of draining. Control filter blocks the order immediately because it leads directly to prohibited state. Contrary to the last example, filters cannot anticipate the blockage because attackers have a high knowledge of the system and the attack occurs at the right moment to prevent anticipation. However, the physical system is not damaged.

Now we consider an attack that replaces any order by the following one $a_{\mathrm{attack2}} = [1\ 1\ 0]$ that opens the two valves V_1 and V_2. At the first cycle, when control filter received this order two alerts are raised. One because control law constraints are violated and the system moves away from the optimal trajectory and the other because child state leads quickly to forbidden state (3 transitions). After filling state, sensors H_0, H_1, and H_2 are activated. When control filter received a_{attack2}, the order is blocked because one of the security rules is not respected. Distinction identifies attack because sequence orders does not respect the control law and a failure on valves does not open these components. Thus, an attack on ICS is detected and identified by the filters.

To conclude, a failure occurs on our system and sensor H_1 breaks down. PLC that sends order a_1 will not receive event H_1 activated but H_2 activated. Indeed, control filter will accept the order waiting for event H_1. The report filter will detect an anomaly by receiving the event H_2 activated whereas event H_1 is expected. Distinction based only on distance cannot be done. Indeed, failure and attack are possible causes of this anomaly. For example, an attacker may inject a false report to fail the control law. In this example, the system is protected from damage but distinction is not possible.

These examples show that type of attacks and hackers knowledge impact performances of filters on anticipation and distinction criteria. Thus, attacks that mimic failures are impossible to distinguish. However, filters protect the system by preventing execution of illicit order that leads the system in prohibited state. To complete our case study, a step of heating is added on the product C just before emptying the tank T_3. Our model of the process becomes a hybrid automaton because a continuous component appears in the system (temperature). In order to monitor the evolution of this variable, the use of distance in the sense of Carcano is necessary. Thus, if the temperature gets close to a critical area (too high for example), then the filter will be able to detect it and stop the heating of the tank. This mechanism completes previous ones when systems have continuous variables.

5.5 Conclusions and Perspectives

5.5.1 Conclusions

Industrial Control Systems are used in many critical infrastructures and application domains to insure productivity and safety. Introduction of Ethernet TCP/IP or other technologies inherited from Information Systems improves on several aspects these ICS. However, IT solutions also bring vulnerabilities that have not been anticipated and fixed. Hence, hackers exploit vulnerabilities to target ICS and severely damage physical systems, goods, and people. This work highlights the need of detection mechanisms to stop bad orders and anticipate deviations. Thus, the main issue is to distinguish an attack from a failure that leads to the scientific problem. Then, a technological obstacle is alighted on the implantation in an existing architecture.

Our approach proposed to take advantages of security (Intrusion Detection System) and safety solutions (Filter Approach) in a mutual reinforcement. Our study offers ICS a solution for level 1 and 0, as the last bulwark for protecting the system. Thus, filters implementation for detecting cyberattacks and methodology for synthetizing these verification blocks are enounced. Integration in ICS architecture is easy and reduces the surface contact for an attacker. Methodology of conception takes advantage of risk assessment for finding critical states as well as necessary parameters and also Petri nets and deterministic-finite automaton models for detection mechanisms. So, any action leading to critical states is blocked by control filter. Notions of distance by computing shortest way to prohibited states are explained to anticipate deviations for discrete events systems and indicate if ICS get closer to prohibited states. Adaptation of detection mechanisms for hybrid systems, with continuous components, has also been highlighted. The application example shows efficiency of our approach to protect these systems.

However, the concept of distance alone does not provide solutions to distinguish a failure from an attack. The concept of trajectory, which follows evolution of distance across time, correlated with previous sequences of states and actions opens new perspectives.

5.5.2 Perspectives

Distinction between attacks and failures is clearly an issue for our work. Indeed, they both lead to a loss of service of the installation but, in case of attacks, hackers will always try to bring the system into a prohibited state in order to make it inoperative. Thus, by studying evolution of distance and sequence of orders sent to the process, our work will be able to identify attacks. To do so, evolution of distance between current and prohibited states is studied in parallel with evolution of the orders sent. Indeed, a system can completely get closer to a prohibit state and respect the control law. For example, the order to open a valve fulfilling a tank respects the control law but brings prohibited state of overflow closer. If sequence orders break control law, a hacker may damage the system. The limit of this approach is the identification of an attack when hackers mimic a failure on the system. An other issue can come from the risk assessment step if prohibited states are not identified or if some parameters are missing to describe the system. Machine learning can be a good solution to endow filters with adaptive capacities. This solution will be studied in future works. However, learning approaches face several issues in the context of cyberattacks, mainly the learning structure and the volume and the quality of data computed.

Moreover, in this paper, filters take into account only combinational constraints, detection mechanism can be improved by adding temporal windows in our process model. In this paper, an approach for protecting critical parts of a process has been developed. It focuses only on the lower levels of the CIM architecture which makes it possible to reduce the complexity of the studied system by targeting only the most critical parts. This implies that not all the system is protected. The question of applicability of this approach on higher levels will be raised, knowing that the complexity will also grow. So that, our approach need to be improved to deal with multiple discrete modes in the case of large scale systems.

However, our approach presents good results for this example and tests on larger examples will be conducted. Finally, the implementation of this solution on real industrial platforms is scheduled for the coming months. Time detection and influence of architecture on detection rate will be studied.

Acknowledgments This research was supported by the Direction Generale de l'Armement (DGA) Maîtrise de l'Information based in Bruz, France.

References

1. K. Stouffer, V. Pillitteri, S. Lightman, M. Abrams, A. Hahn, *Guide to Industrial Control Systems (ICS) Security*, National Institute of Standards and Technology, NIST SP 800-82r2 (2015)
2. J. Clarhaut, N. Dupoty, F. Ebel, J. Hennecart, F. Vicogne, *Cyberdéfense: La sécurité de l'informatique industrielle (domotique, industrie, transports)* Editions (ENI, France, 2015)
3. E.D. Knapp, *Industrial Network Security: Securing Critical Infrastructure Networks for Smart Grid, Scada, and Other Industrial Control Systems*, 2nd edn. (Elsevier, Waltham, 2014)
4. Y. Ashibani, Q.H. Mahmoud, Cyber physical systems security: analysis, challenges and solutions. Comput. Secur. **68**, 81–97 (2017)
5. ODVA, *EtherNet/IP - CIP on Ethernet Technology* (2016)
6. EtherCAT Technology Group, *EtherCAT: The Ethernet Fieldbus* (2012)
7. Y. Fourastier et al., *Pietre-Cambacedes, Cybersécurité des installations industrielles: défendre ses systèmes numériques*. Cépaduès Editions, 2015
8. S. McLaughlin et al., The cybersecurity landscape in industrial control systems. Proc. IEEE **104**(5), 1039–1057 (2016)
9. F. Khorrami, P. Krishnamurthy, R. Karri, Cybersecurity for control systems: a process-aware perspective. IEEE Des. Test **33**(5), 75–83 (2016)
10. RISI - The Repository of Industrial Security Incidents, 09 Sept 2016. [En ligne]. Disponible sur: http://www.risidata.com/Database/event_date/asc. Consulté le: 09 Sept 2016
11. M. Abrams, J. Weiss, *Malicious Control System Cyber Security Attack Case Study - Maroochy Water Services* (Secur. Water Wastewater Syst, Australia 2008)
12. N. Falliere, L.O. Murchu, E. Chien, W32. stuxnet dossier. Symantec Security Response, Version 1.4, févr. (2011)
13. R.M. Lee, M.J. Assante, T. Conway, German steel mill cyber attack, in *SANS ICS 2014* (2014)
14. R.M. Lee, M.J. Assante, T. Conway, Analysis of the Cyber Attack on the Ukrainian Power Grid, in *SANS ICS 2016* (2016)
15. ICS-CERT, ICS-CERT/The Industrial Control Systems Cyber Emergency Response Team, 15 Sept 2016. [En ligne]. Disponible sur: https://ics-cert.us-cert.gov/. Consulté le: 15 Sept 2016
16. M. Caselli, E. Zambon, F. Kargl, Sequence-aware intrusion detection in industrial control systems, in *Proceedings of the 1st ACM Workshop on Cyber-Physical System Security* (New York, NY, 2015), pp. 13–24
17. W. Li, L. Xie, Z. Deng, Z. Wang, False sequential logic attack on SCADA system and its physical impact analysis. Comput. Secur. **58**, 149–159 (2016)

18. Y. Wang, Z. Xu, J. Zhang, L. Xu, H. Wang, G. Gu, SRID: state relation based intrusion detection for False data injection attacks in SCADA, ed. by M. Kutyłowski, J. Vaidya, in Computer Security - ESORICS 2014 (Springer, Heidelberg, 2014), pp. 401–418
19. J. Graham, J. Hieb, J. Naber, Improving cybersecurity for industrial control systems, in *2016 IEEE 25th International Symposium on Industrial Electronics (ISIE)* (2016), pp. 618–623
20. D.-T. Nguyen, *Diagnostic en ligne des systèmes à événements discrets complexes: approche mixte logique/probabiliste*, (Université Grenoble Alpes, Français, 2015)
21. D.E. Denning, An intrusion-detection model. IEEE Trans. Softw. Eng. **13**(2), 222–232 (1987)
22. R. Mitchell, I.-R. Chen, A survey of intrusion detection techniques for cyber-physical systems. ACM Comput. Surv. **46**(4), 1–29 (2014)
23. S. Cheung, B. Dutertre, M. Fong, U. Lindqvist, K. Skinner, A. Valdes, Using model-based intrusion detection for SCADA networks, in *Proceedings of the SCADA security scientific symposium*, vol. 46 (2007), pp. 1–12
24. N. Goldenberg, A. Wool, Accurate modeling of Modbus/TCP for intrusion detection in SCADA systems. Int. J. Crit. Infrastruct. Prot. **6**(2), 63–75 (2013)
25. R.R.R. Barbosa, R. Sadre, A. Pras, Difficulties in modeling SCADA traffic: a comparative analysis, in *Passive and Active Measurement*, vol. 7192 (Berlin, Germany, 2012), pp. 126–135
26. R.R.R. Barbosa, R. Sadre, A. Pras, Flow whitelisting in SCADA networks. Int. J. Crit. Infrastruct. Prot. **6**(3–4), 150–158 (2013)
27. C. Zimmer, B. Bhat, F. Mueller, S. Mohan, Time-based intrusion detection in cyber-physical systems, in *Proceedings of the 1st ACM/IEEE International Conference on Cyber-Physical Systems* (New York, NY, 2010), pp. 109–118
28. C. Bellettini, J.L. Rrushi, A product machine model for anomaly detection of interposition attacks on cyber-physical systems, ed. by S. Jajodia, P. Samarati, S. Cimato, in *Proceedings of The Ifip Tc 11 23rd International Information Security Conference* (Springer, Boston, 2008), pp. 285–300
29. S. McLaughlin, Blocking unsafe behaviors in control systems through static and dynamic policy enforcement, in *Proceedings of the 52nd Annual Design Automation Conference*, (New York, NY, 2015), pp. 55:1–55:6
30. S. Pan, T. H. Morris, U. Adhikari, V. Madani, Causal event graphs cyber-physical system intrusion detection system, in *Proceedings of the Eighth Annual Cyber Security and Information Intelligence Research Workshop* (New York, NY, 2013), pp. 40:1–40:4
31. A. Carcano, A. Coletta, M. Guglielmi, M. Masera, I. Nai Fovino, A. Trombetta, A Multidimensional critical state analysis for detecting intrusions in SCADA systems. IEEE Trans. Ind. Inform. **7**(2), 179–186 (2011)
32. I.N. Fovino, A. Coletta, A. Carcano, M. Masera, Critical state-based filtering system for securing SCADA network protocols. IEEE Trans. Ind. Electron. **59**(10), 3943–3950 (2012)
33. É. Zamaï, Architecture de surveillance-commande pour les systèmes à événements discrets complexes, PhD thesis, Université Paul Sabatier - Toulouse III (1997)
34. M. Combacau, M. Courvoisier, A hierarchical and modular structure for FMS control and monitoring, in *Proceedings [1990]. AI, Simulation and Planning in High Autonomy Systems* (1990), pp. 80–88
35. L.E. Holloway, B.H. Krogh, Monitoring behavioral evolution for on-line fault detection, in *IFAC/IMACS International Conference "Fault Detection, Supervision and Safety for Technical Processes", SAFEPROCESS'91* (Baden Baden, Germany, 1991), pp. 313–319
36. D. Cruette, J.P. Bourey, J.C. Gentina, Hierarchical specification and validation of operating sequences in the context of FMSs. Comput. Integr. Manuf. Syst. **4**(3), 140–156 (1991)
37. J.M. Flaus, *Risk Analysis: Socio–technical and Industrial Systems* (Wiley, Somerset, 2013)
38. ANSSI. Ebios méthode de gestion des risques (2010)
39. G. Zhou, G. Biswas, W. Feng, A comprehensive diagnosis of hybrid systems for discrete and parametric faults using hybrid I/O automata, in *9th IFAC Symp. Fault Detect. Superv. AndSafety Tech. Process. SAFEPROCESS 2015*, vol. 48, issue 21 (2015), pp. 143–149

40. A. Favela, H. Alla, J.M. Flaus, Modeling and analysis of time invariant linear hybrid systems, in *1998 IEEE International Conference on Systems, Man, and Cybernetics, 1998*, vol. 1 (1998), pp. 839–844
41. S. Henry, E. Zamaï, M. Jacomino, Logic control law design for automated manufacturing systems. Eng. Appl. Artif. Intell. **25**(4), 824–836 (2012)
42. S. Genc, S. Lafortune, Predictability of event occurrences in partially-observed discrete-event systems. Automatica **45**(2), 301–311 (2009)
43. J. Chen, R. Kumar, Stochastic failure prognosability of discrete event systems. IEEE Trans. Autom. Control **60**(6), 1570–1581 (2015)

Chapter 6
Active Diagnosis for Switched Systems Using Mealy Machine Modeling

Jeremy Van Gorp, Alessandro Giua, Michael Defoort, and Mohamed Djemaï

6.1 Introduction

Switched systems are systems involving both continuous and discrete dynamics. They can describe a wide range of physical and man-made systems (i.e., power converters, multi-tank systems, transmission systems, etc.). They have been widely studied during the last decade (see, for instance, [15]). Most of the attention has been focused on stability and stabilization problems [2, 14, 15, 18]. In the power electronic field, since the 1950s, power converters are used in traction systems, power supplies, or numerical amplifiers. Among these systems, multicellular converters, which appeared at the beginning of the 1990s are based on the association in series of elementary commutation cells. The *multicellular converter* is an interesting switched system widely studied in the literature on control, observation, and diagnosis. Its structure enables the reduction of the losses due to the commutations of power semiconductors while allowing low cost components. A blocked cell or a blocked switch or the internal components ageing can lead to critical situations for the system if the control law is not broken off or adapted (tolerant control).

J. Van Gorp (✉)
Conservatoire National des Arts et Métiers (CNAM), CEDRIC - LAETITIA, Paris, France
e-mail: jeremy.van_gorp@cnam.fr

A. Giua
Aix Marseille Université, CNRS, ENSAM, Université de Toulon, Marseille, France
DIEE, University of Cagliari, Cagliari, Italy
e-mail: alessandro.giua@lsis.org; giua@diee.unica.it

M. Defoort · M. Djemaï
University of Valenciennes, CNRS, UMR 8201 - LAMIH, Valenciennes, France
e-mail: michael.defoort@univ-valenciennes.fr; mohamed.djemai@univ-valenciennes.fr

M. Sayed-Mouchaweh (ed.), *Diagnosability, Security and Safety of Hybrid Dynamic and Cyber-Physical Systems*, https://doi.org/10.1007/978-3-319-74962-4_6

Occurrence of faults can be extremely detrimental, not only to the equipment and surroundings but also to the human operator if they are not detected and isolated in time. Moreover, usually, a fault tolerant controller [16, 23] cannot be applied if the fault is not isolated, i.e., if the exact nature of the fault that has occurred is not identified. Fault detection and isolation (FDI) have been widely investigated using various methods [8, 11, 12]. Observer-based FDI techniques rely on the estimation of outputs from measurements with the observer in order to detect the fault. The observability and observer design problems for hybrid systems have been studied using different approaches. The Z-observability concept was introduced in [13] to study the observability of some particular classes of hybrid systems. Using a similar approach in [24], it is provided a generalization of observability concepts. Analytical redundancy, i.e., mathematical relationship between measured and estimated variables in order to detect possible faults, can be computed by the analysis of the parity space [9, 29] or using a Bond Graph [17], for instance. However, due to the particular structure of the multicellular converter, the state components are only partially observable for every fixed configuration of the switches. Hybrid observers have been proposed for this system [7, 25, 27, 28] but they cannot be easily applied in real-time to solve the fault observation problem.

Several contributions have also been presented in the discrete event systems (DES) framework. Necessary and sufficient conditions for diagnosability, in the case of multiple failures, are developed both for automata [20] (I-diagnosability) and Petri nets [4, 5]. For DES, the diagnosability analysis and the online diagnosis are computed by a diagnoser where the available measurements are considered as inputs of the diagnoser. It leads to an estimated state which could be either "normal" or "faulty" or "uncertain" after the occurrence of every observable event.

The classical model used in DES diagnosis is finite state machine (FSM) and a system is seen as a spontaneous generator of events. However, in many physical systems, the system evolution is driven by the control input and the diagnosability conditions depend both on the system structure and on the control strategy. Hence, some studies proposed an active diagnosis, using a supervisor, to simultaneously ensure the control and the diagnosability of the system. It is proposed, in [1], an algorithm that controls the system toward diagnosable states when a fault is detected. However, following this approach, the system may cross nondiagnosable regions in order to isolate the fault. In [6, 21], a diagnoser was used to block controllable events that drive the system into nondiagnosable regions. For the multicellular converter, the control law design, satisfying the stability conditions associated with the diagnosability, is complex. Indeed, the unobservable events, related to the system, define uncertain states in the diagnoser and the diagnosability conditions cannot be satisfied. In our approach, the algorithm of [1] is extended. The set of uncertain states, associated to the diagnoser, is partitioned in order to distinguish *uncertain states*, which may be explained by a fault but consistent with the evolution of the nominal model, from *uncertain fault* state where the occurrence of a fault has been detected and a suitable control input can be applied to identify it.

In this chapter, an active diagnosis algorithm for switched systems is proposed. The introduced algorithm in [26] is extended and an experimental validation is developed. Here, we assume that the only control input that drives the evolution of the system is represented by the switching function. This function specifies the active mode. Furthermore, discrete outputs are also available, as a result of each transition between modes, in order to detect and isolate the fault. Under these assumptions, a Mealy Machine (MM), i.e., an automaton with inputs and outputs, may be used to represent the system. Indeed, if suitably selected, an input applied to the MM may be used to steer the diagnoser out of the set of uncertain states, thus improving the detection procedure. In this context, the diagnoser, presented in [21], is re-defined in order to introduce the uncertain states and the uncertain fault states. Some transitions of the automaton, including those corresponding to faults, may occur in the absence of a control input and may be unobservable.

In the nominal situation, the control input is selected by the controller according to a given specification and a diagnoser observes the evolution. Although the state of the diagnoser may be uncertain (i.e., a fault may have or may have not occurred), as long as the observed evolution can be explained by the nominal model, no alarm is generated by the diagnoser. Hence, such a system may be nondiagnosable in the sense of [20]. However, when the diagnoser detects an abnormal behavior, i.e., an evolution that cannot be explained without the occurrence of a fault, an alarm is generated and the control objective becomes the isolation of the fault if necessary. A fault isolating sequence can be determined based on the well-known notion of *homing sequences* defined in testing theory [3].

The study of testing procedure for FSM has been first motivated as fundamental research in computer science [3]. In [10], a fault diagnosis algorithm based on testing was investigated. In [22] the testing theory was applied for diagnosis using Input/Output automata. They consider state faults contrary to our approach where a fault is modeled by an unobservable event on transitions and thus is more general. The problem of determining a synchronizing sequence for interpreted Petri nets, i.e., an input sequence that drives the system to a known state is considered in [19]. In this paper, an adapted algorithm to compute the fault isolating sequences for MMs, and a generic algorithm, for the active diagnosis, are presented. If a corresponding isolating sequence can be computed for each uncertain fault state of the diagnoser, using interconnection between a diagnoser and an online testing algorithm we are able to isolate every fault for switched systems.

The chapter is organized as follows. Section 6.2 deals with the problem formulation and introduces the system and diagnoser modeling. In Sect. 6.3, a testing condition is defined. An algorithm is presented in order to compute the fault isolating sequences. An algorithm combining a MM diagnoser and a testing procedure is also proposed in order to solve the fault diagnosis problem. Simulation results, on the 2-cells converter, and experimentation results, on the 3-cells converter, are presented in Sect. 6.4 to highlight the efficiency of the proposed approach.

6.2 Problem Statement and Modeling

6.2.1 Preliminaries on DES Diagnosis

Hereafter, some definitions from [21] and the diagnoser modeling are reformulated to account for faulty uncertain states. The classical DES approach for diagnosis [20, 21] considers a system modeled by a deterministic finite automaton (DFA):

$$G = (X, \Sigma, \delta, x_0) \tag{6.1}$$

where X is the state set, Σ is the set of events, $\delta : X \times \Sigma \rightarrow X$ is the (partial) transition function and x_0 is the initial state of the system. The state x_0 is assumed to be known.

The model G accounts for the normal and faulty behavior of the system, described by the *prefix-closed language* $L(G)$ generated by G, i.e., a subset of Σ^* where Σ^* denotes the Kleene closure of Σ. The event set Σ is partitioned as $\Sigma = \Sigma_o \cup \Sigma_{uo}$ where Σ_o represents the set of the observable events and Σ_{uo} the unobservable events. The *fault event set* is defined as $\Sigma_f \subseteq \Sigma_{uo}$ and may be partitioned into m different fault classes $\Sigma_f = \Sigma_{f_1} \cup \Sigma_{f_2} \cup \ldots \cup \Sigma_{f_m}$.

Let us re-define [21] the *projection operator* $P : \Sigma^* \rightarrow \Sigma_o^*$ such that:

$$\begin{aligned}
P(\epsilon) &= \epsilon \\
P(\sigma) &= \sigma \qquad\qquad \textit{if } \sigma \in \Sigma_o \\
P(\sigma) &= \epsilon \qquad\qquad \textit{if } \sigma \in \Sigma_{uo} \\
P(s\sigma) &= P(s)P(\sigma) \ \textit{if } s \in \Sigma^*,\ \sigma \in \Sigma
\end{aligned}$$

where ϵ is the empty word. Therefore, P simply erases the unobservable events from a trace. The *inverse projection operator* with codomain in $L(G)$ is the relation $P^{-1} : \Sigma_o^* \rightarrow 2^{L(G)}$ that associates to each word of observable events w the set of traces that may have generated it, i.e., $P^{-1}(w) = \{s \in L(G) \mid P(s) = w\}$. In the following, we will denote by $s \in \Sigma^*$ a trace of events generated by the DFA and by $w = P(s) \in \Sigma_o^*$ an observed word, i.e., the observable projection of a generated trace.

The diagnosis problem for a DFA G consists in determining if, given an observed word $w \in \Sigma_o^*$, a fault has occurred or not, i.e., if a transition labeled with a fault event in $\Sigma_f \subseteq \Sigma_{uo}$ has been fired or not and find the fault class. This may be done using a *diagnoser*, i.e., a DFA on the alphabet of observable events.

Definition 1 Given a DFA G with set of events $\Sigma = \Sigma_o \cup \Sigma_{uo}$ and set of fault events $\Sigma_f = \Sigma_{f_1} \cup \Sigma_{f_2} \cup \ldots \cup \Sigma_{f_m}$. Let $\mathscr{F} = \{F_1, F_2, \ldots, F_m\}$ be the set of labels associated to the fault classes. A diagnoser for the DFA defined by Eq. (6.1) is a DFA

$$Diag(G) = (Y, \Sigma_o, \delta_y, y_0)$$

such that

- $Y \subseteq (X \times \{N\}) \cup (X \times 2^{\mathscr{F}})$, i.e., each state of the diagnoser is a set of pairs

$$y = \{(x_1, \gamma_1), (x_2, \gamma_2), \ldots, (x_k, \gamma_k)\},$$

  where $x_i \in X$ and $\gamma_i = N$ or $\gamma_i \subseteq \mathscr{F}$ (with $\gamma_i \neq \emptyset$), for $i = 1, 2, \ldots, k$. Here N is interpreted as meaning Normal (no fault has occurred), while F_i as meaning that a failure of class F_i has occurred.
- The initial state y_0 of the diagnoser is defined to be $\{(x_0, N), (x_1, \gamma_1), \ldots, (x_k, \gamma_k)\}$, i.e., from a known initial state x_0, if there exist unobservable traces s_i, for $i = 1, \ldots, k$, whose projections are ϵ, the initial state y_0 also contains all pairs (x_i, γ_i) such that $x_i \in X$ is reachable with an unobservable trace s_i and γ_i denotes the fault classes that may have occurred in s_i or N if no fault has occurred in s_i.
- $\delta_y(y_0, w) = y_w$ if and only if

$$\begin{aligned} y_w = \{(x, N) \mid (\exists s \in P^{-1}(w))\ \delta(x_0, s) = x \wedge s \cap \Sigma_f = \emptyset\} \\ \cup\{(x, \gamma_i) \mid (\exists s \in P^{-1}(w))\ \delta(x_0, s) = x\ \wedge i \in \{1, 2, \ldots, m\}, \\ s \cap \Sigma_{f_i} \neq \emptyset \wedge \gamma_i = F_i\}, \end{aligned}$$

  i.e., the execution in $Diag(G)$ of a word w yields a state y_w containing:

  - all pairs (x, N) where x can be reached in G executing a string in $P^{-1}(w)$ that does not contain a fault event;
  - all pairs (x, γ_i) where x can be reached in G executing a string in $P^{-1}(w)$ that contains, for each $\gamma_i \subseteq \mathscr{F}$, a fault event of class Σ_{f_i}.

For each state, $y = \{(x_1, \gamma_1), (x_2, \gamma_2), \ldots (x_k, \gamma_k)\}$ of $Diag(G)$, a diagnosis value $\varphi(y)$ is associated such that:

- $\varphi(y) = N$ (no fault state): if $\gamma_i = N$ for all $i = 1, 2, \ldots, k$,
- $\varphi(y) = U$ (uncertain state): if there exist $i, j \in \{1, 2, \ldots, k\}$ such that $\gamma_i = N$ and $\gamma_j \subseteq \mathscr{F}$,
- $\varphi(y) = F$ (isolated fault state): if $\gamma_i \neq N$ and $\gamma_i = \gamma_j$ for all $i, j = 1, 2, \ldots, k$,
- $\varphi(y) = U_F$ (uncertain fault state): if $\gamma_i \neq N$ for all $i = 1, 2, \ldots, k$ and there exist $i, j = 1, 2, \ldots, k$ such that $\gamma_i \neq \gamma_j$.

Thus, a diagnoser allows one to associate to each observed word w a diagnosis state $\varphi(y_w)$ where $y_w = \delta_y(y_0, w)$ is the state reached in $Diag(G)$ by executing word w from the diagnoser initial state y_0.

Remark 1 Following *Definition* 1, if the diagnosis value is $\varphi(y) = U_F$, it means that the detection of the fault is ensured whereas its isolation is only possible when $\varphi(y) = F$. A fault is not diagnosable if there does not exist a corresponding state in the diagnoser with $\varphi(y) = F$.

The objective of this chapter is to design an algorithm which solves the fault diagnosis problem for a large class of switched systems.

6.2.2 Switched System Modeling

In this chapter, the proposed approach can address the diagnosis problem of a class of switched systems which is generally represented by the following model:

$$\begin{aligned} \dot{\mathfrak{x}}(t) &= \mathfrak{A}_{\eta(t)}(\mathfrak{x}(t), \mathfrak{f}(t)) \\ \mathfrak{O}(t) &= \mathfrak{C}_{\eta(t)}(\mathfrak{x}(t), \mathfrak{f}(t)) \end{aligned} \tag{6.2}$$

where $\mathfrak{x}(t)$ is the continuous state, $\mathfrak{O}(t)$ is the continuous output, $\mathfrak{f}(t)$ is the fault vector, and $\eta(t)$ represents the switching function which is piecewise constant and $\eta(t) : [0, \infty) \rightarrow \{1, 2, \ldots, \mathfrak{N}\}$. $\mathfrak{N}$ denotes the known number of discrete modes or subsystems. In general, function $\eta(t)$ could depend on an external control input and/or the state $\mathfrak{x}(t)$ and/or the fault vector $\mathfrak{f}(t)$. The measured variables are the output signal $\mathfrak{O}(t)$ and eventually the continuous state $\mathfrak{x}(t)$ if it is observable. Here, a fault can be considered on the system parameters, actuators, or sensors. The multiple fault occurrences are not considered. Hereafter, we assume that all continuous variables of system (6.2) can be represented by sets of discrete variables.

In most of the existing studies on the diagnosis in the DES framework, the set of events is only based on one information from the system (input or output signal). The hybrid models allow representing the complex dynamics (continuous and discrete) of a system. It can appear that this class of systems needs a more accurate method for the diagnosis. In order to design a new approach of diagnosis using the formalism of DFA for the class of switched systems, it is interesting to consider the system as a MM. Using this particular modeling, the set of discrete events can be enriched with the combination of input and output signals. An event will be defined by a pair input/output. The idea is to highlight the equivalence between DFA and MM in order to deduce a MM diagnoser using the formalism of DFA previously redefined.

The switched systems can be modeled as MMs, where the input event corresponds to the active mode of the system and the output event to the sensor readings.

Formally a Mealy Machine is a structure:

$$M = (X, I, O, \zeta, \lambda, x_0) \tag{6.3}$$

where X is the set of discrete states, I and O are the set of input and output events, $\zeta : X \times I \rightarrow X$ is the transition function, $\lambda : X \times I \rightarrow O$ is the output function, and x_0 is the initial state of the system.

Here, we consider that the set of input events can be partitioned as $I = I_c \cup I_{uc}$. Events in I_c are *controllable* events, i.e., they denote controlled transitions that are triggered by an external control input. Events in I_{uc} are *uncontrollable* events, i.e., they denote autonomous transitions that may occur without being triggered by an external control input. The set of fault events $I_f = I_{f1} \cup \ldots \cup I_{fm}$ is a subset of I_{uc}. Note that the transition function of a MM is total on the set of controllable input events, i.e., for all $x \in X$ and for all $i \in I_c$, $\zeta(x, i)$ is defined. This means that a controllable input may be applied regardless of the state of the machine. We also assume that the set of output events O may contain the special symbol Ø that denotes transitions whose occurrence does not generate as output a measurable event.

One can easily convert, for the purpose of diagnosis, a MM to an equivalent DFA with the same state set and alphabet $\Sigma = I \times O$. A transition of the MM $\zeta(x, i) = \bar{x}$ with output function $\lambda(x, i) = o$ can be represented in the DFA by a transition $\delta(x, (i, o)) = \bar{x}$. The set of unobservable events of the DFA is $\Sigma_{uo} = I_{uc} \times \{\emptyset\}$, the set of fault events can be redefined as $\Sigma_f = \{I_f \times \{\emptyset\}\}$ and $\Sigma_o = \Sigma \setminus \Sigma_{uo}$. Once a MM has been converted into an equivalent DFA, a diagnoser can be designed to solve the diagnosis problem. Below, an example is given in order to highlight the equivalence between MM and DFA. The corresponding MM diagnoser is illustrated.

Example 1 Consider the MM $M = (X, I, O, \zeta, \lambda, x_0)$ with $X = \{1, 2, 3\}$, $I = \{a, b, \varepsilon_1, \varepsilon_f\}$, $I_c = \{a, b\}$, $I_{uc} = \{\varepsilon_1\}$, $I_f = \{\varepsilon_f\}$, $O = \{o_1, o_2, o_3, \emptyset\}$, $x_0 = \{1\}$, transition and output function:

ζ	a	b	ε_1	ε_f
1	3	2		
2	1	2	1	
3	2	3		1

λ	a	b	ε_1	ε_f
1	o_1	o_2		
2	o_3	o_2	Ø	
3	o_2	o_1		Ø

Using the first line of the left table, one can see that, in the MM M, there exist transitions from the state 1 to states 2 and 3 using input events noted b and a with $\zeta(1, a) = 3$, $\zeta(1, b) = 2$. Associated to the second table, the corresponding output event is represented with $\lambda(1, a) = o_1$, $\lambda(1, b) = o_2$. Using the proposed modeling, an equivalent DFA of this MM can be deduced. Couples (a, o_1) and (b, o_2) are two events of Σ_o in the new representation using the formalism of DFA.

The equivalent DFA is shown in Fig. 6.1(left) where the set of observable events is $\Sigma_o = \{(a, o_1), (a, o_2), (a, o_3), (b, o_1), (b, o_2)\}$, the set of unobservable events is $\Sigma_{uo} = \{(\varepsilon_1, \emptyset), (\varepsilon_f, \emptyset)\}$, and the set of fault events is $\Sigma_{f1} = \{(\varepsilon_f, \emptyset)\}$ (here we have a single fault class). The diagnoser for this DFA is shown in Fig. 6.1(right), where each state y of $Diag(G)$ is labelled with its corresponding diagnosis value $\varphi(y)$ in square brackets.

The objective of the following section is to design an algorithm in order to detect and isolate faults in spite of the presence of uncertain fault states (i.e., $\varphi(y) = U_F$) in the diagnoser.

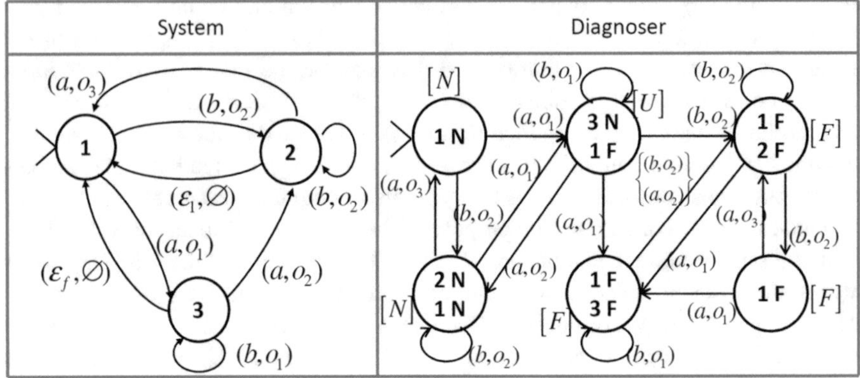

Fig. 6.1 On the left, a DFA G. On the right, its diagnoser automaton $Diag(G)$

6.3 Active Diagnosis

It is assumed that in normal conditions the control inputs of the MM (i.e., the switching sequence of the system) are selected by a controller to satisfy a given objective. In parallel to the controller, a diagnoser is used to detect the evolution of the system. There is no interaction between the diagnoser and the controller when no fault has been detected, i.e., while the diagnoser is in a state with diagnosis value N or U. In such a condition, in fact, the diagnoser behavior may be explained by a nominal evolution and no alarm is generated. However, when a fault has been detected (when $\varphi(y) = U_F$ or F), the control objective is suspended for safety reasons and a fault isolation procedure is applied. Here, the trade-off between the control objective and the active diagnosis is not studied.

In particular, if the diagnoser is in a state F, the fault has been isolated because it is known exactly which fault classes have occurred. On the contrary, when the diagnoser is in one of the uncertain fault states U_F, the control input sequence will be selected on the basis of a testing procedure to design an active diagnoser [21] that isolates the fault identifying the class of the fault that has occurred.

6.3.1 *Testing Condition*

In this subsection, the active diagnosis procedure for the MM defined in Eq. (6.3) is described. It consists in finding a control input sequence which isolates the fault.

In order to design the proposed algorithm, we need to define a function which specifies, for each state $y \in Y$ of the diagnoser and for each control input sequence $\alpha \in I_c$, the set of pairs (y', β) where $y' \in Y$ is the state of the diagnoser reached if $\beta \in O$ has been observed.

Definition 2 Given the diagnoser (*Definition* 1) associated with the DFA equivalent to the MM Eq. (6.3), we define the following function $f : Y \times I_c^* \to 2^{Y \times O^*}$ as follows. For all $y \in Y$ and all $\alpha \in I_c^*$:

$$\begin{aligned} f(y, \alpha) = \{(y', \beta) \mid \delta_y(y, \sigma) = y', \\ \sigma = (i_1, o_1)(i_2, o_2) \ldots (i_k, o_k), \\ \alpha = i_1 i_2 \ldots i_k,\ \beta = o_1 o_2 \ldots o_k\}. \end{aligned} \tag{6.4}$$

Proposition 1 *The input sequence $\alpha \in I_c^*$ isolates the faults from uncertain fault state $y_u \in Y$ such that $\varphi(y_u) = U_F$ if and only if*

$$f(y_u, \alpha) \subseteq \{(y_i, \beta_i) \mid \varphi(y_i) = F\}. \tag{6.5}$$

Proof Obviously, condition (6.5) is a necessary condition for sequence α to isolate the fault. Since the diagnoser is a deterministic automaton, $(y', \beta), (y'', \beta) \in f(y, \alpha)$ implies $y' = y''$, i.e., the state of the diagnoser, reached by applying a given control input sequence α, is perfectly known from the observed output sequence β. This ensures that condition (6.5) is also sufficient.

From *Proposition* 1, an active diagnosability condition for the MM Eq. (6.3) can be deduced.

Proposition 2 *A switched system modeled by a MM Eq. (6.3) is actively diagnosable, using the MM diagnoser (corresponding to Definition 1), if there is at least one control input sequence which verifies Proposition 1 for each uncertain fault state of its diagnoser.*

Remark 2 The above *Proposition* 2 is slightly different from the active diagnosability definitions usually considered in the literature [1, 21]. In this study, a MM modeling is used for the system and its diagnoser in order to highlight input/output transitions and to design an adapted algorithm which solves the active diagnosis problem for the class of switched systems.

The proposed approach is inspired by the notion of homing sequence that is studied in testing theory [3]. A homing sequence is an input sequence that brings a MM (with outputs) from an unknown state to a known state, i.e., after the input sequence is applied by observing the output sequence, one can unambiguously determine the current state of the MM (see [3] for further details). Indeed, our objective consists in finding a control input sequence in the MM diagnoser which isolates the fault or disambiguates the fault class by observing the output sequence.

For a system which satisfies *Proposition* 2, a sequence that isolates the fault can be determined, using the following approach to compute all fault isolating sequences corresponding to the set of uncertain fault states.

6.3.2 Algorithm

Before introducing our proposed algorithm, let us consider the following example.

Example 2 Consider the MM given in Fig. 6.2. There are three different fault classes, i.e., $\Sigma_{f_1} = F1$, $\Sigma_{f_2} = F2$, and $\Sigma_{f_3} = F3$. $X = \{1, 2, 3, 4, 5, 6, 7\}$, $I = \{a, b, c, d, F_1, F_2, F_3\}$ with $I_{f1} = \{F_1\}$, $I_{f2} = \{F_2\}$ and $I_{f3} = \{F_3\}$, $I_{uc} = I_f$, $I_c = \{a, b, c, d\}$, $O = \{1, 2, \emptyset\}$ and $x_0 = \{1\}$. The corresponding diagnoser contains 41 states and it is not detailed here. Our approach consists in applying an algorithm which detects and isolates the fault that has occurred. In the following figures, we have chosen to decompose the diagnoser in steps in order to explain the algorithm. Figures 6.3 and 6.4 illustrate two parts of the diagnoser during its construction in order to achieve the fault detection and isolation. Figure 6.3 presents the *MM diagnoser of detection* in the detection step. It has a unique nominal state $(1\ N, 2\ F1, 3\ F2, 5\ F3)$ since we assume that an uncertain state U in the MM diagnoser is not a faulty situation. This diagnoser shows transitions which allow the fault diagnosis or only detection. It has four uncertain fault states U_F and three isolated fault states F.

Considering the system in Fig. 6.2 with initial state 1, the sequence of observable events $(b, \emptyset)(d, 2)$, for instance, allows detecting a fault but not to isolate it. On the diagnoser (Fig. 6.3), this sequence leads to the uncertain fault state $(6\ F1, 7\ F2)$ with $\varphi(y) = U_F$. When a fault is detected, the nominal control objective is suspended for safety reason.

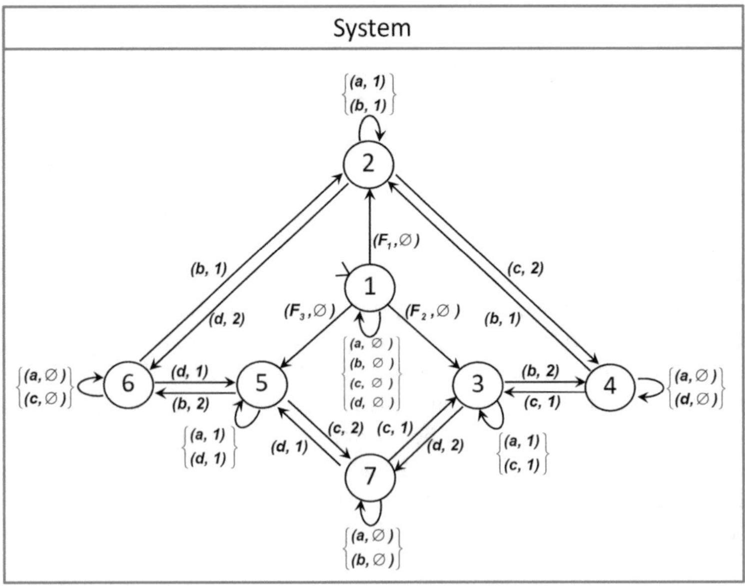

Fig. 6.2 Example of a MM with three fault classes

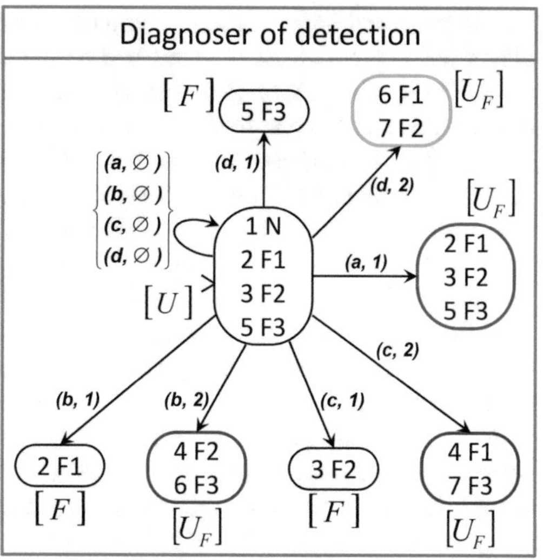

Fig. 6.3 MM diagnoser for the detection step

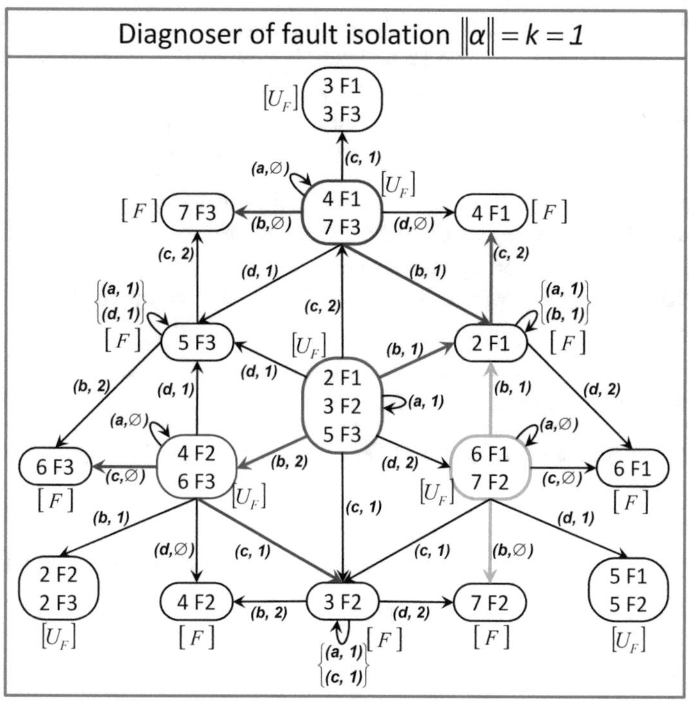

Fig. 6.4 MM active diagnoser for the isolation step, condition (6.5) is satisfied

The proposed approach is to compute a fault isolating sequence from the MM diagnoser. Figure 6.4 presents the *MM active diagnoser* when a sequence $\alpha \in I_c^*$, defined by $k = 1$ event can be observed for the isolation step. It highlights the set of reachable states from the detection step after observation of $\|\alpha\| = k = 1$ event, where $\|.\|$ is the length of a string. Our off-line objective is to analyze this part of the diagnoser in order to find a fault isolating sequence for each uncertain fault state of the *MM diagnoser of detection* (Fig. 6.3). The idea is to increment k while no input sequence $\alpha \in I_c^*$ with $\|\alpha\| = k$ verifying *Proposition* 1 can be found for an uncertain fault state U_F of the detection step.

Following this *MM active diagnoser* Fig. 6.4, all sequences of one event can be tested from each uncertain fault state. Considering condition (6.5) on the diagnoser, the control input event b can be applied as a fault isolating sequence for the states $(4\ F1, 7\ F3)$ (blue state) and $(6\ F1, 7\ F2)$ (green state). Indeed, observing if the corresponding output event is $\emptyset$ or 1, we can isolate the fault $F1$ or $F2$ or $F3$. The event b is not a valid isolating sequence for the uncertain fault states $(4\ F2, 6\ F3)$ (magenta state) and $(2\ F1, 3\ F2, 5\ F3)$ (red state). This sequence does not verify *Proposition* 1 because it leads to uncertain fault states $(4\ F2, 6\ F3)$ or $(2\ F2, 2\ F3)$. The input event c can be taken as a fault isolating sequence for the state $(4\ F2, 6\ F3)$. If the corresponding output event is $\emptyset$, we can isolate the fault $F3$ and if the output event is 1, then the fault $F2$ can be isolated. Following this strategy, the uncertain fault state $(2\ F1, 3\ F2, 5\ F3)$ requires $\|\alpha\| = 2$. Hereafter, the corresponding diagnoser with $k = 2$ is not presented but from Fig. 6.4, we can propose the fault isolating sequence bc whereas bb is not a valid isolating sequence.

The proposed idea is to compute a minimal fault isolating sequence for each uncertain fault states (U_F) of the MM diagnoser in the detection step using the testing theory and based on homing sequences.

Function HomingSequence can be applied off-line to compute all fault isolating sequences.

Function $HomingSequence(Diag(G))$

---

1. **Input:** The diagnoser $Diag(G) = (Y, \Sigma_o, \delta_y, y_0)$
2. Create a set $\mathscr{I}' = \emptyset$
3. **For** all $y_{ui} \in Y$

   3.1. **If** $\varphi(y_{ui}) = U_F$ and $\exists \sigma \in \Sigma_o, \exists y \in Y$ with $\varphi(y) = N$ or U *s.t.* $\delta_y(y, \sigma) = y_{ui}$

      3.1.1. Let $\alpha_{ui} = \varepsilon$ (ε is the empty word)
      3.1.2. Let k=1
      3.1.3. **While** $\alpha_{ui} = \varepsilon$

      - **If** $\exists \alpha \in I_c^*$ *s.t.* $\|\alpha\| = k$ and α verifies *Proposition 1* for the state y_{ui}

        $\alpha_{ui} \leftarrow \alpha$

      - **Else**

        $k \leftarrow k + 1$

End if

End while

3.1.4. $\mathscr{I}' = \mathscr{I}' \cup \{(y_{ui}, \alpha_{y_{ui}})\}$

End if

End for

4. **Output:** $\mathscr{I}' = \{(y_{u1}, \alpha_{y_{u1}}), (y_{u2}, \alpha_{y_{u2}}), \ldots, (y_{ui}, \alpha_{y_{ui}}), \ldots\} \rightarrow$ all pairs combining an uncertain fault state y_{ui} with a minimal fault isolating sequence $\alpha_{y_{ui}}$.

For a system which satisfies *Proposition* 2, *Function HomingSequence* allows finding the set of minimal fault isolating sequences in order to isolate the fault as quickly as possible after its detection. Indeed, in *Function HomingSequence*, step 3.1. verifies that there exists a transition between the uncertain fault state y_{ui} and a state y such that $\varphi(y) = N$ or U. This function is designed in order to increment the size k of the sequence when no sequence can be found for an uncertain fault state y_{ui}. A set of pairs can be proposed in order to combine each uncertain fault state with a fault isolating sequence. If a fault is not diagnosable, this function could return an empty set and the variable k goes to infinity.

According to the example (Fig. 6.2), the *Function HomingSequence* can be applied on the MM diagnoser (Fig. 6.3) in order to find all minimal fault isolating sequences. The following algorithm is an application of the proposed example. The function is illustrated just for 2 uncertain fault states (steps between lines 3.1.4 and 4 correspond to other uncertain fault states of the MM diagnoser, these are not detailed in this chapter in order to simplify notations). The computed output in line 4 is for all uncertain fault states.

Function *HomingSequence*(*Diag*(*G*)) applied on the MM diagnoser (Fig. 6.3)

---

1. **Input:** The diagnoser *Diag*(*G*) corresponding to the MM Fig. 6.2
2. Create a set $\mathscr{I}' = \emptyset$
3. **For** $y_{ui} = (6\,F1, 7\,F2)$ (or $y_{ui} = (4\,F1, 7\,F3)$)

3.1. $\varphi(y_{ui}) = U_F$ and $\exists\sigma = (d, 2)$ (or $\exists\sigma = (c, 2)$), $\exists y = (1\,N, 2\,F1, 3\,F2, 5\,F3)$ with $\varphi(y) = U$ such that $\delta_y(y, \sigma) = y_{ui}$

3.1.1. Let $\alpha_{ui} = \varepsilon$
3.1.2. Let k=1
3.1.3. **While** $\alpha_{ui} = \varepsilon$

- $\exists\alpha = b$ such that $\|b\| = 1$ and b verifies *Proposition* 1 for the state $(6\,F1, 7\,F2)$ (or $(4\,F1, 7\,F3)$)

  $\alpha_{ui} \leftarrow b$

End while

3.1.4. $\mathscr{I}' = \mathscr{I}' \cup \{((6\,F1, 7\,F2), b)\}$ (or $\cup\{((4\,F1, 7\,F3), b)\}$)

⋮

4. **Output:** $\mathscr{I}' = \{((6\ F1, 7\ F2), b), ((4\ F1, 7\ F3), b), ((4\ F2, 6\ F3), c), ((2\ F1, 3\ F2, 5\ F3), bc)\}$

The proposed MM active diagnoser algorithm can be summarized by *Algorithm* 1.

Algorithm 1 Active diagnoser

1. Compute $\mathscr{I} = HomingSequence(Diag(G))$
2. **Loop**
   - 2.1. Nominal control of the system (defined according to the control objective)
   - 2.2. Follow the occurred events (i, o) in the MM active diagnoser
   - 2.3. **If** a fault is detected ($\varphi(y) = U_F$ or F)
     - 2.3.1. Stop the control objective
     - 2.3.2. **If** $\varphi(y) = F$
       - The fault is isolated using the MM diagnoser
     - 2.3.3. **Else**
       - Apply the homing sequence α_y corresponding to the pair $(y, \alpha_y) \in \mathscr{I}$
       - Follow the occurred events (i, o) in the MM diagnoser in order to reach a final state $y_f \in Y$ such that $\varphi(y_f) = F$ and the fault is isolated

       End if
     - 2.3.4. **STOP**

     End if

   End loop

Following *Algorithm* 1, in the first step, all minimal fault isolating sequences are computed off-line using *Function Homing Sequence* for each uncertain fault state of the MM diagnoser computed for the detection step. In the second step, the nominal control can be applied and the MM diagnoser follows the occurred events (i, o) (the diagnosis value $\varphi(y)$ can be equal to N, U, U_F or F). If a fault is detected, the control objective is broken off. If the diagnosis value $\varphi(y) = F$, the fault class is isolated and the algorithm is ended. If the fault is only detected (i.e., $\varphi(y) = U_F$), then corresponding fault isolating sequence can be applied in order to achieve the diagnosis objective.

6.4 Application to the Multicellular Converter

In this section, the proposed diagnosis algorithm is applied to the multicellular converter. The details of the algorithm are presented with simulation results using a 2-cells converter (4 modes). Experimental results on a 3-cells converter (8 modes)

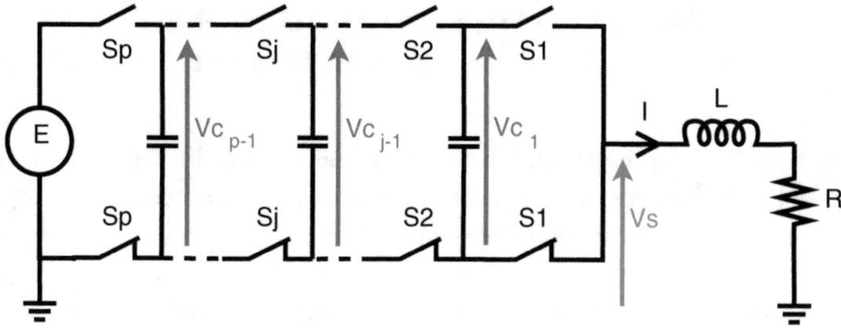

Fig. 6.5 Multicellular converter associated to an inductive load

highlight the effectiveness of the proposed approach and show that the algorithm can be generalized for this class of switched system and applied in real time.

6.4.1 Multicellular Converter Modeling

The multicellular converter is based on the combination of p elementary cells of commutation. The current flows from the source E toward the output through the different switches. The converter shows, by its structure illustrated in Fig. 6.5, a hybrid behavior due to the discrete variables, i.e., switches. Note that because of the presence of $(p - 1)$ floating capacitors, there are also continuous variables, i.e., currents and voltages.

The dynamics of the converter, with a load consisting in a resistance R and an inductance L, can be expressed by the following differential equations:

$$\begin{cases} \dot{I} &= -\frac{R}{L}I + \frac{E}{L}S_p - \sum_{j=1}^{p-1} \frac{V_{cj}}{L}(S_{j+1} - S_j) \\ \dot{V}_{c_j} &= \frac{I}{c_j}(S_{j+1} - S_j), \quad j = 1, \ldots, p-1 \end{cases} \tag{6.6}$$

where I is the load current, c_j is the capacitance, V_{c_j} is the voltage in the j-th capacitor, and E is the voltage of the source. Here, it is assumed that only the output voltage V_s can be measured:

$$V_s = ES_p - \sum_{j=1}^{p-1} V_{c_j}(S_{j+1} - S_j) \tag{6.7}$$

Each commutation cell is controlled by the binary signal $S_j \in \{0, 1\}$. Signal $S_j = 1$ means that the upper switch of the j-th cell is "on" and the lower switch is "off" whereas $S_j = 0$ means that the upper switch is "off" and the lower switch is "on."

Remark 3 System defined by (6.6) and (6.7) is not observable in the classical sense. Indeed, if $\forall j \in \{1, \ldots, p\}$, $S_j = 0$ or $S_j = 1$, then the internal voltages V_{c_j} cannot be estimated.

It is important to highlight that in order to standardize the industrial production, the electrical switches constraints should be similar in each cell. This requirement implies a unique voltage switch constraint of $\frac{E}{p}$. Thus, the discrete control laws, which determine the evolution of the control signals S_j, ensure the simultaneous regulation of the load current and capacitor voltages such that:

$$V_{c_j,ref} = j\frac{E}{p}, \quad \forall j \in \{1, \ldots, p\} \tag{6.8}$$

A driver applies the control strategy on the switches of each cell (see Fig. 6.6(left) for the 2-cells converter). $[S_1, \ldots, S_p]^T \in \{0, 1\}^p$ is a boolean vector describing the configuration or mode of the system.

Assuming that the control law is computed using a PWM module (Fig. 6.6(left)), the switching sequence, which depends on the desired load current, is known. Since the transient period is very short, one can only consider the steady state value for each mode. Therefore, the hybrid control strategy is defined by 2^p modes. It creates a stairs behavior of the output voltage, i.e., $V_s \in \{0, \frac{E}{p}, \frac{2E}{p}, \ldots, E\}$. In order to reduce the harmonic contents and the switching losses of semiconductors during the different commutations, the control limits the variation of the output voltage to $\frac{E}{p}$. Indeed, the control operates one cell at once.

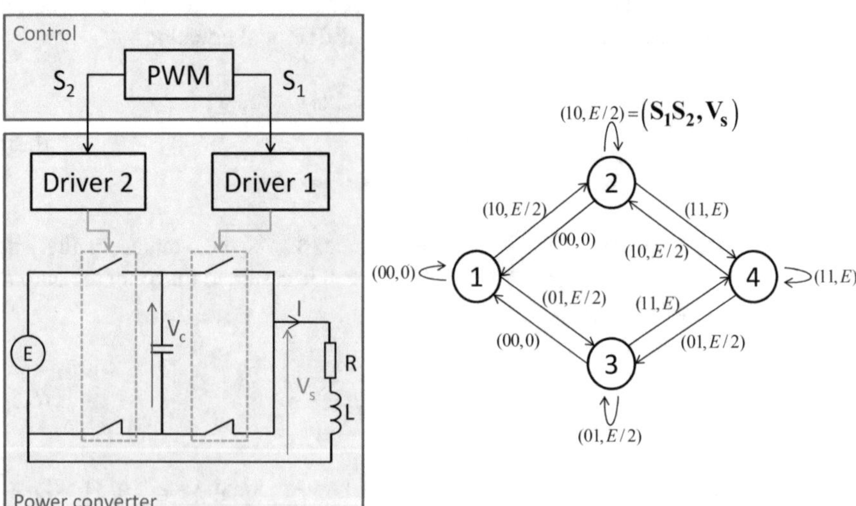

Fig. 6.6 Topology of a 2-cells converter with a PWM based control and the corresponding MM in its nominal behavior

6.4.2 *Active Fault Diagnosis for a 2-Cells Converter*

Without loss of generality, we consider the case $p = 2$ in order to simplify notations. Anyway, the proposed approach can be easily applied for any p.

6.4.2.1 2-Cells Converter Modeling

Figure 6.6 depicts the topology of the 2-cells converter associated to an inductive load and its corresponding MM for the nominal modes, where the control signals S_1S_2 represent the input events and the discrete values, associated to V_s, are the output set.

The model of the 2-cells converter involves that the reference voltage of the capacitor is such that $V_c = \frac{E}{2}$ and the output voltage is defined as $V_s \in \{0, \frac{E}{2}, E\}$ when the transient is ignored.

In this work, only faults which occur on a commutation cell are considered. It is possible that a commutation cell is blocked due to a faulty driver. For the 2-cells converter, four faults can be defined. The fault event set is $\Sigma_f = f_1 \cup f_2 \cup \bar{f}_1 \cup \bar{f}_2$, where f_j (resp. $\bar{f}_j$) indicates that the j-cell is blocked in $S_j = 1$ (resp. $S_j = 0$). The fault states are denoted according to the corresponding nominal state. For instance, the fault state $2\bar{f}_2$ is the equivalent state of 2 in the presence of fault $\bar{f}_2$.

Figure 6.7 shows the MM representation of the 2-cells converter. The output set is $O = \{\emptyset, 0, 1, 2\}$ and corresponds to Table 6.1. The output set represents the output voltage variations. The input set is $I = \{\varepsilon_f, s_1s_2, \bar{s}_1s_2, s_1\bar{s}_2, \bar{s}_1\bar{s}_2\}$ with $I_{uc} = \{\varepsilon_f\}$. s_j

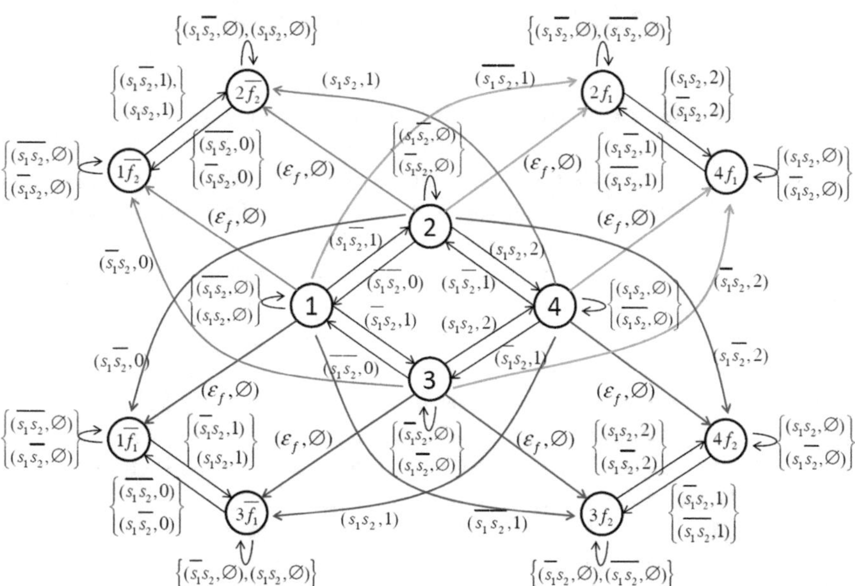

Fig. 6.7 MM modeling for the 2-cells converter considering $((S_2, S_1), V_s\ variation)$ as the observable quantity

Table 6.1 Output voltage variations and the output set for the 2-cells converter

$O = \{\emptyset, 0, 1, 2\}$	V_s *variation*
$\emptyset$	*No variation*
0	$E/2$ *to* 0
1	0 *or* E *to* $E/2$
2	$E/2$ *to* E

Table 6.2 Observable faults associated with the fault classes

Fault events	Classes
$(\bar{s_1}s_2, 2)$	f_1
$(s_1\bar{s_2}, 2)$	f_2
$(\bar{s_1}\bar{s_2}, 1)$	f_1, f_2
$(s_1s_2, 1)$	$\bar{f_1}, \bar{f_2}$
$(\bar{s_1}s_2, 0)$	$\bar{f_2}$
$(s_1\bar{s_2}, 0)$	$\bar{f_1}$

(resp. $\bar{s_j}$) indicates a control law $S_j = 1$ (resp. $S_j = 0$). Each transition edge is labeled with the values of the input and output. The system has unobservable faults, noted by pair $(\varepsilon_f, \emptyset)$.

Remark 4 The MM of the converter (given in Fig. 6.7) contains observable faults based on physical considerations of the system between the input and output (linked to the output value V_s). An expert can associate these faults with the different fault classes. Observable faults represented by the events associated with their fault classes are given in Table 6.2.

The MM modeling allows taking into account the change in sensor readings when a same control is applied. It improves the fault detection procedure.

6.4.2.2 Algorithm Associated with the 2-Cells Converter

Figure 6.8 shows the diagnoser corresponding to the 2-cells converter, modeled by its equivalent DFA and assuming that the control is broken off if a fault is detected. Each state of the diagnoser is a set of pairs (x_i, y_i) where $x_i \in X$ and $y_i \in \{N, f_1, \bar{f_1}, f_2, \bar{f_2}\}$. It should be pointed out that it has two uncertain fault states, $(2\bar{f_2}, 3\bar{f_1})$ and $(2f_1, 3f_2)$. Indeed, if the state of the system is, for instance, 1 (or 4), a fault event $(\bar{s_1}\bar{s_2}, 1)$ (or $(s_1s_2, 1)$) enables to detect a fault but does not enable to isolate it. Using the proposed diagnoser, the states $4f_1$, $4f_2$, $1\bar{f_1}$, and $1\bar{f_2}$ can be directly isolated using the observations $(\bar{s_1}s_2, 0)$, $(s_1\bar{s_2}, 0)$, $(\bar{s_1}s_2, 2)$, and $(s_1\bar{s_2}, 2)$ (see Fig. 6.8). By a classical approach [21], from the state of the system 2 or 3, the observations $(\bar{s_1}\bar{s_2}, \emptyset)$ and $(s_1s_2, \emptyset)$ also lead to the fault diagnosis. Therefore, a fault can always be detected but may not directly be isolated.

Associated to the MM diagnoser, a fault isolating sequence can be computed, using *Function HomingSequence*, to eliminate the uncertainty between states $(2\bar{f_2}, 3\bar{f_1})$ and $(2f_1, 3f_2)$ (see Fig. 6.9). The input event $(\bar{s_1}s_2) \in I_c^*$ satisfying

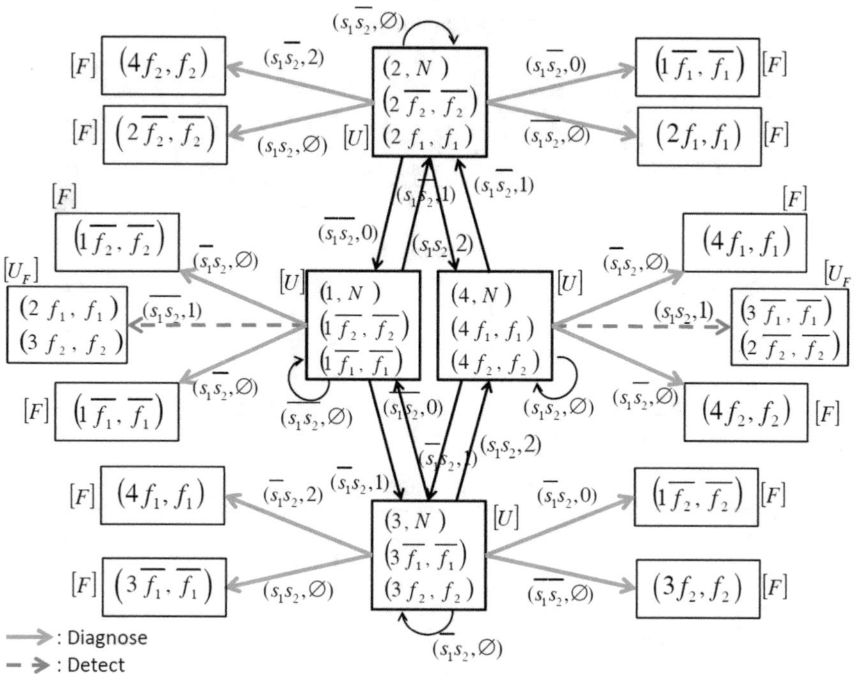

Fig. 6.8 Part of the diagnoser associated to the 2-cells converter, considering that the control is broken off if a fault can be detected

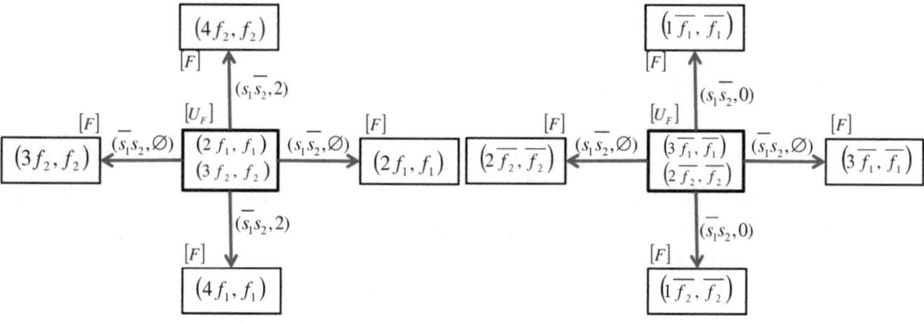

Fig. 6.9 Homing sequences allowing faults isolation (i.e., $(s_1\bar{s}_2)$ and $(\bar{s}_1 s_2)$ with $\|\alpha\| = 1$)

condition (6.5) can be a fault isolating sequence for the system (the input event $(s_1\bar{s}_2) \in I_c^*$ can be also used).

Remark 5 The diagnoser given in Fig. 6.8 cannot isolate a fault if the initial state x_0 is unknown. Here, it is considered that the initial conditions of the system are known and the initial mode is without fault. The initial mode corresponds to the mode without control (all $S_j = 0$) (see Fig. 6.5) and will be defined with mode 1.

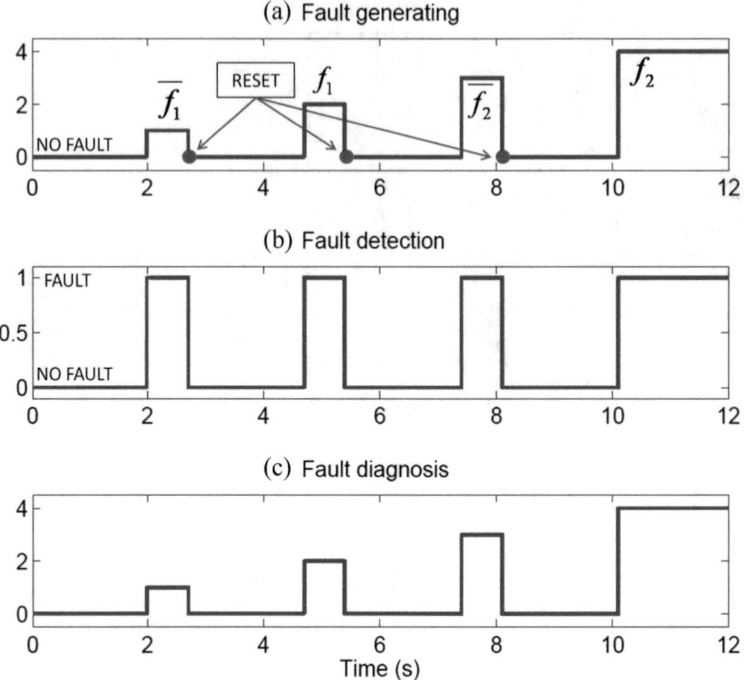

Fig. 6.10 Fault detection and isolation, using the proposed active diagnosis algorithm. (**a**) Fault evolution. (**b**) Fault detection using the proposed diagnoser. (**c**) Isolation using the diagnoser and the homing sequences given in Figs. 6.8 and 6.9

6.4.2.3 Simulation Results

In this section, some simulations are carried out to show the effectiveness of the proposed approach. Equations (6.6) and (6.7) are written using Matlab/Simulink, a PWM module controls the 2-cells converter and a Stateflow module is used to model the DFA. The parameters used in the simulation are as follows:

$$E = 60\,\text{V},\ c = 400\,\mu\text{F},\ R = 200\,\Omega,\ L = 0.1\,\text{H}$$

Figure 6.10a depicts the evolution of faults. In order to highlight the efficiency of the diagnoser, the simulation takes into account all kind of faults $\{f_1, f_2, \bar{f_1}, \bar{f_2}\}$. Figure 6.10b highlights the fault detection and Fig. 6.10c illustrates the fault diagnosis using the proposed strategy. Indeed, a reset of the system is realized between each fault. The state is re-initialized at $x_0 = [V_{cref}, I_{ref}]^t = [30, 0.2]^t$ and the mode is 1. Figure 6.11 shows the evolution of the mode of the DFA.

One can see, in Fig. 6.10, that the diagnoser, using the MM representation, fulfils the objective, i.e., the faulty modes are well detected and isolated. In Fig. 6.11, one can note that faults $\bar{f_1}$ and $\bar{f_2}$ are identified using the proposed fault isolating

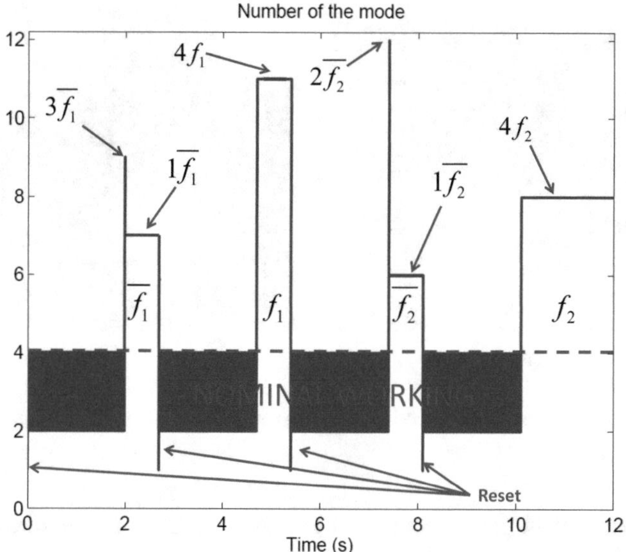

Fig. 6.11 Mode commutations (nominal and faulty)

sequence. Indeed, these faults generate an uncertain fault state in the diagnoser. Using the testing theory, a sequence is applied, among the fault isolating sequences given in Fig. 6.9, i.e., $(s_1\bar{s}_2)$ or $(\bar{s}_1 s_2)$. This sequence depends on the uncertain state of the diagnoser. It enables to eliminate the uncertain states and isolate the corresponding fault.

6.4.3 *Active Fault Diagnosis for a 3-Cells Converter*

In order to highlight the performance of the proposed active diagnosis, we have also performed some experimental validations.

6.4.3.1 Experimental Setup

To demonstrate the effectiveness of the proposed strategy, experimental investigations have been realized on a test bench which consists of a 3-cells converter. The schematic view of the overall platform is shown in Fig. 6.12a. The experimental setup (see Fig. 6.12b) is described as follows:

- The power block is composed of a 3-cells converter with three legs. The nominal bench characteristics, obtained after identification, are: $c_1 = 40 \times 10^{-6}$ F, $c_2 = 40 \times 10^{-6}$ F, $E = 60$ V.

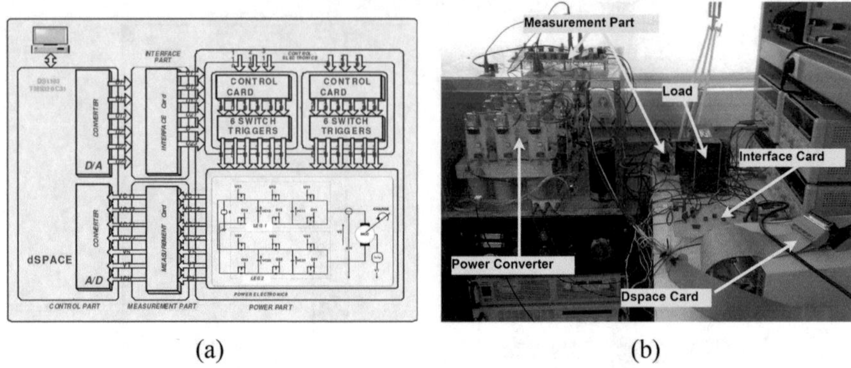

Fig. 6.12 Schematic view of the overall platform (**a**). A photography of the experimental setup (**b**)

- The measurement part is composed of voltage sensors to measure the voltage across the floating capacitors and a current transductor to measure the load current. A low pass filter has been added.
- The computer is equipped with Mathworks software and an interface Dspace card DSP1103, based on a floating point DSP (TMS320C31) with ControlDesk software in order to visualize the state during the experiment. In order to obtain the best resolution, the minimum sampling period for the Dspace has been chosen, i.e. $T_{ech} = 7 \times 10^{-5}$ s.
- The three control inputs, designed by the proposed scheme, are computed and delivered by the interface Dspace card. An interface card allows to protect, by insulation, the DSP of the power electronics.
- The load is composed of an inductance and a resistance: $R = 200\,\Omega,\ L = 1\,\text{H}$.

6.4.3.2 3-Cells Converter Modeling

Figure 6.13 depicts the MM of the 3-cells converter associated to an inductive load for the nominal modes.

The model of the 3-cells converter involves that the reference voltages of the capacitors are such that $V_{c_1 ref} = \frac{E}{3}$ and $V_{c_2 ref} = \frac{2E}{3}$. The output voltage is defined as $V_s \in \{0, \frac{E}{3}, \frac{2E}{3}, E\}$ (considering the system in the steady state). Similarly with the model of the 2-cells converter, the fault event set may be defined with 6 fault classes $\Sigma_f = f_1 \cup f_2 \cup f_3 \cup \bar{f_1} \cup \bar{f_2} \cup \bar{f_3}$ (associated to each cells of the converter). The output set is $O = \{\emptyset, 0, 1, 2, 3\}$ and corresponds to Table 6.3. The input set is $I = \{\varepsilon_f, s_1s_2s_3, \bar{s_1}s_2s_3, s_1\bar{s_2}s_3, \bar{s_1}\bar{s_2}s_3, s_1s_2\bar{s_3}, \bar{s_1}s_2\bar{s_3}, s_1\bar{s_2}\bar{s_3}, \bar{s_1}\bar{s_2}\bar{s_3}\}$. The initial conditions of the system are defined by $x_0 = [V_{c_1}, V_{c_2}, I]^T = [0, 0, 0]^T$ and the initial mode is 1.

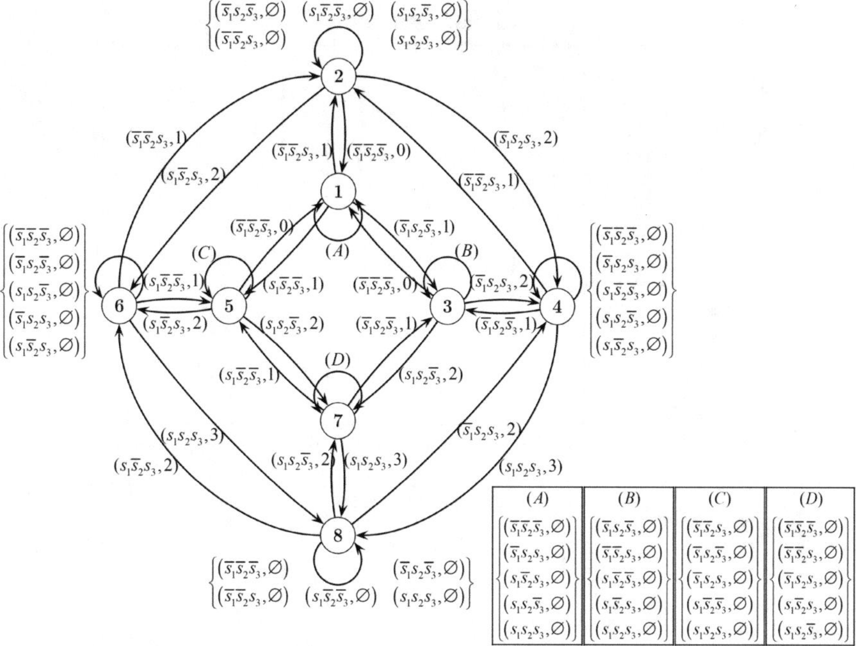

Fig. 6.13 Nominal MM of the 3-cells converter (without faults)

Table 6.3 Output voltage variations and the output set for the 3-cells converter

$O = \{\emptyset, 0, 1, 2, 3\}$	V_s *variation*
$\emptyset$	*No variation*
0	$E/3$ *to* 0
1	0 *or* $2E/3$ *to* $E/3$
2	$E/3$ *or* E *to* $2E/3$
3	$2E/3$ *to* E

Remark 6 If the fault $\bar{f_1}$ or $\bar{f_2}$ or $\bar{f_3}$ occurs, then $V_s \in \{0, \frac{E}{3}, \frac{2E}{3}\}$. If the fault f_1 or f_2 or f_3 occurs, then $V_s \in \{\frac{E}{3}, \frac{2E}{3}, E\}$. When a fault occurs (when a cell is blocked), the system becomes similar to the 2-cells converter (4 modes).

Remark 7 This work considers the system in steady state. During the experimentation, there is a transient period to fulfil the control objective $x_{ref} = [V_{c_1 ref}, V_{c_2 ref}, I_{ref}]^T = [20, 40, 0, 17]^T$. Therefore, after each reset, the system is re-initialized at x_0 and a delay, corresponding to its transient time, is considered on the active diagnosis procedure.

In this paper, the diagnoser, associated to the 3-cells converter, is not detailed in order to simplify notations. The diagnosis algorithm follows the same procedure than the 2-cells converter. Some experimental results are carried out to show that the approach can be generalized for this class of systems and applied in real time.

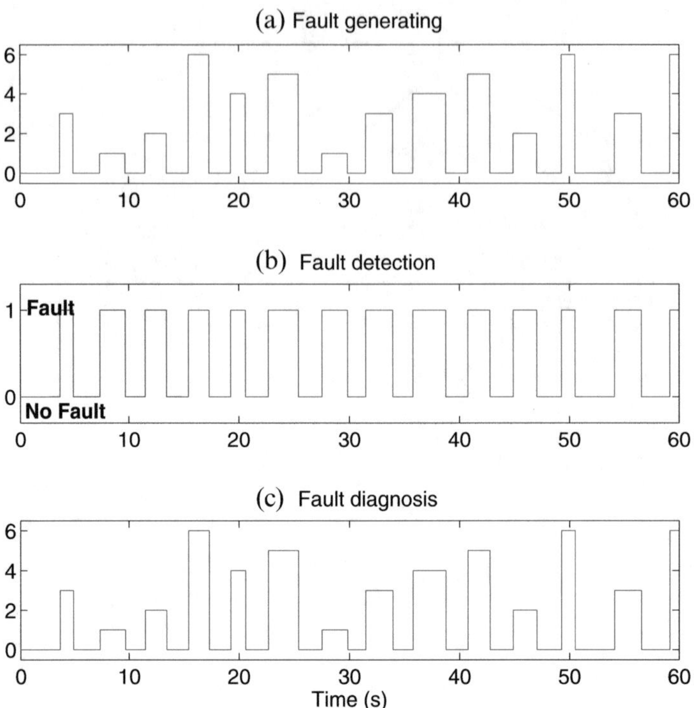

Fig. 6.14 Fault detection and isolation, using the proposed active diagnosis algorithm for the 3-cells converter. (**a**) Fault evolution. (**b**) Fault detection using the proposed diagnoser. (**c**) Isolation using the diagnoser and the homing sequences

6.4.3.3 Experimental Results

Figure 6.14a depicts the evolution of faults. In order to highlight the efficiency of the approach in real time, the experimentation takes into account all kind of faults $\{f_1, f_2, f_3, \bar{f}_1, \bar{f}_2, \bar{f}_3\}$. The faults are manually generated in order to interact with the control. A reset of the system is realized between each fault (see Fig. 6.15b). Figures 6.15 and 6.16 show, respectively, the evolution of the actual mode of the DFA and the state evolution of the converter. For each fault class, the diagnoser is initialized and the control ensures the state regulation. In Fig. 6.16, the nominal working of the converter between each generated fault is illustrated. When a fault is detected, the control is broken off and a fault isolating sequence can be applied in order to isolate it.

One can see, in Fig. 6.14, that the diagnoser, using the MM representation, fulfils the objective, i.e., the faulty modes are well detected and isolated. In Fig. 6.15, one can note that faults are identified by the same approach as the 2-cells converter and using the fault isolating sequences.

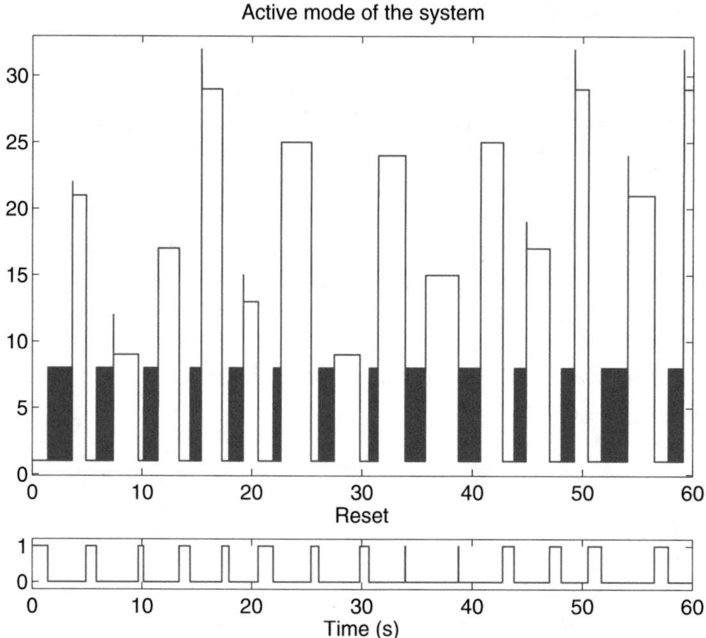

Fig. 6.15 Mode commutations (nominal and faulty) for the 3-cells converter

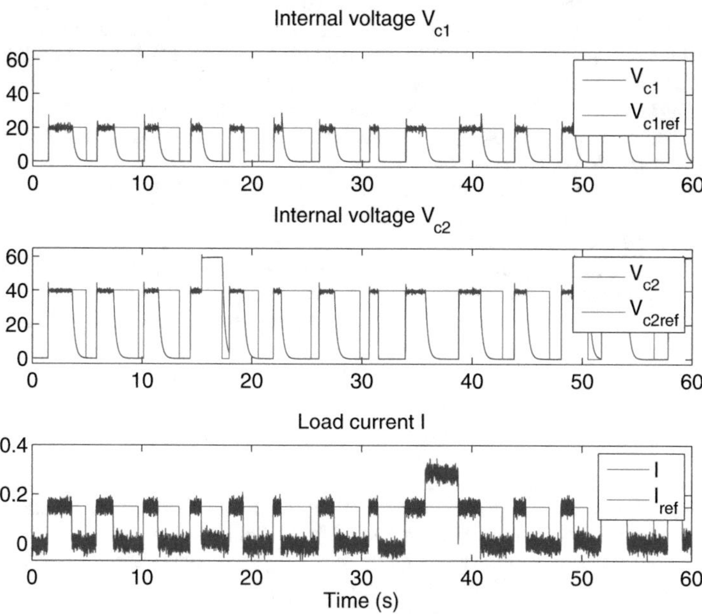

Fig. 6.16 Evolution of the state and the reference for the 3-cells converter

6.5 Conclusion

An active diagnosis for a class of switched systems which may not satisfy the diagnosability conditions is designed. A Mealy Machine modeling is used to define an appropriate diagnoser which reduces the uncertain state subset. Some diagnosability conditions of faults are deduced using this representation. If the MM diagnoser satisfies these conditions, an algorithm combining the proposed diagnoser and a testing procedure can be used in order to solve the fault diagnosis problem. A study on the cascade multicellular converter is carried out to detect and isolate faulty cells. Simulation results, on the 2-cells converter, are detailed and highlight the effectiveness of the proposed algorithm. Experimental results, on the 3-cells converter, show that the approach can be generalized for this class of switched system and applied in real time.

References

1. M. Bayoudh, L. Travé-Massuyès, An algorithm for active diagnosis of hybrid systems casted in the DES framework, in *2nd IFAC Workshop on Dependable Control of Discrete Systems* (2009), pp. 329–334
2. M.S. Branicky, Multiple Lyapunov functions and other analysis tools for switched and hybrid systems. IEEE Trans. Autom. Control **43**(4), 475–482 (1998)
3. M. Broy, B. Jonsson, J.-P. Katoen, *Model-Based Testing of Reactive Systems*, ed. by M. Leucker, A. Pretschner. Lecture Notes in Computer Science, vol. 3472 (Springer, Berlin, 2005)
4. M.P. Cabasino, A. Giua, S. Lafortune, C. Seatzu, A new approach for diagnosability analysis of Petri nets using verifier nets. IEEE Trans. Autom. Control **57**(12), 3104–3117 (2012)
5. M.P. Cabasino, A. Giua, C. Seatzu, Diagnosis using labeled Petri nets with silent or undistinguishable fault events. IEEE Trans. Syst. Man Cybern. A **43**(2), 345–355 (2013)
6. M. Daigle, G. Biswas, Improving diagnosability of hybrid systems through active diagnosis, in *Safeprocess09* (2009), pp. 217–222
7. M. Defoort, J. Van Gorp, M. Djemaï, K. Veluvolu, Hybrid observer for switched linear systems with unknown inputs, in *7th IEEE Conference on Industrial Electronics and Applications* (2012), pp. 594–599
8. P.M. Franck, Fault diagnosis in dynamic systems using analytical and knowledge-based redundancy-a survey and some new results. Automatica **26**(3), 459–474 (1990)
9. J. Gertler, Fault detection and isolation using parity relations. Control. Eng. Pract. **5**(5), 653–661 (1997)
10. Q. Guo, R.M. Hierons, M. Harman, K. Derderian, Heuristics for fault diagnosing when testing from finite state machines. J. Softw. Test. Verif. Reliab. **17**(1), 41–57 (2007)
11. I. Hwang, S. Kim, Y. Kim, C.E. Seah, A survey of fault detection, isolation, and reconfiguration methods. IEEE Trans. Control Syst. Technol. **18**(3), 636–653 (2010)
12. R. Isermann, *Fault Diagnosis of Technical Process-Applications* (Springer, Heidelberg, 2006)
13. W. Kang, J.P. Barbot, L. Xu, *On the Observability of Nonlinear and Switched Systems*. Lecture Notes in Control and Information Sciences (Springer Berlin, 2009)
14. D. Liberzon, *Switching in Systems and Control*. Systems and Control: Foundations and Applications (Birkhäuser, Boston, MA, 2003)
15. H. Lin, P.J. Antsaklis, Stability and stabilizability of switched linear systems: a survey of recent results. IEEE Trans. Autom. Control **54**(2), 308–322 (2009)

16. L. Maharjan, T. Yamagishi, H. Akagi, J. Asakura, Fault-tolerant operation of a battery-energy-storage system based on a multilevel cascade PWM converter with star configuration. IEEE Trans. Power Electron. **25**(9), 2386–2396 (2010)
17. K. Medjaher, J. Andrews, C.H. Bérenguer, L. Jackson (eds.), A bond graph model-based fault detection and isolation, in *Maintenance Modelling and Applications. Chapter 6 : Fault Diagnostics* (Det Norske Veritas, Akershus, 2011), pp. 503–512
18. S. Pettersson, B. Lennartson, Hybrid system stability and robustness verification using linear matrix inequalities. Int. J. Control **75**(16–17), 1335–1355 (2002)
19. M. Pocci, I. Demongodin, N. Giambiasi, A. Giua, Testing experiments on synchronized Petri nets. IEEE Trans. Autom. Sci. Eng. **11**(1), 125–138 (2014)
20. M. Sampath, R. Sengupta, S. Lafortune, K. Sinnamohideen, D.C. Teneketzis, Failure diagnosis using discrete-event models. IEEE Trans. Control Syst. Technol. **4**(2), 105–124 (1996)
21. M. Sampath, S. Lafortune, D. Teneketzis, Active diagnosis of discrete-event systems. IEEE Trans. Autom. Control **43**(7), 908–929 (1998)
22. M. Schmidt, J. Lunze, Active diagnosis of deterministic I/O automata, in *4th IFAC Workshop on Dependable Control of Discrete Systems* (2013), pp. 79–84
23. W. Song, A.Q. Huang, Fault-tolerant design and control strategy for cascaded H-bridge multilevel converter-based STATCOM. IEEE Trans. Ind. Electron. **57**(8), 2700–2708 (2010)
24. A. Tanwani, D. Liberzon, Invertibility of switched nonlinear systems. Automatica **46**(12), 1962–1973 (2010)
25. J. Van Gorp, M. Defoort, M. Djemai, N. Manamanni, Hybrid observer for the multicellular converter, in *Proceedings IFAC ADHS 12* (2012), pp. 259–264
26. J. Van Gorp, A. Giua, M. Defoort, M. Djemai, Active diagnosis for a class of switched systems, in *IEEE 52nd Annual Conference on Decision and Control* (2013), pp. 5003–5008
27. J. Van Gorp, M. Defoort, K. Veluvolu, M. Djemai, Hybrid sliding mode observer for switched linear systems with unknown inputs. J. Franklin Inst. **351**(7), 3987–4008 (2014)
28. J. Van Gorp, M. Defoort, M. Djemai, K. Veluvolu, Fault detection based on higher-order sliding mode observer for a class of switched linear systems. IET Control Theory Appl. **9**(15), 2249–2256 (2015)
29. S. Yoon, S. Kim, J. Bae, Y. Kim, E. Kim, Experimental evaluation of fault diagnosis in a skew-configured UAV sensor system. Control Eng. Pract. **19**(2), 158–173 (2011)

Chapter 7
Secure Diagnosability of Hybrid Dynamical Systems

Gabriella Fiore, Elena De Santis, and Maria Domenica Di Benedetto

7.1 Introduction

Hybrid systems are heterogeneous dynamical systems characterized by the interaction of continuous and discrete dynamics. Hybrid systems provide a powerful modeling framework to deal with a great variety of applications (such as smart grids, automotive and air traffic management systems, unmanned vehicles, and many others) where physical processes (described by means of differential or difference equations) and computational and communication components (modeled as discrete systems) are tightly interconnected. All the above-mentioned applications are safety critical, in the sense that their failure can cause irreparable damage to the physical systems being controlled and to the people who depend on it [2, 3]. Even when the disruption of these complex systems is not life threatening for people, it could have a large impact on society, by causing large direct and indirect economic losses. For these reasons, the study of security issues for hybrid systems is presently one of the most significant challenges. In this respect, the observability and diagnosability properties of a hybrid system play an important role. In fact, they are essential in characterizing the possibility of identifying the system's hybrid state, and in particular the occurrence of some specific states that may correspond to a malfunctioning of the system due to a fault or an attack.

One of the most important challenges when dealing with security for hybrid systems is to provide countermeasures to the aim of increasing the resilience of the system with respect to malicious attacks. One possible strategy is being able

G. Fiore (✉) · E. De Santis · M. D. Di Benedetto
University of L'Aquila, Department of Information Engineering, Computer Science and Mathematics (DISIM), Center of Excellence DEWS, L'Aquila, Italy
e-mail: gabriella.fiore@univaq.it; elena.desantis@univaq.it; mariadomenica.dibenedetto@univaq.it

M. Sayed-Mouchaweh (ed.), *Diagnosability, Security and Safety of Hybrid Dynamic and Cyber-Physical Systems*, https://doi.org/10.1007/978-3-319-74962-4_7

to reconstruct the true state of the system despite the presence of an adversarial attacker. For this reason, the main focus of this chapter is on investigating under which conditions the hybrid system's internal state can be correctly estimated, despite the presence of attacks, and to provide efficient algorithms to perform this estimation.

The state can be reconstructed instantaneously or within a finite time interval. More precisely, observability corresponds to the possibility of determining the current discrete state of the process as well as the continuous one, on the basis of the observed output information. Diagnosability, a property that is closely related to observability but is more general, corresponds to the possibility of detecting the occurrence of particular subsets of hybrid states, for example faulty states, on the basis of the observations, within a finite time interval.

Diagnosability has been extensively studied for Finite State Machines (FSMs) in the last two decades, with different approaches depending on the model, on the available output information, and on the objective for which state reconstruction is needed (see [18, 19, 24, 27, 37] to name a few). Recently, decentralized and distributed approaches to diagnosability have also been investigated [25, 26, 36]. An exhaustive review of the state of the art of diagnosis methods for discrete event systems can be found in the survey [38] and in [28]. Due to the lack of a common formalism, in [8] the authors propose a unifying framework where observability and diagnosability are defined and characterized with respect to a subset of the state space for finite state systems. For this class of systems, time flow is not taken into account, so that the results obtained for FSMs cannot, in general, be applied directly to hybrid systems where the discrete evolution interacts with the continuous one.

In the hybrid domain, checking the diagnosability property and determining its decidability and computational complexity are difficult issues and several problems remain open [7]. Some efforts in this direction have been made in [1, 9–12, 21, 23]. In [21] and [23] the authors formulate the hybrid diagnosis problem as a model selection problem, to detect and isolate a component fault causing a deviation from the nominal system operation. In [1] continuous dynamics of the hybrid system are abstracted by defining a set of signature events associated to mode signature changes, which are used to properly enrich the underlying discrete event system. Thanks to this abstraction, diagnosability analysis of the hybrid system is cast into a purely discrete event framework and hybrid diagnosability conditions are provided. A similar approach is used in [11] and [12], where the event signatures are generated on the basis of the distinguishability between continuous dynamics associated to the discrete states, when the underlying original discrete event system is not diagnosable. However, the distinguishability notion is not explicitly characterized. In [35] distinguishability between continuous dynamics is exploited, and an online diagnoser is proposed. In [10] an abstraction procedure is presented, which verifies the diagnosability of a hybrid automaton by means of an equivalent durational graph (a special subclass of timed automata [34]). Instead, in [9] a trajectory-based abstraction is proposed to take into account also the presence of measurement uncertainty.

In this chapter, we propose a formal definition of diagnosability for hybrid systems, and we characterize this property in the case where the available information may be corrupted by an external attacker. We also provide an abstracting procedure that can be used to determine if a hybrid system is diagnosable. More specifically, our procedure combines the available continuous and discrete information of the hybrid dynamical system and obtains an abstracted FSM $\mathscr{M}$. Then, by checking the diagnosability of $\mathscr{M}$ it is possible to infer the diagnosability of the original hybrid system.

The detection of the occurrence of some particular hybrid states is based on the observed output information, which consists of a continuous output and a discrete one. We suppose that the continuous output information is measured by sensors that monitor the system and send information to the estimator, which has to reconstruct the hybrid state of the system. Sensor measurements are exchanged by means of a wireless communication network and may be compromised by a malicious attacker. Our purpose is to investigate how adversarial attacks on the observed output information may affect the estimation of the hybrid state. In this chapter, we make use of a model-based approach to investigate hybrid system's diagnosability. Other techniques have been proposed for this purpose, which are based on machine learning algorithms (as described, e.g., in [29]), but the comparison between data-based and model-based approaches is out of scope of this chapter.

This chapter is organized as follows. In Sect. 7.2 we introduce the secure state estimation problem for dynamical systems. In Sect. 7.3 we define the hybrid system model and describe the property of distinguishing between any two discrete states of the hybrid system, on the basis of the continuous output information only. In Sect. 7.4 we focus on the discrete structure of the hybrid system. In particular, as the diagnosability of the hybrid system is investigated by means of an abstracted discrete event system, we provide a general description of FSMs as well as a characterization of diagnosability for this class of systems. In Sect. 7.5 we provide a formal definition of diagnosability for hybrid systems, by considering the general case where the available information may be corrupted by an external attacker. We describe the abstracting procedure that allows inferring diagnosability of the hybrid system, and provide an example. Finally, we illustrate sufficient conditions for the hybrid system to be diagnosable.

Notation In this chapter we use the following notation. The symbols $\mathbb{N}$, $\mathbb{R}$ denote the set of nonnegative integer, and real numbers, respectively. I indicates the identity matrix, $\mathbf{0}$ indicates the null matrix of proper dimensions (which can be trivially deduced by the context). Given a vector $x \in \mathbb{R}^n$, $\text{supp}(x)$ is its support, that is the set of indexes of the non-zero elements of x; $\|x\|_0$ is the cardinality of $\text{supp}(x)$, that is the number of non-zero elements of x. The vector $x \in \mathbb{R}^n$ is said to be s-sparse if $\|x\|_0 \leq s$. $\mathbb{S}_s^n$ indicates the set containing all the s-sparse vectors $x_i \in \mathbb{R}^n$ such that $\|x_i\|_0 \leq s$. Given the function $y : \mathbb{N} \to \mathbb{R}^p$, $y|_{[t_0,t_0+T-1]}$ is the collection of T samples of y, i.e. $y|_{[t_0,t_0+T-1]} = (y(t_0)^\top \; y(t_0+1)^\top \; \cdots \; y(t_0+T-1)^\top)^\top$. The function y is said to be cyclic s-sparse if, given a set $\Gamma \subset \{1, \ldots, p\}$, such that $|\Gamma| = s$, $y(t) \in \mathbb{S}_s^p$ and $\text{supp}(y(t)) \subseteq \Gamma$, for all $t \in \mathbb{N}$. $\mathbb{CS}_s^{pT}$ is the set containing all the cyclic s-sparse vectors $y \in \mathbb{R}^{pT}$. Given a matrix $M \in \mathbb{R}^{n \times m}$ and a set $\Gamma \subset \{1, \ldots, n\}$, we denote by $\overline{M}_\Gamma \in \mathbb{R}^{(n-|\Gamma|)\times m}$ the matrix obtained from M by removing the rows whose indexes are contained in Γ. For a set $Y \in X$, the symbol $\overline{Y}$ denotes the complement of Y in X, i.e. $\overline{Y} = \{x \in X : x \notin Y\}$. ϵ indicates the null event. For a string σ, $|\sigma|$ denotes its length; $\sigma(i)$ denotes the i-th element, $i \in \{1, \ldots, |\sigma|\}$; $\sigma|_{[a,b]}$ indicates the string $\sigma(a)\sigma(a+1)\cdots\sigma(b)$. $P(\sigma)$ is the projection of the string σ, that is the string obtained from σ by erasing the symbol ϵ.

7.2 The Problem of Secure State Estimation

Given a physical process, we consider the scenario illustrated in Fig. 7.1, where sensor measurements are sent to the controller through a wireless communication network. The controller estimates the state of the system and, based on this estimation, sends the control signal to the actuators. We assume that sensor measurements may be compromised by an external malicious attacker, and we consider both the case in which the attacker compromises the sensor nodes and the case in which the attacker affects the communication links between sensors and controller, for example by spoofing sensor measurements or launching deception attacks [33]. We suppose that the attack is not represented by a specific model, but it is assumed to be unbounded and influencing only a small subset of sensors, that is, the attack intensity may be unbounded but it is sparse. More precisely, when the attacker has the ability to compromise s nodes on a set of p devices, we define it as a s-sparse attack. We assume that the actual number of nodes under attack is unknown, but an upper bound is known, that is, $s \le \bar{s} \ll p$. This assumption is motivated by the fact that it is reasonable to consider that, in a real system, the attacker cannot reach the whole set of monitoring devices. We also assume that the set of attacked nodes is unknown, but fixed over time. This is compatible with the assumption that the attacker does not have arbitrary access to the whole set of devices.

The problem of estimating the internal state of a system, when sensors can be corrupted by a malicious attacker, is called “secure state estimation” problem [13].

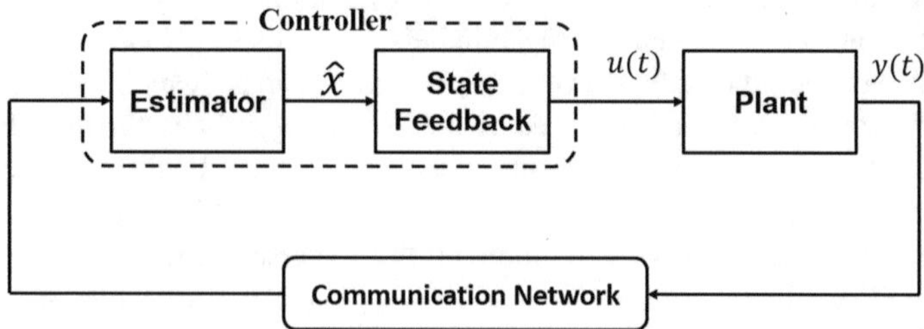

Fig. 7.1 Conceptual block diagram of the control system. Sensor measurements $y(t)$ are exchanged by means of a communication network, $\hat{x}$ indicates the estimate of the state

Recent results on security for dynamical systems focus on this case. In [14] the authors propose a method to estimate the state of a linear time invariant system when a fixed set of sensors and actuators is corrupted by deception attacks. They prove that, if the number of corrupted nodes is smaller than a certain threshold, then it is possible to exactly recover the internal state of the system by means of an algorithm derived from compressed sensing technique and error correction over the reals. A computationally efficient version of the algorithm is presented in [30], where a notion of strong observability is introduced as well as a recursive algorithm that estimates the state despite the presence of the attack. In order to overcome the limitations imposed by the combinatorial nature of the problem, in [32] the authors formulate the problem as a satisfiability one, and propose a sound and complete algorithm based on the Satisfiability Modulo Theory paradigm. This approach is extended to nonlinear differentially flat systems in [31]. The same assumption on the sparsity of the attack signal is made in [5] and [16], but here the authors consider a more general case where the set of attacked nodes can change over time. A similar approach is used in [6] where a continuous time linear system is considered.

Within this framework, the corrupted discrete-time linear dynamical system can be described as follows:

$$\begin{aligned} x(t+1) &= Ax(t) + Bu(t) \\ y(t) &= Cx(t) + w(t) \end{aligned} \tag{7.1}$$

where $t \in \mathbb{N}$, $x(t) \in \mathbb{R}^n$ is the system's state, $y(t) \in \mathbb{R}^p$ is the output signal, $u(t) \in \mathbb{R}^m$ is the input signal, and $w(t) \in \mathbb{S}_s^p$ is the s-sparse attack vector on sensor measurements. In the papers mentioned above the authors assume that the malicious attacker has only access to a subset of sensors $K_w \subset \{1, \dots, p\}$, meaning that the set of attacked nodes is fixed over time (but unknown). It is only assumed that the number or attacked sensors is bounded above by $\bar{s}$.

Assumption 1 *The attack on sensor measurements is cyclic s-sparse (for brevity, s-sparse).*

Roughly speaking, Assumption 1 means that we know that the set of attacked sensors has bounded cardinality (that is, $|K_w| \leq s < p$), but we do not know which nodes are actually compromised. Let $w_k(t)$ denote the k-th component of $w(t) \in \mathbb{R}^p$, $k \in \{1, \dots, p\}$ (i.e., the component of $w(t)$ corresponding to the k-th sensor), at time $t \in \mathbb{N}$. If $k \notin K_w$, then $w_k(t) = 0$ for all $t \in \mathbb{N}$ and the k-th sensor is said to be secure (i.e., not attacked). If $k \in K_w$, then $w_k(t)$ can assume any value and this corresponds to the case in which the attacker has access to the k-th sensor.

All of the above-mentioned works are concerned with the state estimation for linear or nonlinear systems and cannot be directly applied to hybrid systems. To the best of our knowledge, we provide in this chapter the first contribution to the study of the secure state estimation problem for hybrid systems, which we introduce in the next section.

7.3 Hybrid Dynamical Systems

In this section we first introduce the general model of a linear hybrid dynamical system (see [7, 20] for additional details), whose continuous output may be corrupted by a sparse attack. Then, we provide conditions for the distinguishability, only based on the continuous output information, between any two discrete states of the hybrid system, and for the detection of a transition between them.

7.3.1 Definition of a Hybrid Dynamical System

Definition 1 A linear hybrid dynamical system (LH-system) is a tuple

$$\mathscr{H} = (\Xi, \Xi_0, Y, h, S, E, G, R, \delta, \Delta) \tag{7.2}$$

in which:

- $\Xi = Q \times \mathbb{R}^n$ is the hybrid state space, where the finite set $Q = \{1, \ldots, N\}$ is the discrete state space, $\mathbb{R}^n$ is the continuous state space. The hybrid state is $\xi = (q, x)$ with $q \in Q$ the discrete state (also called mode or location), and $x \in \mathbb{R}^n$ the continuous state.
- $\Xi_0 = Q_0 \times \mathbb{R}^n \subset \Xi$ is the set of initial hybrid states, Q_0 is the set of initial discrete states.
- $Y = Y_d \times \mathbb{R}^p$ is the hybrid output space, where the finite set Y_d is the discrete output space, $\mathbb{R}^p$ is the continuous output space.
- $h : Q \to Y_d$ is the discrete output function.
- S is a function which associates a linear dynamical system $S(q)$ to each discrete state $q \in Q$. $S(q)$ (also indicated as S_q) is described by the following equations:

$$\begin{aligned} x(t+1) &= A_q x(t) + B_q u(t) \\ y(t) &= C_q x(t) + w(t) \end{aligned} \tag{7.3}$$

  where $t \in \mathbb{N}$, $x(t) \in \mathbb{R}^n$ is the (continuous) state of the system, $y(t) \in \mathbb{R}^p$ is the (continuous) output signal, $u(t) \in \mathbb{R}^m$ is the (continuous) input signal, $w(t) \in \mathbb{S}_s^p$ is the s-sparse attack vector on sensor measurements, $A_q \in \mathbb{R}^{n \times n}$, $B_q \in \mathbb{R}^{n \times m}$, $C_q \in \mathbb{R}^{p \times n}$, for all $q \in Q$.
- $E \subset Q \times Q$ is the set of admissible discrete transitions. A transition (also called switching) from the discrete state $i \in Q$ to $j \in Q$ is indicated by the pair (i, j).
- $G : E \to 2^{\mathbb{R}^n}$ is the function which associates to the transition $e \in E$ a linear subspace $G(e) \subseteq \mathbb{R}^n$, called guard set.
- $R : E \times \mathbb{R}^n \to \mathbb{R}^n$ is the linear reset function. $R(e, x) = R_e x$, with $R_e \in \mathbb{R}^{n \times n}$.
- $\delta : Q \to \mathbb{R}^+$ is the function which associates to $q \in Q$ the minimum dwell time $\delta(q)$.

- $\Delta : Q \to \mathbb{R}^+ \cup \{\infty\}$ is the function which associates to $q \in Q$ the maximum dwell time $\Delta(q)$.

To refer to the evolution in time of an LH-system we resort on the concepts of hybrid time set and execution, as follows (see [20] and [7]).

A hybrid time basis τ is defined as a finite or infinite sequence of time intervals $[t_{k-1}, t_k)$, $k = 1, \ldots, \text{card}(\tau)$, with $t_k > t_{k-1}$ and $t_0 = 0$. Given a hybrid time basis τ, time instants t_k are called switching times. Let $\text{card}(\tau) = L$, a time basis can be:

- finite, if L is finite and $t_L \neq \infty$;
- infinite, if either $L = \infty$ or $t_L = \infty$;
- Zeno, if $L = \infty$ and $t_L \neq \infty$.

Throughout this chapter we assume that the LH-system $\mathscr{H}$ is non-Zeno, that is, for each finite time interval, a finite number of switching occurs. We also indicate by $\mathscr{T}$ the set of all non-Zeno hybrid time bases, and by $\mathscr{U}$ the set of admissible control inputs.

Definition 2 ([7]) An execution of an LH-system is a tuple $\chi = (\xi_0, \tau, u, \xi)$ in which:

- $\xi_0 = (q_0, x_0) \in \Xi_0$ is the initial hybrid state;
- $\tau \in \mathscr{T}$ is the hybrid time basis. The switching times are such that the minimum and maximum dwell time constraints are satisfied in each discrete state, that is:

$$\delta\left(q(t_{k-1})\right) \leq t_k - t_{k-1} \leq \Delta\left(q(t_{k-1})\right), \; k = 1, \ldots, L;$$

- $u \in \mathscr{U}$ is the continuous control input;
- $\xi : \mathbb{N} \to \Xi$ is the hybrid state evolution, defined as:

$$\begin{aligned} \xi(t_0) &= \xi_0 \\ \xi(t) &= (q(t), x(t)), \; t \in [t_0, t_L) \end{aligned}$$

  where $q : \mathbb{N} \to Q$ is the discrete state evolution, and $x : \mathbb{N} \to \mathbb{R}^n$ is the continuous state evolution. At time $t \in [t_{k-1}, t_k)$, $q(t) = q(t_{k-1})$, $x(t)$ is the unique solution of Eq. (7.3) for the dynamical system $S(q(t_{k-1}))$, with initial time t_{k-1}, initial state $x(t_{k-1})$ and control law $u|_{[t_{k-1},t)}$, $(q(t_{k-1}), q(t_k)) \in E$, $x(t_k) = R_{q(t_k),q(t_{k+1})}x(t_k^-)$, and $x(t_k^-) \in G\left((q(t_k), q(t_{k+1}))\right)$.

Given a state execution χ, the output function $\eta : \mathbb{N} \to Y$ is defined as:

$$\eta(t) = (y_d(t), y(t))$$

where $y(t)$ is the continuous component of the output, and $y_d(t)$ is the discrete component of the output such that:

$$y_d(t_k) = h(q(t_k)), k = 0, \dots, L-1$$
$$y_d(t) = \epsilon, \forall t \in \mathbb{N} : t \neq t_k.$$

The evolution in time of the discrete state of the LH-system is described by the triple (q_0, τ, q).

7.3.2 Secure Mode Distinguishability

Reconstructing the discrete mode of an LH-system corresponds to understanding which continuous dynamical system is evolving. This can be done either by using only the discrete output information or only the continuous output information, or by using mixed information, both discrete and continuous. When the discrete information is not sufficient to identify the discrete state, the possibility of distinguishing between two continuous dynamical systems on the basis of the continuous output information is needed. Therefore, in this section we introduce the distinguishability property between two dynamical systems S_i and S_j, $(i, j) \in Q \times Q$ of $\mathscr{H}$. This will be instrumental in providing conditions for the diagnosability of the LH-system (further details can be found in [15] and [17]).

Let y_i, $i \in Q$, be the continuous output evolution when the dynamical system S_i is active with initial state $x(0) = x_{0i}$, and let $\mathscr{U}$ be the set of all input functions $u : \mathbb{N} \to \mathbb{R}^m$. A generic input sequence $u|_{[0,\tau-1)}$ is any input sequence that belongs to a dense subset of the set $\mathbb{R}^{m\tau}$, equipped with the L_∞ norm.

Definition 3 Two linear systems S_i and S_j, $(i, j) \in Q \times Q$, are securely distinguishable with respect to generic inputs and for all s-sparse attacks on sensors (shortly, s-securely distinguishable), if there exists $\tau \in \mathbb{N}$ such that $y_i|_{[0,\tau-1]} \neq y_j|_{[0,\tau-1]}$, for any pair of initial states x_{0i} and x_{0j}, for any pair of s-sparse attack vectors $w_i|_{[0,\tau-1]} \in \mathbb{CS}_s^{\tau p}$ and $w_j|_{[0,\tau-1]} \in \mathbb{CS}_s^{\tau p}$, and for any generic input sequence $u|_{[0,\tau-1)}$, $u \in \mathscr{U}$. The linear systems S_i and S_j are called s-securely indistinguishable if they are not s-securely distinguishable.

In order to characterize the s-secure distinguishability property, we consider the augmented linear system S_{ij} depicted in Fig. 7.2, which is fully described by triple (A_{ij}, B_{ij}, C_{ij}), such that:

$$A_{ij} = \begin{bmatrix} A_i & \mathbf{0} \\ \mathbf{0} & A_j \end{bmatrix}, \quad B_{ij} = \begin{bmatrix} B_i \\ B_j \end{bmatrix}, \quad C_{ij} = \begin{bmatrix} C_i & -C_j \end{bmatrix} \tag{7.4}$$

with $A_{ij} \in \mathbb{R}^{2n\times 2n}$, $B_{ij} \in \mathbb{R}^{2n\times m}$, $C_{ij} \in \mathbb{R}^{p\times 2n}$. The following matrices are also associated with the augmented system S_{ij}:

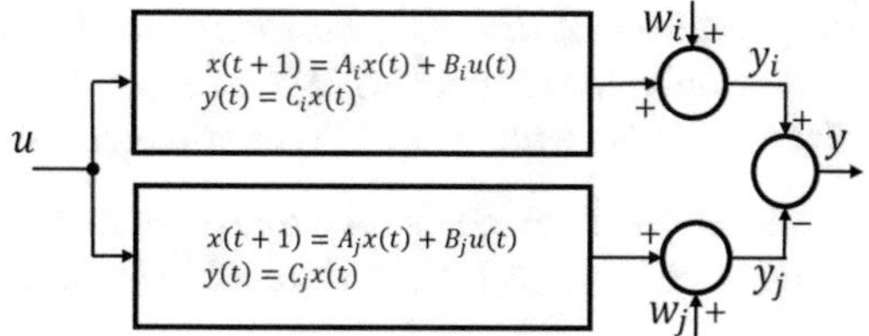

Fig. 7.2 Attack on sensors and secure actuators: the augmented controlled linear system S_{ij}

$$O_{ij} = \begin{bmatrix} C_{ij} \\ C_{ij}A_{ij} \\ \vdots \\ C_{ij}A_{ij}^{2n-1} \end{bmatrix} = \begin{bmatrix} O_i^{(2n-1)} & -O_j^{(2n-1)} \end{bmatrix},$$

$$M_{ij} = \begin{bmatrix} \mathbf{0} & \mathbf{0} & \cdots & \mathbf{0} \\ C_{ij}B_{ij} & \mathbf{0} & \cdots & \mathbf{0} \\ \vdots & \vdots & \cdots & \vdots \\ C_{ij}A_{ij}^{2n-2}B_{ij} & C_{ij}A_{ij}^{2n-3}B_{ij} & \cdots & C_{ij}B_{ij} \end{bmatrix} \tag{7.5}$$

where $\mathbf{0} \in \mathbb{R}^{p\times m}$ is the null matrix, and $M_{ij} \in \mathbb{R}^{2np\times(2n-1)m}$. $O_{ij} \in \mathbb{R}^{2np\times 2n}$ is the $2n$ steps-observability matrix for the augmented system S_{ij}, and it is made up of the $2n$ steps-observability matrices $O_i^{(2n-1)}$ and $O_j^{(2n-1)}$ for the linear systems S_i and S_j, respectively.

Given the set $\Gamma \subset \{1,\ldots,p\}$, $|\Gamma| \leq 2s$, let $M_{ij,\overline{\Gamma}} \in \mathbb{R}^{2n(p-2s)\times(2n-1)m}$ be the matrix obtained by the triples $(A_i, B_i, \overline{C}_{i,\Gamma})$ and $(A_j, B_j, \overline{C}_{j,\Gamma})$.

Theorem 1 *Two linear systems S_i and S_j, $(i,j) \in Q\times Q$, are s-securely distinguishable with respect to generic inputs and for all s-sparse attacks on sensors if and only if for any set Γ, with $\Gamma \subseteq \{1,\ldots,\gamma\}$, $|\Gamma| \leq 2s$, the matrix $M_{ij,\Gamma} \neq \mathbf{0}$.*

In the following we provide conditions to detect the occurrence of a transition between two discrete modes of the LH-system, by using only the (corrupted) continuous output information (the control signals are assumed to be secure).

Definition 4 A discrete transition $e = (i,j) \in E$ with $G(e) = \mathbb{R}^n$ is said to be $\overline{x}$-observable, $\overline{x} \in \mathbb{R}^n$, if there exists $d \in \mathbb{N}$ such that, for any switching time $t_k \in \mathbb{N}$, for any pair of s-sparse attack vectors $w_i|_{[t_k,t_k+d]} \in \mathbb{CS}_s^{dp}$ and $w_j|_{[t_k,t_k+d]} \in \mathbb{CS}_s^{dp}$, and for any generic input sequence $u|_{[t_k,t_k+d)}$, with $u \in \mathcal{U}$, $y_i|_{[t_k,t_k+d]} \neq y_j|_{[t_k,t_k+d]}$. The transition is said to be observable if it is $\overline{x}$-observable for any $\overline{x} \in \mathbb{R}^n$.

Since we are considering the transition between mode i and mode j, then $y_i|_{[t_k,t_k+d]}$ represents the continuous output of the linear system S_i with initial state $x_i(t_k) = x(t_k^-)$, whereas $y_j|_{[t_k,t_k+d]}$ represents the continuous output of the linear system S_j with initial state $x_j(t_k) = R_{ij}x(t_k^-)$. $x(t_k^-)$ indicates the continuous state before the transition takes place, R_{ij} denotes the reset map.

Proposition 1 *A discrete transition $e = (i, j) \in E$ with $G(e) = \mathbb{R}^n$ is observable if and only if the pair (S_i, S_j) is s-securely distinguishable (that is, securely distinguishable with respect to generic inputs and for all s-sparse attacks).*

A weaker condition holds for a transition to be $\overline{x}$-observable. In this case, the pair of linear systems (S_i, S_j) is not required to be distinguishable.

Proposition 2 *A discrete transition $e = (i, j) \in E$ with $G(e) = \mathbb{R}^n$ is $\overline{x}$-observable if and only if the following holds:*

$$\overline{x} \notin \ker(\overline{O}_{i,\Gamma} - \overline{O}_{j,\Gamma} R_{ij})$$

for any set Γ such that $|\Gamma| \leq 2s$.

7.4 Finite State Systems

It is possible to associate an FSM to an LH-system, by extracting the dependence of the discrete dynamics from its continuous evolution. Therefore, in this section we provide some background on FSMs. Moreover, we recall the diagnosability notion for FSMs as defined in [8].

7.4.1 Background on FSMs

Definition 5 A Finite State Machine (FSM) is a tuple

$$M = (Q, Q_0, Y_d, h, E) \tag{7.6}$$

in which Q is the finite set of states, $Q_0 \subset Q$ is the set of initial states, Y_d is the finite set of outputs, $h : Q \to Y_d$ is the output function, $E \subset Q \times Q$ is the transition relation.

For a state $i \in Q$, we can define the set of its successors $succ(i) = \{j \in Q : (i, j) \in E\}$ and the set of its predecessors $pre(i) = \{j \in Q : (j, i) \in E\}$. In this chapter, we make the following standard assumption:

Assumption 2 (Liveness) *$succ(i) \neq \emptyset$, for all $i \in Q$.*

The discrete behavior of a linear hybrid system $\mathscr{H} = (\Xi, \Xi_0, Y, h, S, E, G, R, \delta, \Delta)$, defined as in Eq. (7.2), is usually represented by means of the nondeterministic FSM $M = (Q, Q_0, Y_d, h, E)$. Given a time basis τ with $\text{card}(\tau) = L$, and the evolution in time (q_0, τ, q) of the discrete state of $\mathscr{H}$, the corresponding state evolution of the associated M is described by a string $\sigma(k) = q(t_{k-1}), k = 1, \ldots, L$.

In particular, a state execution of an FSM M is any finite or infinite string σ with symbols in Q which satisfies:

$$\begin{aligned} \sigma(1) &\in Q \\ \sigma(k+1) &\in succ(\sigma(k)),\ k = 1, \ldots, |\sigma| - 1 \end{aligned} \tag{7.7}$$

We denote by Q^* the set of all finite and infinite strings with symbols in Q, by $\mathscr{X}^*$ the set of all state executions, by $\mathscr{X}^*_{Q_0}$ the set of state executions $\sigma \in \mathscr{X}^*$ with $\sigma(1) \in Q_0$, by $\mathscr{X}$ the set of infinite state executions with $\sigma(1) \in Q_0$, thus $\mathscr{X} \subset \mathscr{X}^* \subset Q^*$. Given a set $\Psi \subset Q$, $\mathscr{X}^\Psi$ is the set of finite state executions $\sigma \in \mathscr{X}^* \backslash \mathscr{X}$ with last symbol in Ψ.

We indicate by $\mathbf{h} : \mathscr{X}^* \rightarrow (Y_d \backslash \{\epsilon\})^*$ the function which associates the corresponding output string to a state execution of M (ϵ indicates the null event). That is, given a state execution $\sigma \in \mathscr{X}^*$, and the corresponding output string $s = h(\sigma(1)) \cdots h(\sigma(|\sigma|))$,

$$\mathbf{h}(\sigma) = P(s) \tag{7.8}$$

if σ is finite, otherwise $\mathbf{h}(\sigma) = P(s_\infty)$, where s_∞ is an infinite string recursively defined as:

$$\begin{aligned} s_1 &= h(\sigma(1)) \\ s_{k+1} &= s_k h(\sigma(k+1)),\ k = 1, 2, \ldots \end{aligned} \tag{7.9}$$

Given the state execution $\sigma \in \mathscr{X}^*_{Q_0}$, $\mathbf{h}^{-1}(\mathbf{h}(\sigma)) = \{\widehat{\sigma} \in \mathscr{X}^*_{Q_0} : \mathbf{h}(\widehat{\sigma}) = \mathbf{h}(\sigma)\}$.

Let $Q' \subset Q$ be given, $reach(Q')$ is the set of states which can be reached starting from Q', that is:

$$reach(Q') = \{i \in Q : (\sigma(1) \in Q') \wedge (\sigma(k) = i) \wedge (\sigma \in \mathscr{X}^*)\}. \tag{7.10}$$

7.4.2 *Definitions of Diagnosability for FSMs*

In this section, we consider FSMs as in Definition 5, in which $h : Q \rightarrow Y_d$.

Given a set $Q_c \subset Q$, we investigate the possibility of inferring from the output execution, that the state belongs to the set Q_c at some step during the execution of the FSM. The set Q_c may represent any set of states of interest (including faulty and unsafe states).

For a state execution $\sigma \in \mathscr{X}$, two cases are possible:

1. $\sigma(k) \in Q_c$, for some $k \in \mathbb{N}$,
2. $\sigma(k) \notin Q_c$, for any $k \in \mathbb{N}$.

If the first condition holds, let k_σ denote the minimum value of $k \in \mathbb{N}$ such that $\sigma(k) \in Q_c$, that is:

$$k_\sigma = k \in \mathbb{N} : (\sigma(k) \in Q_c) \wedge (k = 1 \vee \sigma(l) \notin Q_c, \forall\, l \in [1, k-1]) \tag{7.11}$$

If the second condition holds, then $k_\sigma = \infty$.

Definition 6 The FSM M is diagnosable with respect to a set $Q_c \subset Q$ (shortly, Q_c–diagnosable) if there exists $T \in \mathbb{N}$ such that, for any string $\sigma \in \mathscr{X}$ with finite k_σ, any string $\widehat{\sigma} \in \mathbf{h}^{-1}(\mathbf{h}(\sigma|_{[1,k_\sigma+T]}))$ is such that $\widehat{\sigma}(k) \in Q_c$.

We can also propose an equivalent definition of Q_c-diagnosability based on the existence of a diagnoser, as follows:

Definition 7 The FSM M is diagnosable with respect to a set $Q_c \subset Q$ (shortly, Q_c-diagnosable) if there exist $T \in \mathbb{N}$ and a function $\mathscr{D} : (Y_d \backslash \{\epsilon\})^* \rightarrow \{0, 1\}$ called diagnoser such that

i) if $(\sigma(\hat{k}) \in Q_c) \wedge (\hat{k} = 1 \vee \sigma(k) \notin Q_c, \forall k \in [1, \hat{k}-1])$, then $\mathscr{D}(\mathbf{h}(\sigma|_{[1,\hat{k}+T]})) = 1$;
ii) if $\mathscr{D}(\mathbf{h}(\sigma|_{[1,k]})) = 1 \wedge ((k = 1) \vee (\mathscr{D}(\mathbf{h}(\sigma|_{[1,l]})) = 0, \forall l \in [0, k-1])$, then $\sigma(\hat{k}) \in Q_c$ for some $\hat{k} \in [\max\{1, k-T\}, k]$.

The following notions are instrumental to recall the result on Q_c-diagnosability of FSM M, as stated in [8]. Throughout the rest of this section we make the following assumption.

Assumption 3 $\epsilon \notin Y_d$.

We define the following symmetric sets:

$$\Pi = \{(i,j) \in Q \times Q : h(i) = h(j)\} \tag{7.12}$$

and

$$\Theta = \{(i,j) \in Q \times Q : i = j\} \tag{7.13}$$

We can now recall the indistinguishability notions between state trajectories of M.

Definition 8 Two state trajectories $\sigma_1 \in \mathscr{X}^*$ and $\sigma_2 \in \mathscr{X}^*$ are called indistinguishable if $\mathbf{h}(\sigma_1) = \mathbf{h}(\sigma_2)$. The pair $(i,j) \in \Pi$ is k–forward distinguishable if there exist $\sigma_1 \in \mathscr{X}_{\{i\}}$ and $\sigma_2 \in \mathscr{X}_{\{j\}}$ such that $|\sigma_1| = |\sigma_2| = k$ and $\mathbf{h}(\sigma_1) = \mathbf{h}(\sigma_2)$. The pair $(i,j) \in \Sigma \subset \Pi$ is k-backward distinguishable in Σ if there exist $\sigma_1 \in \mathscr{X}^{\{i\}}$ and $\sigma_2 \in \mathscr{X}^{\{j\}}$ such that $|\sigma_1| = |\sigma_2| = k$, $\sigma_1(l) \in \Sigma$, $\sigma_2(l) \in \Sigma$, for all $l \in [1,k]$ and $\mathbf{h}(\sigma_1) = \mathbf{h}(\sigma_2)$.

To characterize the diagnosability property for FSMs, we define the following subsets (see [8] for additional details and for recursive algorithms to compute these sets):

1. $S^* \subset \Pi$ is the maximal set of pairs $(i,j) \in \Pi$ such that there exist two indistinguishable state executions $\sigma_1 \in \mathscr{X}^{\{i\}} \cap \mathscr{X}_{Q_0}$;
2. $F^* \subset \Pi$ is the maximal set of pairs $(i,j) \in \Pi$ which are k-forward indistinguishable, for any $k \in \mathbb{N}, k \geq 1$;
3. Λ_k is the set of pairs $(i,j) \in S^*$, with $i \in Q_c$ and $j \in \overline{Q_c}$ (or vice versa $i \in \overline{Q_c}$ and $j \in Q_c$) for which there exist two indistinguishable state trajectories $\sigma_1 \in \mathscr{X}_{\{i\}}$ and $\sigma_2 \in \mathscr{X}_{\{i\}}$, $|\sigma_1| = |\sigma_2| = k$, such that $\sigma_2(l) \in \overline{Q_c}$, for all $l \in [1,k]$ (or, conversely $\sigma_1(l) \in \overline{Q_c}$, for all $l \in [1,k]$);
4. $\Lambda^* \subset (F^* \cap S^*)$ is the set of pairs $(i,j) \in S^*$, such that for all $\bar{k} \in \mathbb{N}$, there exists $k \geq \bar{k}$ such that $(i,k) \in \Lambda_k$.

Given the FSM $M = (Q, Q_0, Y_d, h, E)$, we define the FSM $\widetilde{M} = (Q, Q_0, Y_d, h, \widetilde{E})$, where $(i,j) \in \widetilde{E}$ if and only if $(i,j) \in E$ and $i \notin Q_c$. Let $\widetilde{S}^*$ be the set of pairs reachable from Q_0 with two indistinguishable state evolutions, computed for $\widetilde{M}$.

The following theorem will be instrumental in proving the main result of this chapter.

Theorem 2 ([8]) *The FSM M is Q_c-diagnosable if and only if* $\widetilde{S}^* \cap \Lambda^* = \emptyset$.

Remark 1 The set $\widetilde{S}^* \cap \Lambda^*$ is the set of pairs $(i,j) \in \widetilde{S}^*$ such that:

- only one of the two states i or j belongs to Q_c,
- i and j are the ending states of a pair of indistinguishable state executions of the FSM $\widetilde{M}$, with initial state in Q_0, such that one of these executions never crosses the set Q_c,
- i and j are the initial states of a pair of arbitrarily long indistinguishable state executions of the FSM M, such that one of these executions never crosses the set Q_c.

7.5 Diagnosability of LH-Systems

In this section, we define and characterize the diagnosability property for LH-systems. We provide a formal definition of diagnosability of an LH-system. Then, we propose an abstracting procedure that can be used to determine if a hybrid system is diagnosable. More specifically, our procedure combines the available continuous and discrete information of the LH-system $\mathscr{H}$ to obtain an abstracted FSM $\mathscr{M}$. Then, by checking the diagnosability of $\mathscr{M}$ it is possible to infer the diagnosability of $\mathscr{H}$. Finally, sufficient diagnosability conditions are established.

7.5.1 Definitions of Diagnosability

Given a linear hybrid system $\mathscr{H}$ as in (7.2), let $\mathscr{U}$ and $\mathscr{Y}$ denote the collection of all input and hybrid output sequences of $\mathscr{H}$, respectively. In this section we make the following assumptions.

Assumption 4 $\delta(q) \geq \delta_{\min}$, *for all* $q \in Q$.

Assumption 5 $\Delta(q) \neq \infty$, *for all* $q \in Q$.

Definition 9 The LH-system $\mathscr{H}$ is s-securely diagnosable with respect to $\Omega \subset \Xi$ (shortly, s-securely Ω-diagnosable) if there exists $T \in \mathbb{N}$ and a function $\mathscr{D}$: $\left(\mathscr{U} \times \mathscr{Y} \times \mathbb{S}_s^p\right) \to \{0, 1\}$, called the diagnoser, such that

i) if

$$\xi\left(\widehat{t}\right) \in \Omega \wedge \left(\widehat{t} = 0 \vee \left(\xi\left(t\right) \notin \Omega, \forall t \in \left[0, \widehat{t} - 1\right], \widehat{t} > 0\right)\right)$$

then $\mathscr{D}\left(u|_{[0,\hat{t}+T-1]}, \eta|_{[0,\hat{t}+T]}\right) = 1$, with $\eta|_{[0,\hat{t}+T]} = (y_d|_{[0,\hat{t}+T]}, y|_{[0,\hat{t}+T]} + w|_{[0,\hat{t}+T]})$, for any generic input sequence $u|_{[0,\hat{t}+T-1]}$, with $u \in \mathscr{U}$, and for any attack sequence $w|_{[0,\hat{t}+T]} \in \mathbb{CS}_s^{(\hat{t}+T)p}$;

ii) if for any generic input sequence $u|_{[0,t-1]}$, with $u \in \mathscr{U}$, and for any attack sequence $w|_{[0,t]} \in \mathbb{CS}_s^{tp}$, $\mathscr{D}\left(u|_{[0,t-1]}, \eta|_{[0,t]}\right) = 1$ and

$$\left(t = 0 \vee \left(\mathscr{D}\left(u|_{[0,t'-1]}, \eta|_{[0,t']}\right) = 0, \forall t' \in [0, t-1], t > 0\right)\right)$$

then $\xi\left(\widehat{t}\right) \in \Omega$, for some $\widehat{t} \in [\max\{0, t - T\}, t]$.

The previous definition extends the notion of Ω-diagnosability presented in Definition 7 for FSMs to the class of LH-systems. The more general case where the continuous output information may be corrupted by s-sparse attacks is taken into account here.

The case where the sensor measurements are secure is considered in the following definition.

Definition 10 The LH-system $\mathscr{H}$ is diagnosable with respect to $\Omega \subset \Xi$ (shortly, Ω-diagnosable) if there exist $T \in \mathbb{N}$ and a function $\mathscr{D} : (\mathscr{U} \times \mathscr{Y}) \to \{0, 1\}$, called the diagnoser, such that

i) if

$$\xi\left(\widehat{t}\right) \in \Omega \wedge \left(\widehat{t} = 0 \vee \left(\xi\left(t\right) \notin \Omega, \forall t \in \left[0, \widehat{t} - 1\right], \widehat{t} > 0\right)\right)$$

then $\mathscr{D}\left(u|_{[0,\hat{t}+T-1]}, \eta|_{[0,\hat{t}+T]}\right) = 1$, with $\eta|_{[0,\hat{t}+T]} = (y_d|_{[0,\hat{t}+T]}, y|_{[0,\hat{t}+T]})$, for any generic input sequence $u|_{[0,\hat{t}+T-1]}$, with $u \in \mathscr{U}$;

ii) if for any generic input sequence $u|_{[0,t-1]}$, with $u \in \mathscr{U}$, $\mathscr{D}\left(u|_{[0,t-1]}, \eta|_{[0,t]}\right) = 1$ and

$$\left(t = 0 \vee \left(\mathscr{D}\left(u|_{[0,t'-1]}, \eta|_{[0,t']}\right) = 0, \forall t' \in [0, t-1], t > 0\right)\right)$$

then $\xi\left(\widehat{t}\right) \in \Omega$, for some $\widehat{t} \in [\max\{0, t-T\}, t]$.

In [8] some diagnosability definitions for FSMs are proposed, which are parametric with respect to the delay required for the detection of a critical state, and the uncertainty in the determination of the time step at which the crossing event $\sigma(k) \in \Omega$ occurs. This parametric approach can be extended to the case of LH-systems, but we leave this extension for future work.

7.5.2 Abstracting Procedure

In this section, we provide a detailed description of an abstracting procedure that, starting from an LH-system $\mathscr{H}$, derives an LH-system having purely discrete output information, which is equivalent to $\mathscr{H}$ with respect to diagnosability.

Let $\sim$ indicate the s-secure indistinguishability relation on the set of discrete modes Q, as defined in Sect. 7.3.2. We define the equivalence class of a state $i \in Q$ as the set $\{j \in Q \mid S_j \sim S_i\}$. The set of equivalence classes induces a partition of Q, which is called the quotient space of Q by the relation $\sim$ and is denoted by $Q/\sim$. To each equivalence class, we can associate a label in a certain set of labels $Y_\sim$.

The abstracting procedure consists of the following three main phases.

Step 0. Given the linear hybrid system $\mathscr{H}$ as in Eq. (7.2), define the quotient space $Q/\sim$ and associate a label (in the set $Y_\sim$) to any equivalence class.

Step 1. $\mathscr{H} \to \mathscr{H}^{(1)}$, where $\mathscr{H}^{(1)}$ is a linear hybrid system in which the initial states have no predecessors.

Step 2. $\mathscr{H}^{(1)} \to \mathscr{H}^{(2)}$, where $\mathscr{H}^{(2)}$ has additional outputs associated to discrete transitions.

Step 3. $\mathscr{H}^{(2)} \to \mathscr{H}^{(3)}$, in which $\mathscr{H}^{(3)}$ is a linear hybrid system with purely discrete information.

7.5.2.1 Step 0

In the first step, given the linear hybrid system $\mathscr{H}$, we derive the quotient space induced by the s-secure indistinguishability relation. This phase can be represented by the following algorithm.

procedure QUOTIENT SPACE(Q)
  Initialize Nclass $= 0$, class $= zeros(N, N)$
  while $Q \neq \emptyset$ **do**
    Nclass $++$, col $= 1$
    for $i \in Q$ **do**
      class[Nclass, 1] $= i$

```
            Q = Q\i
            for j ∈ Q do
                Check the s-secure distinguishability property
                if M_{ij,Γ} = 0, ∀ Γ then
                    class[Nclass, col] = j
                    Q = Q\j
                end if
            end for
        end for
    end while
    Return class
end procedure
```

Then, we associate a label in the set $Y_{\sim}$ to each equivalence class.

7.5.2.2 Step 1

Let the LH-system $\mathcal{H}$ as in Eq. (7.2) be given. Consider the associated FSM $M = (Q, Q_0, Y_d, h, E)$, and apply Algorithm 0, as described in section "From Mealy to Moore" in the Appendix, to obtain the FSM $M^{(1)} = (Q^{(1)}, Q_0^{(1)}, Y_d, h^{(1)}, E^{(1)})$, in which the set of initial states has no predecessors, and the injection map $g : Q \to \mathbb{N}$ is defined in Eq. (7.26). Based on the FSM $M^{(1)}$, the following LH-system is defined:

$$\mathcal{H}^{(1)} = (\Xi^{(1)}, \Xi_0^{(1)}, Y, h^{(1)}, S^{(1)}, E^{(1)}, G^{(1)}, R^{(1)}, \delta^{(1)}, \Delta^{(1)}) \tag{7.14}$$

in which:

- $\Xi^{(1)} = Q^{(1)} \times \mathbb{R}^n$
- $\Xi_0^{(1)} = Q_0^{(1)} \times \mathbb{R}^n$
- $S^{(1)}(i) = S(i),\ S^{(1)}(g(i)) = S(i),\ i \in Q$
- $G^{(1)}((i,j)) = G((i,j)),\ G^{(1)}((g(i),j)) = G((i,j)),\ (i,j) \in E$
- $R^{(1)}((i,j)) = R((i,j)),\ R^{(1)}((g(i),j)) = R((i,j)),\ (i,j) \in E$
- $\delta^{(1)}(i) = \delta(i),\ \delta^{(1)}(g(i)) = \delta(i),\ i \in Q$
- $\Delta^{(1)}(i) = \Delta(i),\ \Delta^{(1)}(g(i)) = \Delta(i),\ i \in Q$

Proposition 3 *The LH-system $\mathcal{H}$ is s-securely Ω-diagnosable if and only if $\mathcal{H}^{(1)}$ is s-securely Ω-diagnosable.*

Proof Directly follows from Algorithm 0.

7.5.2.3 Step 2

Throughout the rest of this chapter we make the following assumption.

Assumption 6 *We assume that the LH-system $\mathscr{H}$ has initial states with no predecessors, i.e.* $pre(i) = \emptyset,\ \forall\, i \in Q_0$.

Assumption 6 is made for the ease of notation, to consider $\mathscr{H}^{(1)} = \mathscr{H}$. It is motivated by the equivalence with respect to diagnosability between the LH-systems $\mathscr{H}$ and $\mathscr{H}^{(1)}$ stated in Proposition 3.

In this section we describe how to obtain the LH-system $\mathscr{H}^{(2)}$, in which an additional binary output is associated to each discrete transition. In particular, a binary signal with logical value 1 is associated if the transition can be detected from the continuous or discrete output information; otherwise, a binary signal with logical value 0 is associated. This additional signal associated to each transition can be modeled as a discrete input, therefore $\mathscr{H}^{(2)}$ can be described as:

$$\mathscr{H}^{(2)} = (\Xi, \Xi_0, W_d = \{1,0\}, Y^{(2)}, h^{(2)}, S^{(2)}, E^{(2)}, G, R, \delta, \Delta) \tag{7.15}$$

where $Y^{(2)} = Y_d^{(2)} \times \{\mathbf{0}\}$, $Y_d^{(2)} = (Y_d \cup Y_\sim \cup \{1,0\})^*$, $W_d = \{1,0\}$ is the finite discrete input space (with input symbol being associated to the discrete transition), $h^{(2)} : (Q \cup W_d) \rightarrow Y_d^{(2)}$. $E^{(2)} \subset Q \times W_d \times Q$ is the set of admissible discrete transitions (a transition from $i \in Q$ to $j \in Q$ determined by an event $\omega \in W_d$ is indicated by the triple (i, ω, j)).

Concerning the discrete output function, the following holds:

$$h^{(2)}(\omega) = \omega,\ \omega \in W_d$$

$$h^{(2)}(i) = \begin{cases} c(i) \circ h(i), & i \in Q \wedge h(i) \neq \epsilon \\ c(i) \circ 0, & i \in Q \wedge h(i) = \epsilon \end{cases} \tag{7.16}$$

where $c(i) \in Y_\sim$ indicates the label associated to the equivalence class to which the discrete state $i \in Q$ belongs, and the symbol $\circ$ represents the concatenation between two strings.

The set $E^{(2)}$ is constructed with the following procedure:

procedure SET $E^{(2)}(\mathscr{H})$
  Initialize $E^{(2)} = \emptyset$
  for $(i, j) \in E$ **do**
    if $h(j) = \epsilon$ **then**
      Case 1 The pair $(S(i), S(j))$ is s-securely distinguishable.
      According to Proposition 1 in Sect. 7.3.2 the transition (i, j) is observable, thus $E^{(2)} = E^{(2)} \cup \{(i, 1, j)\}$.
      Case 2 The pair $(S(i), S(j))$ is s-securely indistinguishable and there exists a set Γ, $|\Gamma| \leq 2s$, such that $(\overline{O}_{i,\Gamma} = \overline{O}_{j,\Gamma} R_{ij})$.
      According to Proposition 2 in Sect. 7.3.2 the transition (i, j) can not be detected for any generic input function and for any value of the continuous state at the switching time instant, thus
      $E^{(2)} = E^{(2)} \cup \{(i, 0, j)\}$.

Case 3 The pair $(S(i), S(j))$ is s-securely indistinguishable but $(\overline{O}_{i,\Gamma} \neq \overline{O}_{j,\Gamma} R_{ij})$ for any set Γ, $|\Gamma| \leq 2s$.
According to Proposition 2 in Sect. 7.3.2 the transition (i, j) can be detected depending on the state at the switching time, (that is $x(t_k^-) \notin \ker(\overline{O}_{i,\Gamma} - \overline{O}_{j,\Gamma} R_{ij})$, for any set Γ, $|\Gamma| \leq 2s$), thus
$E^{(2)} = E^{(2)} \cup \{(i, 0, j) \cup \{(i, 1, j)\}$.
else $E^{(2)} = E^{(2)} \cup \{(i, 1, j)\}$
end if
end for
end procedure

Remark 2 $\mathscr{H}^{(2)}$ is an LH-system system with purely discrete information. In fact, the continuous output information which allows to distinguish between the continuous dynamics is remodeled as a discrete output associated to the discrete transitions (therefore $\mathscr{H}^{(2)}$ is modeled by a Mealy Machine).

As $\mathscr{H}^{(2)}$ and $\mathscr{H}$ are not equivalent with respect to the s-secure Ω-diagnosability property, further steps are required.

7.5.2.4 Step 3

The discrete behavior of $\mathscr{H}^{(2)}$ is described by the Mealy Machine $M^{(2)}$, which can be transformed (by using Algorithm 1 described in the Appendix) into a Moore Machine $M'^{(3)} = (Q^{(3)}, Q_0^{(3)}, Y_d^{(3)}, h'^{(3)}, E^{(3)})$, in which no information is associated to discrete transitions.

From the finite state machine $M'^{(3)}$, the LH-system

$$\mathscr{H}'^{(3)} = (\Xi^{(3)}, \Xi_0^{(3)}, Y^{(3)}, h'^{(3)}, S^{(3)}, E^{(3)}, G^{(3)}, R^{(3)}, \delta^{(3)}, \Delta^{(3)}) \tag{7.17}$$

is derived, with:

- $\Xi^{(3)} = Q^{(3)} \times \mathbb{R}^n$
- $\Xi_0^{(3)} = Q_0^{(3)} \times \mathbb{R}^n$
- $Y^{(3)} = Y_d^{(3)} \times \mathbb{R}^p$
- $S^{(3)}(i) = S(f^{-1}(i))$
- $G^{(3)}(i, j) = G(f^{-1}(i), f^{-1}(j))$
- $R^{(3)}(i, j) = R(f^{-1}(i), f^{-1}(j))$
- $\delta^{(3)} = \delta(f^{-1}(i))$
- $\Delta^{(3)} = \Delta(f^{-1}(i))$

where $f : Q \to 2^{\mathbf{Q}}$ is the point to set mapping defined in Algorithm 1 in the Appendix. Recalling that $c(i) \in Y_\sim$ indicates the label associated to the equivalence class to which the discrete state $i \in Q^{(3)}$ belongs, the discrete output function $h'^{(3)}$ is defined as:

- for $i \in Q_0^{(3)}$

$$h'^{(3)}(i) = \begin{cases} c(i) \circ 0, & i \in Q_0^{(3)} \wedge h(i) = \epsilon \\ c(i) \circ h(i), & i \in Q_0^{(3)} \wedge h(i) \neq \epsilon \end{cases} \tag{7.18}$$

- for $i \in Q^{(3)} \backslash Q_0^{(3)}$

$$h'^{(3)}(i) = \begin{cases} t(i) \circ c(i) \circ 0, & i \in Q_0^{(3)} \wedge h(i) = \epsilon \\ t(i) \circ c(i) \circ h(i), & i \in Q_0^{(3)} \wedge h(i) \neq \epsilon \end{cases} \tag{7.19}$$

where $t(i)$ is a symbol representing the possibility to detect a transition $(j, i) \in E^{(3)}$ by using the continuous output information. In particular:

$$t(i) = \begin{cases} 0, & \text{if}(j, i) \in E^{(3)} \text{cannot be detected for any } j \in pre(i) \\ 1, & \text{if}(j, i) \in E^{(3)} \text{can be detected for any } j \in pre(i) \end{cases} \tag{7.20}$$

From the LH-system $\mathscr{H}'^{(3)}$, the LH-system

$$\mathscr{H}^{(3)} = (\Xi^{(3)}, \Xi_0^{(3)}, Y^{(3)}, h^{(3)}, S^{(3)}, E^{(3)}, G^{(3)}, R^{(3)}, \delta^{(3)}, \Delta^{(3)}) \tag{7.21}$$

is derived, by only reformulating the discrete output function $h'^{(3)}$ into $h^{(3)}$. For the ease of notation, let ψ_i indicate the string $h'^{(3)}(i)$, and let $\psi_i(k)$ indicate the k-th symbol of the string ψ_i (e.g., $\psi_i(1)$ is the first symbol of the string $h'^{(3)}(i)$). The discrete output function $h^{(3)} : Q^{(3)} \rightarrow Y^{(3)}$ is defined as in the following:

- for $i \in Q_0^{(3)}$

$$h^{(3)}(i) = \begin{cases} \psi_i = c(i) \circ h(i) & \text{if } \psi(2) \neq 0 \\ \psi_i(1) = c(i) & \text{if } \psi(2) = 0 \end{cases} \tag{7.22}$$

- for $i \in Q^{(3)} \backslash Q_0^{(3)}$

$$h^{(3)}(i) = \begin{cases} \psi_i(2) \circ \psi_i(3) = c(i) \circ h(i) & \text{if } \psi(3) \neq 0 \\ \psi_i(2) = c(i) & \text{if } \psi(1) = 1 \wedge \psi(3) = 0 \\ \epsilon & \text{if } \psi(1) = 0 \wedge \psi(3) = 0 \end{cases} \tag{7.23}$$

7.5.2.5 Abstracting Procedure: An Example

In this section we propose an example to better understand the abstracting procedure described in Sect. 7.5.2. We consider a dynamical network made up of n nodes, where each node updates its state $x_i \in \mathbb{R}$, $i = 1, \ldots, n$ on the basis of the states of its neighbors and other m external nodes providing an external input. We also assume to measure the state of p nodes. The network topological structure can be represented by an undirected graph $\mathscr{G} = (\mathscr{V}, \mathscr{E})$, where $\mathscr{V} = \{1, \ldots, n\}$ is the set of nodes, and $\mathscr{E}$ is the set of edges. The discrete time collective dynamics of the network can be written as:

$$\begin{aligned} x(t+1) &= -Lx(t) + Bu(t) \\ y(t) &= Cx(t) \end{aligned} \tag{7.24}$$

where L is the Laplacian induced by the graph $\mathscr{G} = (\mathscr{V}, \mathscr{E})$ (see [22] for further information), $x \in \mathbb{R}^n$, $u \in \mathbb{R}^m$, $y \in \mathbb{R}^p$.

Given the nominal dynamics of the network, a node or link disconnection changes the network's topology, thus changing the network collective dynamics, too. We assume that some disconnections can be directly measured (by means of a discrete label representing the active or inactive state of a certain link/node), and some of them cannot be measured (therefore the null event ϵ can be associated to them).

We can represent this scenario by means of an LH-system, in which each discrete state $q \in Q$ is associated to a particular network topology. A discrete label belonging to the discrete output set Y_d is associated to each discrete state, as described above. Moreover, a linear dynamical system S_q is associated to each discrete state $q \in Q$, as follows:

$$\begin{aligned} x(t+1) &= -L_q x(t) + B_q u(t) \\ y(t) &= C_q x(t) + w(t) \end{aligned} \tag{7.25}$$

where $w(t) \in \mathbb{S}_s^p$ represents an s-sparse attack vector on continuous output measurements.

As an example, let us assume that all the network's topologies can be represented by the LH-system in Fig. 7.3 Let $\sim$ indicate the s-secure indistinguishability relation on the set of discrete modes Q, as defined in Sect. 7.3.2. We define the equivalence class of a state $i \in Q$ as the set $\{j \in Q \,|\, S_j \sim S_i\}$. The set of equivalence classes induces a partition of Q, which is called the quotient space of Q by the relation $\sim$ and is denoted by $Q/\sim$. In this case, we assume that $Q/\sim = \{(1, 2, 3), (4, 5)\}$. To each equivalence class, we can associate a label in the set of labels $Y_\sim = \{\alpha, \beta\}$.

The LH-system obtained as a result of the first step of the proposed abstracting procedure is represented in Fig. 7.4.

The second step of the abstracting procedure associates an additional binary output to each discrete transition $e = (i, j) \in E$, based on the s-secure distinguishability property between S_i and S_j and the possibility to detect the transition $e = (i, j) \in E$, as described in Sect. 7.5.2. At the end of Step 2 the LH-system in Fig. 7.5 is obtained.

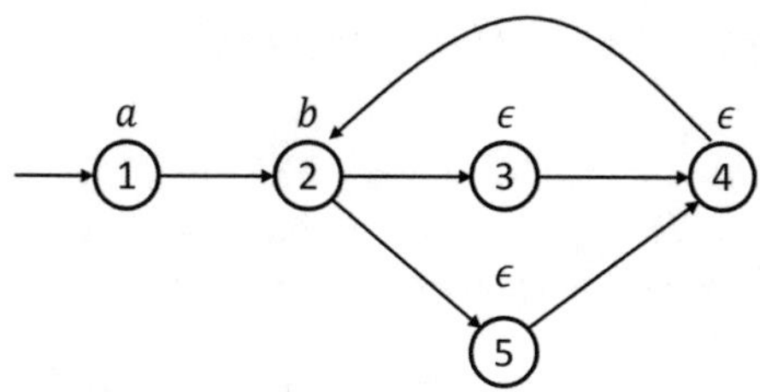

Fig. 7.3 Example: LH-system representing all the network's topologies. Topologies associated to S_1 and S_2 are measured by means of discrete labels $\{a, b\} \in Y_d$, whereas topologies associated to S_3, S_4, and S_5 cannot be measured

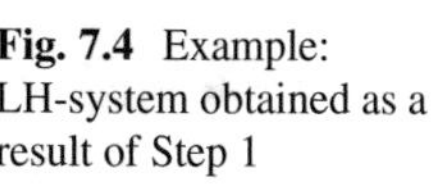
Fig. 7.4 Example: LH-system obtained as a result of Step 1

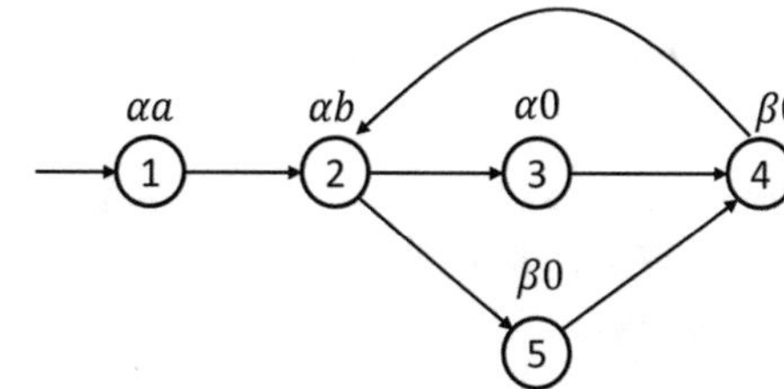

Fig. 7.5 Example: LH-system obtained as a result of Step 2

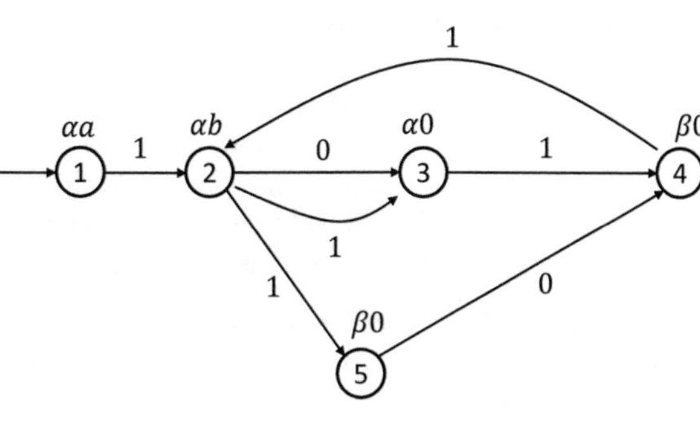

Fig. 7.6 Example: LH-system obtained as a result of Step 3

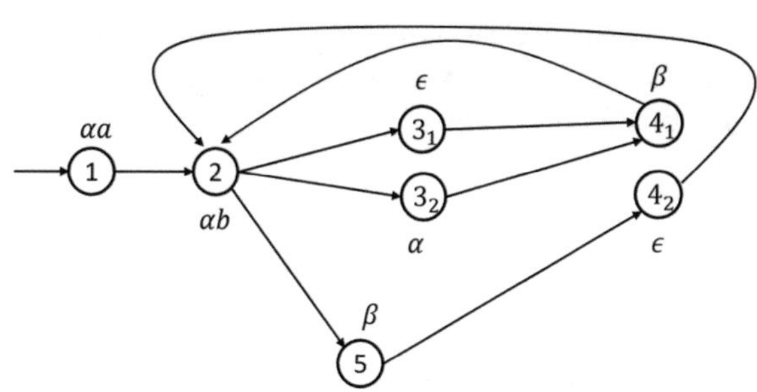

The third step of the abstracting procedure transforms the Mealy Machine obtained at the end of the second step into a Moore Machine, and further manipulates the output labels, as shown in Fig. 7.6.

7.5.3 Checking the Secure Diagnosability Property

Thanks to the abstraction algorithm proposed in Sect. 7.5.2, it is possible to investigate the s-secure Ω-diagnosability of the initial LH-system $\mathcal{H}$ by means of the LH-system $\mathcal{H}^{(3)}$, which has purely discrete output (only associated to discrete states).

Let $\mathcal{M}$ be the FSM associated to $\mathcal{H}^{(3)}$, that is $\mathcal{M} = (Q^{(3)}, Q_0^{(3)}, Y_d^{(3)}, h^{(3)}, E^{(3)})$. In this section we make the following assumption.

Assumption 7 *$\Omega = Q_c \times \mathbb{R}^n \subset \Xi$, where $Q_c \subset Q$ is a set of discrete states of interest.*

Theorem 3 *Suppose Assumptions 4, 5, 7 hold. If $\mathcal{M}$ is Q_c-diagnosable ($Q_c \subset Q^{(3)}$), then $\mathcal{H}$ is s-securely Ω-diagnosable with $\Omega = Q_c \times \mathbb{R}^n$.*

Proof If the FSM $\mathcal{M}$ is Q_c-diagnosable, $Q_c \subset Q^{(3)}$, there exists $T \in \mathbb{N}$ such that the discrete output information $\mathbf{h}(\sigma|_{[1,\hat{k}]})$ allows to infer if $\sigma(k) \in Q_c$ for some $k \in [\max\{1, \hat{k}-T\}, \hat{k}]$. Therefore $\mathcal{H}^{(3)}$ is Ω-diagnosable with $\Omega = Q_c \times \mathbb{R}^n$, and the same property holds for $\mathcal{H}'^{(3)}$. $\mathcal{H}'^{(3)}$ is obtained from $\mathcal{H}^{(2)}$ by applying Algorithm 1 in the Appendix to obtain a Moore Machine, therefore to a state $q^{(2)}(k) = i$ corresponds a discrete state $q^{(3)}(k) \in f(i)$. Thus, Ω-diagnosability of $\mathcal{H}'^{(3)}$ implies s-secure Ω-diagnosability of $\mathcal{H}^{(2)}$ as $\mathcal{H}^{(2)}$ encodes also the information related to the possibility of s-securely distinguishing between any two discrete states (i, j) belonging to the same equivalence class or to detect a transition between them. s-secure Ω-diagnosability of $\mathcal{H}^{(2)}$ implies s-secure Ω-diagnosability of $\mathcal{H}$ as for any execution of $\mathcal{H}$ there is an execution of $\mathcal{H}^{(2)}$ such that $q^{(2)}(k) = q(k)$.

Theorem 3 provides a sufficient condition to infer the s-secure Ω-diagnosability, $\Omega = Q_c \times \mathbb{R}^n$ of the LH-system $\mathcal{H}$, based on the Q_c-diagnosability of the FSM $\mathcal{M}$ abstracted from $\mathcal{H}$ by means of the procedure described in Sect. 7.5.2. Actually, in this result, we are not exploiting the information related to the elapsed time. Therefore, less conservative conditions could be derived by taking into account this additional information.

7.6 Conclusions

In this chapter we introduced a formal definition of diagnosability for LH-systems by extending the one provided in [8] for FSMs. We also defined diagnosability for LH-systems when the continuous output information may be corrupted by an adversarial attacker. Furthermore, we proposed a procedure to obtain an abstracted FSM from the original hybrid system, by combining the available continuous and discrete information. Sufficient conditions for the diagnosability of the given LH-system can then be obtained on the basis of the diagnosability conditions for the abstracted FSM. In our model, the continuous output information may be corrupted by an external malicious attacker. This is motivated by the fact that, since hybrid systems are an important mathematical paradigm to deal with a great variety of safety critical applications, investigating security issues for hybrid systems is particularly relevant.

Appendix

From Mealy to Moore

Depending on the nature of the discrete output signal, two variants of FSMs could be used [4]:

- Moore Machines: the output function associates a discrete output to each discrete state, no output is associated to discrete transitions;
- Mealy Machines: the output function associates a discrete output to each discrete transition, no output is associated to discrete states.

It is possible to transform a Mealy Machine into a Moore Machine, and there exists a vast literature dealing with this transformation (see [7] for a detailed description on this topic). For the sake of clarity, we recap the procedure proposed in [7].

First, we recall the procedure called Algorithm 0 that transforms a finite state machine $M = (Q, Q_0, Y, h, E)$ into a finite state machine $M^{(1)} = (Q^{(1)}, Q_0^{(1)}, Y, h^{(1)}, E^{(1)})$, in which the set of initial states has no predecessors (i.e., $pre(i) = \emptyset,\ \forall\, i \in Q_0^{(1)}$). Algorithm 0 is described in detail in [7].

Given the set $P = \{i \in Q_0 \,:\, pre(i) \neq \emptyset\}$, we define the injection map $g : Q \rightarrow \mathbb{N}$, where:

$$
\begin{aligned}
g(i) &= i, \qquad & i \in Q \backslash P \\
g(i) &\in \mathbb{N} \backslash Q, \qquad & i \in P
\end{aligned}
\tag{7.26}
$$

Roughly speaking, Algorithm 0 splits any state $i \in P$ in the pair of states i and $g(i)$ (which can assume any arbitrary value in $\mathbb{N} \backslash Q$).

procedure ALGORITHM 0(M)
$Q^{(1)} = Q \cup \{g(i),\, i \in P\}$
$Q_0^{(1)} = (Q_0 \backslash P) \cup \{g(i),\, i \in P\}$
$h^{(1)}(i) = h(i);\ h^{(1)}(g(i)) = h(i),\, i \in Q$
$E^{(1)} = E \cup \{(g(i), h) \,:\, i \in P,\, (i, h) \in E\}$
end procedure

We now describe the procedure called Algorithm 1 which transforms a Mealy FSM $M = (Q, Q_0, W_d, Y_d, h, E)$ into a Moore FSM $M' = (Q', Q_0, W_d, Y_d', h', E')$ in which $h'(w) = \epsilon$ for any transition $(i, w, j) \in E$, $w \in W_d$. Each state of the finite state machine M is decomposed in a certain number of new states, any transition ending in each of these new states shares the same discrete input $w \in W_d$. Without loss of generality, we assume now that the FSM M is such that $pre(i) = \emptyset,\ \forall\, i \in Q_0$ (otherwise, Algorithm 0 can be used to obtain this case).

First, for any state $i \in Q \backslash Q_0$, we define the subset of discrete inputs $W_i \subset W_d$, such that the following holds:

- for any $j \in pre(i)$, there exists $w \in W_i$ such that $(j, w, i) \in E$,
- for any $w \in W_i$, there exists $j \in Q$ such that $(j, w, i) \in E$.

Let μ_i be the cardinality of the subset W_i. We define a point to set mapping $f : Q \rightarrow 2^{\mathbf{Q}}$, where $\mathbf{Q} = \{i_h : i \in Q,\ h = 1, 2, \ldots\}$ and:

$$f(i) = \begin{cases} \{i_h,\ h = 1, \ldots, \mu_i\}, & i \in Q \backslash Q_0 \\ \{i\}, & i \in Q_0 \end{cases} \tag{7.27}$$

For a given $s \in \mathbf{Q}, f^{-1}(s)$ indicates the state $i \in Q$ such that $s \in f(i)$.

Lastly, the discrete output space Y'_d of the FSM M' is defined as in the following: $Y'_d = \{h(w) \circ s,\ w \in W_d,\ s \in Y_d\}$.

procedure ALGORITHM 1(M)
  Initialize $Q' = Q,\ E' = E$
  for $i \in Q$ **do**
    $Q' = (Q' \backslash \{i\}) \cup f(i)$
    $E' = E' \backslash \{e \in E' : (e = (j, w, i)) \vee (e = (i, w, j))\}$
    $E' = E' \cup \{(j, w, i_h) : (j, w, i) \in E,\ h = 1, \ldots, \mu_i\}$
    $E' = E' \cup \{(i_h, w, j) : (i, w, j) \in E,\ h = 1, \ldots, \mu_i\}$
  end for
  for $s \in Q'$ **do**
    $h'(s) = h(w) \circ h(f^{-1}(s)),\ w : (j, w, s) \in E'$
  end for
end procedure

References

1. M. Bayoudh, L. Travé-Massuyès, Diagnosability analysis of hybrid systems cast in a discrete-event framework. Discret. Event Dyn. Syst. **24**(3), 309–338 (2014)
2. A.A. Cárdenas, S. Amin, S. Sastry, Research challenges for the security of control systems, in *Proceedings of the 3rd Conference on Hot Topics in Security, HOTSEC'08*, Berkeley, CA. USENIX Association (2008), pp. 6:1–6:6
3. A.A. Cárdenas, S. Amin, S. Sastry, Secure control: towards survivable Cyber-Physical Systems, in *2008 The 28th International Conference on Distributed Computing Systems Workshops* (2008), pp. 495–500
4. C. Cassandras, S. Lafortune, *Introduction to Discrete Event Systems*. SpringerLink Engineering (Springer, Berlin, 2009)
5. Y.H. Chang, Q. Hu, C.J. Tomlin, Secure estimation based Kalman filter for Cyber-Physical Systems against adversarial attacks. *CoRR*, abs/1512.03853v2 (2016)

6. M.S. Chong, M. Wakaiki, J.P. Hespanha, Observability of linear systems under adversarial attacks, in *American Control Conference (ACC)* (2015), pp. 2439–2444
7. E. De Santis, M.D. Di Benedetto, Observability of hybrid dynamical systems. Found. Trends Syst. Control **3**(4), 363–540 (2016)
8. E. De Santis, M.D. Di Benedetto, Observability and diagnosability of finite state systems: a unifying framework. Automatica **81**, 115–122 (2017)
9. Y. Deng, A. D'Innocenzo, M.D. Di Benedetto, S. Di Gennaro, A.A. Julius, Verification of hybrid automata diagnosability with measurement uncertainty. IEEE Trans. Autom. Control **61**(4), 982–993 (2016)
10. M.D. Di Benedetto, S. Di Gennaro, A. D'Innocenzo, Verification of hybrid automata diagnosability by abstraction. IEEE Trans. Autom. Control **56**(9), 2050–2061 (2011)
11. O. Diene, E.R. Silva, M.V. Moreira, Analysis and verification of the diagnosability of hybrid systems, in *53rd IEEE Conference on Decision and Control* (2014), pp. 1–6
12. O. Diene, M.V. Moreira, V.R. Alvarez, E.R. Silva, Computational methods for diagnosability verification of hybrid systems, in *2015 IEEE Conference on Control Applications (CCA)* (2015), pp. 382–387
13. H. Fawzi, P. Tabuada, S. Diggavi, Secure state-estimation for dynamical systems under active adversaries, in *2011 49th Annual Allerton Conference on Communication, Control, and Computing (Allerton)* (2011), pp. 337–344
14. H. Fawzi, P. Tabuada, S. Diggavi, Secure estimation and control for Cyber-Physical Systems under adversarial attacks. IEEE Trans. Autom. Control **59**(6), 1454–1467 (2014)
15. G. Fiore, Secure state estimation for cyber-physical systems. PhD thesis, University of L'Aquila, Department of Information Engineering, Computer Science and Mathematics, 2017
16. G. Fiore, Y.H. Chang, Q. Hu, M.D. Di Benedetto, C.J. Tomlin, Secure state estimation for cyber physical systems with sparse malicious packet drops, in *2017 American Control Conference (ACC)* (2017), pp. 1898–1903
17. G. Fiore, E. De Santis, M.D. Di Benedetto, Secure mode distinguishability for switching systems subject to sparse attacks. IFAC-PapersOnLine **50**(1), 9361–9366 (2017). https://doi.org/10.1016/j.ifacol.2017.08.1442
18. P.M. Frank, Fault diagnosis in dynamic systems using analytical and knowledge-based redundancy. Automatica **26**(3), 459–474 (1990)
19. F. Lin, Diagnosability of discrete event systems and its applications. Discret. Event Dyn. Syst. **4**(2), 197–212 (1994)
20. J. Lygeros, K.H. Johansson, S.N. Simic, J. Zhang, S.S. Sastry, Dynamical properties of hybrid automata. IEEE Trans. Autom. Control **48**(1), 2–17 (2003)
21. S. McIlraith, G. Biswas, D. Clancy, V. Gupta, *Hybrid Systems Diagnosis* (Springer, Berlin, 2000), pp. 282–295
22. M. Mesbahi, M. Egerstedt, *Graph Theoretic Methods in Multiagent Networks* (Princeton University Press, Princeton, 2010)
23. S. Narasimhan, G. Biswas, Model-based diagnosis of hybrid systems. IEEE Trans. Syst. Man Cybern. Part A Syst. Hum. **37**(3), 348–361 (2007)
24. A. Paoli, S. Lafortune, Safe diagnosability for fault-tolerant supervision of discrete-event systems. Automatica **41**(8), 1335–1347 (2005)
25. Y. Pencolé, Diagnosability analysis of distributed discrete event systems, in *Proceedings of the 16th European Conference on Artificial Intelligence, ECAI'04, Amsterdam* (IOS Press, Amsterdam, 2004), pp. 38–42
26. W. Qiu, R. Kumar, Decentralized failure diagnosis of discrete event systems. IEEE Trans. Syst. Man Cybern. Part A Syst. Hum. **36**(2), 384–395 (2006)
27. M. Sampath, R. Sengupta, S. Lafortune, K. Sinnamohideen, D. Teneketzis, Diagnosability of discrete-event systems. IEEE Trans. Autom. Control **40**(9), 1555–1575 (1995)
28. M. Sayed-Mouchaweh, *Discrete Event Systems: Diagnosis and Diagnosability* (Springer Science & Business Media, New York, 2014)
29. M. Sayed-Mouchaweh, E. Lughofer, *Learning in Non-stationary Environments: Methods and Applications* (Springer Science & Business Media, New York, 2012)

30. Y. Shoukry, P. Tabuada, Event-triggered state observers for sparse sensor noise/attacks. IEEE Trans. Autom. Control **61**(8), 2079–2091 (2016)
31. Y. Shoukry, P. Nuzzo, N. Bezzo, A.L. Sangiovanni-Vincentelli, S.A. Seshia, P. Tabuada, Secure state reconstruction in differentially flat systems under sensor attacks using satisfiability modulo theory solving, in *2015 54th IEEE Conference on Decision and Control CDC)* (2015), pp. 3804–3809
32. Y. Shoukry, M. Chong, M. Wakaiki, P. Nuzzo, A.L. Sangiovanni-Vincentelli, S.A. Seshia, J.P. Hespanha, P. Tabuada, SMT-based observer design for Cyber-Physical Systems under sensor attacks, in *2016 ACM/IEEE 7th International Conference on Cyber-Physical Systems (ICCPS)* (2016), pp. 1–10
33. A. Teixeira, I. Shames, H. Sandberg, K.H. Johansson, A secure control framework for resource-limited adversaries. Automatica **51**, 135–148 (2015)
34. S. Tripakis, *Fault Diagnosis for Timed Automata* (Springer, Berlin, 2002), pp. 205–221
35. J. Vento, L. Travé-Massuyès, V. Puig, R. Sarrate, An incremental hybrid system diagnoser automaton enhanced by discernibility properties. IEEE Trans. Syst. Man Cybern. Syst. **45**(5), 788–804 (2015)
36. L. Ye, P. Dague, An optimized algorithm of general distributed diagnosability analysis for modular structures. IEEE Trans. Autom. Control **62**(4), 1768–1780 (2017)
37. T.-S. Yoo, S. Lafortune, Polynomial-time verification of diagnosability of partially observed discrete-event systems. IEEE Trans. Autom. Control **47**(9), 1491–1495 (2002)
38. J. Zaytoon, S. Lafortune, Overview of fault diagnosis methods for discrete event systems. Annu. Rev. Control. **37**(2), 308–320 (2013)

Chapter 8
Diagnosis in Cyber-Physical Systems with Fault Protection Assemblies

Ajay Chhokra, Abhishek Dubey, Nagabhushan Mahadevan, Saqib Hasan, and Gabor Karsai

8.1 Introduction

The Smart Electric Grid is a CPS: it consists of networks of physical components (including generation and transmission subsystems) interacting with cyber components (e.g., intelligent sensors, communication networks, computational and control software). Reliable operation of such CPS is critical. Therefore, these systems are equipped with specialized protection devices that remove the faulty component from the system. However, if there are failures in the fault protection units, this leads to a situation where an incorrect local mitigation in a subsystem results in a larger fault cascade, leading to a blackout. This phenomenon was observed in the recent blackouts [1], where tripping of some lines by the relays (protection devices) overloaded some other parts of the system. These secondary overloaded components were again isolated by pre-defined protection schemes, leading to tertiary effects and so on. This domino effect got disseminated into the whole system, pushing it towards total collapse.

The ultimate challenge in doing fault diagnosis in these cyber-physical systems is to handle the complexity: the sheer size, large number of components, anomalies, and failure modes. Furthermore, the subsystems are often heterogenous and the typical approach is to try and understand the interactions among them, even if the subsystems are from different domains. In the past, we have used the high-level concept to model the interaction between the subsystems—(1) observable degradations,

A. Chhokra (✉) · A. Dubey · N. Mahadevan · S. Hasan · G. Karsai
Institute for Software Integrated Systems, Vanderbilt University, Nashville, 37212, TN, USA
e-mail: ajay.d.chhokra@vanderbilt.edu; abhishek.dubey@vanderbilt.edu; nag.mahadevan@Vanderbilt.Edu; saqib.hasan@vanderbilt.edu; gabor.karsai@Vanderbilt.Edu

M. Sayed-Mouchaweh (ed.), *Diagnosability, Security and Safety of Hybrid Dynamic and Cyber-Physical Systems*, https://doi.org/10.1007/978-3-319-74962-4_8

anomalies, discrepancies caused by failure modes, (2) their propagation, and (3) their temporal evolution towards system-level fault (effects). This approach called Timed Fault Propagation Graphs (TFPG) has been applied to avionics systems, fuel assemblies, and software component assemblies [2, 3] and is based on a discrete-event model that captures the causal and temporal relationships between failure modes (causes) and discrepancies (effects) in a system, thereby modeling the failure cascades while taking into account propagation constraints imposed by operating modes and timing delays. In this graphical model, nodes represent failure modes and discrepancies, edges represent the direction of causality, and attributes of edges capture the conditions (mode and temporal delays) under which the edge is active. The model-based fault diagnostics reasoner receives observations in the form of time-stamped alarms that indicate whether a discrepancy is present and, using abductive reasoning, generates a set of hypotheses about the failure modes that could explain the observed fault signature, i.e. the fault effects.

However, the approach of failure diagnosis with timed fault propagation graphs does not deal with the built-in automatic fault-protection mechanisms of the system. Such local fault protection components are designed to mask the effect of failures and thereby arrest the fault cascades. Additionally, these fault protection components introduce failure modes that are specific to the operation or lack of operation of the protection components. A classical TFPG model is not well suited for capturing the specializations that are introduced by the inherent fault protection mechanisms built into the system. For example, in power systems, there is already a fault-detection/protection system (relays and breakers) that autonomously protects elements of the network. Any protection operation performed by these systems can fall into one of these categories: (a) correct and thereby isolate the area where the fault occurred, (b) incorrect: fires incorrectly when it is not supposed to, (c) backup: accounting for lack of firing of another protection system, or (d) consequence of a previous firing which was incorrect when considering its effect on the global or regional system stability. In effect, the failure can be introduced by the physical components of the power system (e.g. cables) as well as components of the fault-protection system (e.g., breakers/sensors). Furthermore, the autonomous fault protection mechanism changes the network topology automatically (i.e., changes the mode of the system).

To solve this problem, we have developed an extension of TFPG called Temporal Causal Diagrams (TCDs). A TCD model is a behavioral augmentation of Temporal Fault Propagation Graphs (TFPGs) that can efficiently model fault propagation in various domains. The TCD-based diagnosis system is hierarchical. The lower level uses local discrete event diagnosers, called *Observers*, which are generated from the behavior specification of fault management controllers. A higher level reasoner produces system level hypotheses based upon the output of local observers. The approach does not involve complex real-time computations with high-fidelity models, but reasons using efficient graph algorithms to explain the observed anomalies. This approach is applicable to CPS that include supervisory controllers that arrest fault propagation based upon local information without considering system-wide effects. To explain TCD we use examples from power system domain.

The paper is organized as follows, Sect. 8.2 describes the background and literature review followed by brief explanation of cascade phenomenon caused by misoperation of fault management assemblies in power systems (Sect. 8.2.3). Section 8.3 gives an overview of our approach and describes the TCD modeling formalism and diagnosis methodology in detail. The fault diagnosis approach is described in the context of a power system example in Sect. 8.4, followed by concluding remarks in Sect. 8.5.

8.2 Background

8.2.1 Diagnosis in CPS

Diagnostic reasoning techniques share a common process in which the system is continuously monitored and the observed behavior is compared with the expected one to detect abnormal conditions. In many industrial systems, diagnosis is limited to signal monitoring and fault identification via threshold logic, e.g., detecting if a sensor reading deviates from its nominal value. Failure propagation is modeled by capturing the qualitative association between sensor signals in the system for a number of different fault scenarios. Typically, such associations correspond to relations used by human experts in detecting and isolating faults. This approach has been effectively used for many complex engineering systems. Common industrial diagnosis methods include fault trees [4–7], cause-consequence diagrams [8, 9], diagnosis dictionaries [10], and expert systems [11, 12].

Model-based diagnosis (see [13–15] and the references therein), on the other hand, compares observations from the real system with the predictions from a model. Analytical models such as state equations [16], finite state machines [17], hidden Markov models [18], and predicate/temporal logic [19] are used to describe the nominal system behavior. In the presence of a fault, the observed behavior of the system deviates from the nominal behavior expected by the model. The associated discrepancies can then be used to detect, isolate, and identify the fault depending on the type of model and methods used. In consistency-based diagnosis the behavior of the system is predicted using a nominal system model and then compared with observations of the actual behavior of the system to obtain the minimal set of faulty component that is consistent with the observations and the nominal model. Consistency-based diagnosis was introduced in a logical framework in [19] and was later extended in [20]. The approach has been applied to develop diagnosis algorithms for causal systems [21, 22] and temporal causal systems [23, 24].

The diagnosis approach presented here is conceptually related to the temporal causal network approach presented in [24]. However, we focus on incremental reasoning and diagnosis robustness with respect to sensor failures. The causal model presented in this paper is based on the timed failure propagation graph

(TFPG) introduced in [25, 26]. The TFPG model is closely related to fault models presented in [27–29] and used for an integrated fault diagnosis and process control system [30]. The TFPG model was extended in [31] to include mode dependency constraints on the propagation links, which can then be used to handle failure scenarios in hybrid and switching systems. TFPG modeling and reasoning tool has been developed and used successfully in an integrated fault diagnoses and process control system [30].

Additionally, the temporal aspects of the TFPG model are closely related to the domain theoretic notion of temporal dependency proposed in [32]. However, there are several major differences between the two approaches. In particular, TFPG-based diagnosis implements a real-time incremental reasoning approach that can handle multiple failures including sensor/alarm faults. In addition, the underlying TFPG model can represent a general form of temporal and logical dependency that directly incorporates the dynamics of multi-mode systems.

8.2.2 Diagnosis in Power Systems

Since power systems is our example domain, we now present a brief review of fault diagnosis approaches in that domain, which can be categorized into three main branches based on their underlying technique: expert systems [33–36], artificial neural networks [37–40], and analytical model based optimization [41–44]. In addition, approaches based on Petri nets [45] and cause-effect Bayesian networks [46–50] have also been proposed. Expert systems are one of the earliest techniques proposed to address the failure diagnosis problem in power systems. A comprehensive survey of such knowledge-based approaches is available in [51]. The expert systems, in general, suffer from limitation imposed due to the maintenance of the knowledge database and slow response time. Moreover, expert system based approaches are known to produce wrong hypothesis in presence of missing and/or spurious alarms. Artificial neural networks (ANNs) are adaptive systems inspired by biological systems. These approaches, in general, suffer from convergence problems. Further, the ANNs have to be retrained whenever there is a change in network topology as the weights are dependent upon the structure of the power system. A number of model-based analytical methods have been devised over the years for diagnosing failures by generating optimal failure hypotheses that best explain all the events and anomalies. However, these techniques rely heavily on critical and computationally expensive tasks such as the selection of an objective function, development of exact mathematical models for system actions and protective schemes, which greatly influence the accuracy of the failure diagnosis.

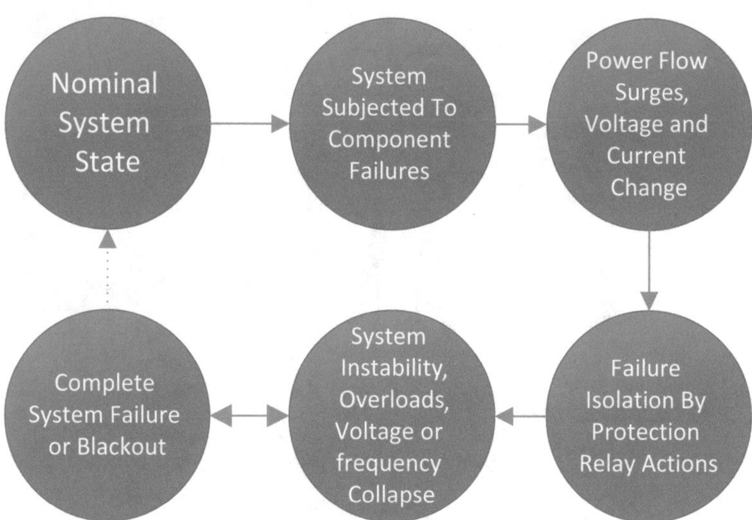

Fig. 8.1 Typical blackout progression in power systems

8.2.3 *Cascade Phenomenon: When Fault Management Controllers Misoperate*

Cascading failures in networked systems are defined as the set of independent events that trigger a sequence of dependent events. Such cascading failures in power grids successively weaken the system by increasing stress on other components and sometimes lead to major blackouts. According to North American Reliability Corporation (NERC), a cascading outage is defined as an uncontrolled loss of any system facilities or load, whether because of thermal overload, voltage collapse, or loss of synchronism, as a result of fault isolation.

Figure 8.1 shows a typical blackout scenario in power systems. The nominal system is subjected to failures from physical and cyber components. These failure modes change the voltage and current at different buses. Fast acting protection devices (relays) react to these changes based on predefined strategies. While these actions are intended to isolate the faulty components and arrest fault propagation, they could have unintended secondary effects such as branch overloads, voltage and/or frequency collapse that can cause instability in the system. A new set of protection elements react to arrest these secondary effects. These secondary actions may cause different tertiary effects and the cycle continues until the system reaches a blackout or there are no more consequences of protective actions.

A simple example of cascading phenomenon using a standard IEEE 14 bus system is shown in Fig. 8.2. It is a simple approximation of American electric power system as of 1960s [52, 53]. The system consists of 14 buses, 5 generators, 11 loads,

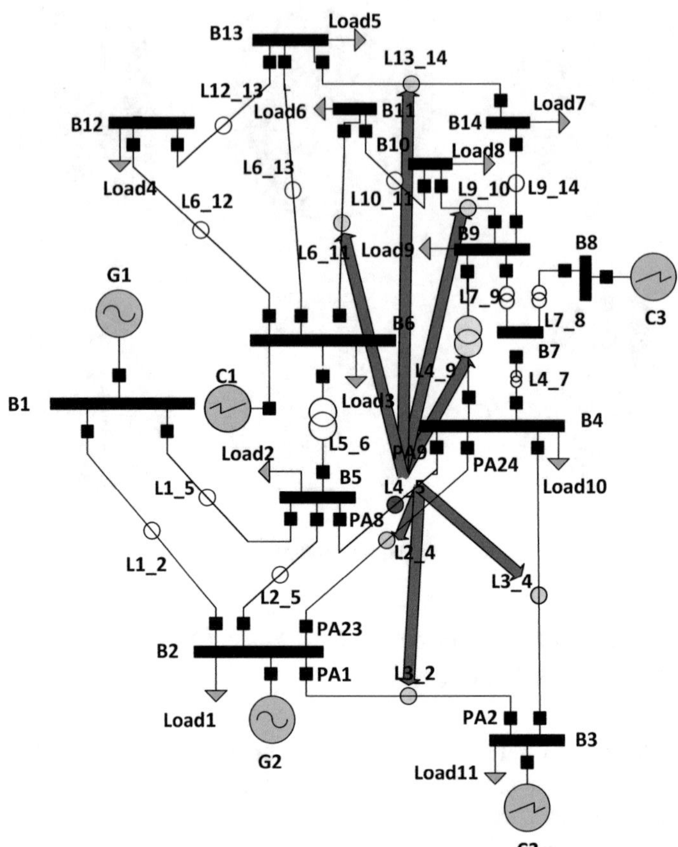

Fig. 8.2 Cascade progression in IEEE 14 Bus system with initial outage in *L4_5* leading to outages in *L3_4*, *L2_4*, and *L2_3* followed by outages in *L6_11*, *L13_14*, *L9_10*, *L4_9*, and ultimately leading to blackout

and 20 branches (transmission lines and transformers). Reference [52] provides bus and branch data in IEEE common data format [54] for creating OpenDSS [55] simulation models. A three phase to ground phase fault is injected in line, *L4_5*. The fault is isolated by tripping the line. This control actions of protection devices lead to overloading of lines *L3_4*, *L2_4*, and *L2_3*. These overloads are removed by tripping these lines. The removal of these secondary effects leads to overloads in lines *L6_11*, *L13_14*, *L9_10*, *L4_9*. The removal of these overloaded branches de-energizes more than 40% of the total system load and is considered as catastrophic event or blackout.

8.3 Temporal Causal Diagrams (TCD)

TCDs are discrete models that appropriately model failure modes, anomalies, and their propagation in both physical and cyber systems. TCD is a combination of Temporal Fault Propagation Graphs (TFPGs) and Time Triggered Automata (TTAs). TFPG based models and reasoning schemes have been used in the past to diagnose faults in physical systems including industrial plants [56, 57], aerospace systems [3], and software systems [58].

However, in cyber-physical systems, there are discrete controllers that try to arrest the failure effect if detected. These protection devices can cause system reconfiguration by instructing actuators to change their state. These devices can also have faults that alter their response to the detection of failure effects and control commands. TFPG based reasoning schemes are not very effective in accounting for faults in both physical system and their corresponding protection assembly (i.e., anomaly detectors, mode detectors, actuators). Failure diagnosis of protection devices is critical for cyber-physical systems, where realistic assessment of fault propagation is not possible without accounting for the behavior of the deployed sensors, controllers, and actuators. The second component of the TCD model, TTA is responsible for modeling the behavior of discrete components in both faulty and non-faulty modes.

TCD framework consists of hierarchical event-driven reasoning engines as shown in Fig. 8.3. The diagnosis system consists of multiple local diagnosers, called *Observers* that track the behavior of protection devices and estimate the presence of failures in both physical and cyber infrastructure (fault management controllers are often implemented in software). These estimates are then passed to a system level reasoner that creates system level hypotheses temporally consistent with the fault propagation graph. The observable events in the case of power transmission system are commands sent by relays to breakers, messages sent by relays to each other, state change of breakers, physical fault detection alarms, etc. The following sections describe the modeling formalism of TCD, which includes an extension to TFPG.

8.3.1 Extending TFPG with Non-deterministic Semantics

A temporal fault propagation graph is a labeled directed graph where nodes are either failure modes or discrepancies. Discrepancies are the failure effects, some of which may be observable. Edges in TFPG represent the causality of the fault propagation and edge labels capture operating modes in which the failure effect can propagate over the edge, as well as a time-interval by which the failure effect could be delayed (see Fig. 8.4). Classically, the diagnostic reasoner of TFPG assumed the correct knowledge of the system modes is always available. However, in the context of self-correcting cyber-physical systems such as power grids, the system mode or

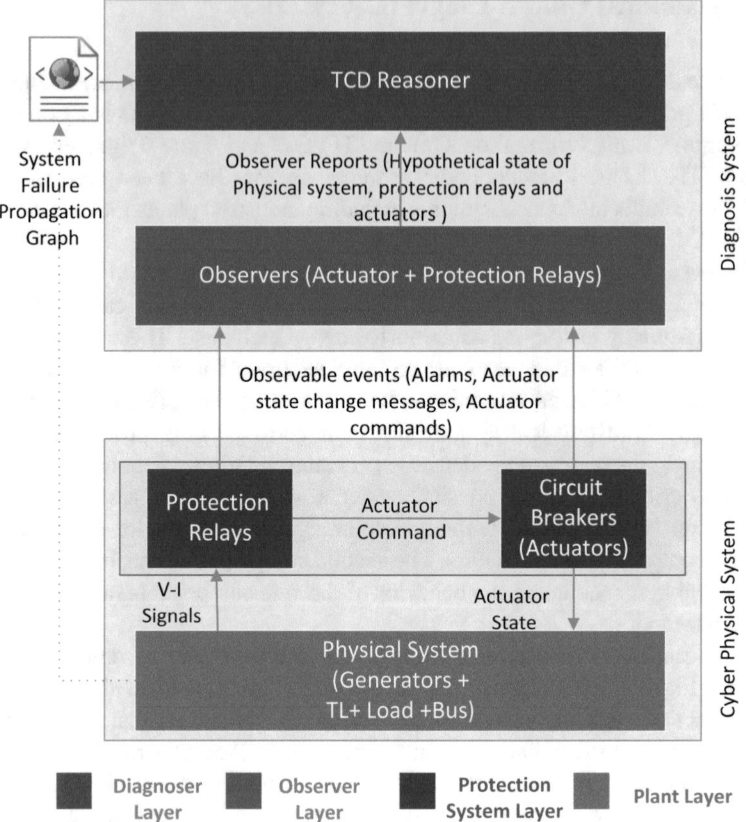

Fig. 8.3 The block-diagram of the Temporal Causal Diagram Diagnosis Framework in the Context of Power Systems

operating conditions depend upon the state of sources, sinks, and the topology of the system. Identification of all operating conditions, i.e. unique system modes is computationally very expensive. In this paper, we use the system topology dictated by the state of the actuators to map an operating condition (i.e., mode) to the fault propagation. However, while such a constraint imposed due to topology of the system is deemed necessary to identify when a fault will not propagate, it is not sufficient to state that the failures will propagate. So we need to extend the TFPG language with an additional map that associates uncertainty to failure edges.

Formally, the extended TFPG is represented as a tuple $\{F_{physical}, D_{physical}, E, M, ET, EM, ND\}$, where

- $F_{physical}$ is a nonempty set of fault nodes in physical system. A fault node can be in two states either present denoted by ON state or absent represented by OFF state. A fault node represents a failure mode of the system or a component, and

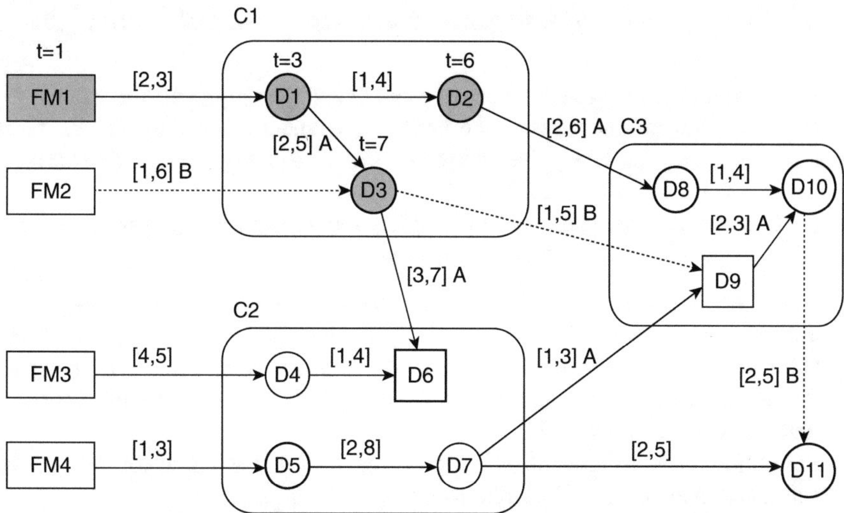

Fig. 8.4 TFPG Model with Failure Modes (FM), Discrepancies (D), and fault propagation links (edges). Labels on edges indicate delay (min,max) values and modal dependencies (letters)

its state represents whether the failure mode is present or not. In the subsequent discussion we will use the terms fault node and failure mode interchangeably.

- $D_{physical}$ is a nonempty set of discrepancy nodes related to fault effects of physical faults.
- $E \subseteq V \times V$ is a set of edges connecting the set of all nodes $V = F_{physical} \cup D_{physical}$.
- M is a nonempty set of system modes. At each time instance t the system can be in only one mode.
- $ET : E \rightarrow I$ is a map that associates every edge in E a time interval $[t_{\min}, t_{\max}] \in I$ that represents the minimum and maximum time for fault propagation over the edge.
- $EM : E \rightarrow M$ is a map that associates every edge in E with a set of modes in M when the edge is active. For any edge $e \in E$ that is not mode-dependent (i.e., active in all modes), $EM(e) = \varnothing$.
- $ND : E \rightarrow \{True, False\}$ is a map that associates an edge, $e \in E$ to *True* or *False*, where *True* implies the propagation along the edge, e **Will** happen, whereas *False* implies the propagation is uncertain and **Can** happen. The destination node of any uncertain edge is referred to as secondary discrepancy while primary discrepancy implies a certain edge. These labels are defined with respect to edges as same discrepancy can act as a destination node of both uncertain and certain edge.

8.3.2 Modeling the Behavior of Fault Management Controllers

The TCD framework relies on the use of an extended time triggered automaton [59] to model the interaction between the fault management controllers and the plant model (TFPG model). Then, given these behaviors we can synthesize the observers that are used in diagnosis step.

Mathematically, the extended time triggered automaton is represented as tuple $(\Sigma, Q, q_0, Q_m, F_{cyber}, D_{cyber}, \mathbb{M}, \alpha(F), \Phi, T)$.[1]

- **Event Set**: Σ is a finite set of events that consists of observable and unobservable events partitioned as $\Sigma = \Sigma_{obs} \cup \Sigma_{unobs}$ such that $\Sigma_{obs} \cap \Sigma_{unobs} = \phi$. Observable events are alarms, commands, and messages exchanged between discrete components, whereas unobservable events are related to introduction of faults in system components.
- **Locations**: Q is a finite set of locations. $q_0 \in Q$ is the initial location of the automaton and $Q_m \subset Q$ is a finite set of marked locations.
- **Discrepancy Set**: D_{cyber} is a finite set of discrepancies associated with the component behavior, partitioned into the sets of observable and unobservable discrepancies.
- **Failure Mode Set**: F_{cyber} is a finite set of unobservable failure modes associated with the component. Similar to a fault node in TFPG, failure mode also has ON and OFF states. δ_t is a function defined over $F_{cyber} \times \mathbb{R}_+$ that maps a failure mode $f \in F_{cyber}$ at time $t \in \mathbb{R}_+$ to *True* if the state of failure mode is ON and to *False* if the state is OFF.
- **Failure Mode Constraints**: $\alpha(F_{cyber})$ represents the set of all constraints defined over members of set F_{cyber}. An individual failure mode constraint, $\omega_t \in \alpha(F_{cyber})$, is a Boolean expression defined inductively as

$$\omega_t := \delta_t(f) \quad | \quad \neg\delta_t(f) \quad | \quad \omega_{1,t} \quad \wedge \quad \omega_{2,t} \tag{8.1}$$

  where $f \in F_{cyber}$ is a failure mode and ω_1, ω_2 are failure mode constraints. A failure mode constraint is True if the Boolean expression is evaluated to be True and False otherwise.
- **Timing Constraints**: Φ is a set of timing constraints defined as $\Phi = [n], (n) | n \in \mathbb{N}_+$, where $[n]$ denotes instantaneous constraints and (n) represents periodic constraints. The timing constraints specify a pattern of time points at which the automaton checks for events and failure node constraints. For instance, periodic constraint, (4), on any outgoing transition from the current state forces the automaton to periodically look for events specified by the edge, every 4 units of time whereas in the case of instantaneous constraint, [4], automaton checks only once.
- **Mode Map**: $\mathbb{M} : Q \rightarrow 2^m$ is a function that maps location $q \in Q$ to mode $m \in M$ defined in the fault propagation graph.

[1] The extension includes sets of failure modes and failure mode guards.

- **Edge**: $T \subset Q \times p(\Sigma) \times \Phi \times \alpha(F_{cyber}) \times p(\Sigma) \times Q$ is a finite set of edges. An edge represents a transition between any two locations. The activation conditions of an edge depend upon the timing, failure mode constraints, and an input event. For example, an edge $< q_1, \sigma_1, [n], \delta(f_1) \wedge \neg\delta(f_2), \sigma_2, q_2 >$ represents a transition from location q_1 to q_2 with an instantaneous time constraint of n units of time and failure mode constraint $\delta(f_1) \wedge \neg\delta(f_2) \in \alpha(F_{cyber})$ defined over the failure modes $f_1, f_2 \in F_{cyber}$. $\sigma_1 \in \Sigma$ is the required input event for this transition to be valid. $\sigma_2 \in \Sigma$ represents the event generated when the transition is taken. Syntactically, a transition is represented as *Event(timing constraint){failure constraint}/Event.* If no event is mentioned, then the transition is valid only if the failure mode constraint evaluates to true as per the timing constraints.

8.3.3 Observers for Postulating the Failures of Controllers

Observers are discrete, finite state machines that consume events produced by their respective tracked devices in order to diagnose faults in their behaviors. There exist a number of approaches for generating discrete diagnosers for dynamic systems based on [60] and [61]. However, the observers presented here are created manually. The events produced by the various observers fall into two categories; an estimation of a state change in discrete components, and a discrepancy detection. The detected anomalies and the local estimate of the state of different components in the plant and protection layer are passed by the observer to the next layer for system level diagnosis.

8.3.4 Combined Diagnosis and Reasoning Approach

The TCD reasoner relies on the fault propagation graph and the output of various observers to hypothesize about the anomalies observed in the system.[2] The reasoner attempts to explain the observations in terms of consistency relationship between the states of the nodes and edges in the fault propagation graph. The states of a node in a fault propagation graph can be categorized as *Physical* (Actual), *Observed*, and *Hypothetical* state [57].

- *Physical state* corresponds to the actual state of the nodes and edges.
- An *Observed state* is the same as the *Physical state*, but defined only for observable nodes.
- A *Hypothetical state* is an estimate of the node's physical state and the time since the last state change happened by the TCD reasoner.

[2]In order to relate to the alarms generated by observers with the failure graph few modifications are performed. The alarms signaled by relays are replaced by their corresponding observers.

Every reasoner hypothesis $h_f \in HSet_t$ consists of a map, $HNode_t$ that associates to every node in the failure graph an evaluation, (*ON*, *OFF*) and time estimate (t_1, t_2). The time estimate (t_1, t_2) denotes the earliest and latest time estimates for the state changes of node v, i.e. from ON to OFF or vice versa. The structure of a hypothesis is described as follows: Hypothesis is a tuple, where elements are related based on temporal consistency. Formally, hypothesis $h_f=\{f, terl, tlat, S, C, I, M, E, U\}$ where:

- $f \in F_{physical}$ is a physical failure mode projected by the hypothesis, h_f and F is the set of physical failure modes defined in Sect. 8.3.1. We are using single physical fault hypothesis which lists only one fault per element of the physical system along with multiple faults in protection system.
- $S \subseteq F_{cyber}$ is a set of faults active in the system. These faults are related to components in the protection system layer as defined in Sect. 8.3.2.
- The interval $[terl, tlat]$ is the estimated earliest and the latest time during which the failure mode f could have been activated. The time estimate for protection layer faults is not supported in the current implementation.
- $C \subseteq D_{physical}$ is the set of discrepancies that are consistent with the hypothesis h_f, where $D_{physical}$ is the set of physical discrepancies described in Sect. 8.3.1. These discrepancies are referred to as consistent discrepancies. We partition the set C into two disjoint subsets, $C1$, $C2$ where $C1$ consists of primary discrepancies and $C2$ contains secondary discrepancies. A discrepancy d w.r.t hypotheses h_f is called primary if the fault propagation linking the discrepancy, d, is certain otherwise it's termed secondary as defined in Sect. 8.3.1.
- $E \subseteq D_{physical}$ is the set of discrepancies which are expected to be activated in the future according to h_f. This set is also partitioned into $E1$ and $E2$ that contain primary and secondary discrepancies, respectively.
- $M \subseteq D_{physical}$ is the set of discrepancies that are missing according to the hypothesis h_f, i.e. alarms related to these discrepancies should have been signaled. This set is also composed of two disjoint sets $M1$ and $M2$ based on primary and secondary discrepancies.
- $I \subseteq D_{physical}$ is the set of discrepancies that are inconsistent with the hypothesis h_f. These are the discrepancies that are in the domain of f but cannot be explained in the current mode.
- $U \subseteq D_{physical}$ is the set of discrepancies which are not explained by this hypothesis h_f as there is no fault propagation link between $d \in U$ and $s \in f \cup S \cup C$, i.e. the discrepancy is not in the domain of f.

For every scenario, the reasoner creates one special hypothesis (conservative), **H0** that associates a spurious detection fault with each of the triggered alarms.

The quality of the generated hypotheses is measured based on four metrics defined as follows:

- ***Plausibility***: It is a measure of the degree to which a given hypothesis explains the current fault and its failure signature. Mathematically, it's defined as

$$Plausibility = \frac{|C1| + |C2|}{|C1| + |C2| + |M1| + |I|}$$

- ***Robustness***: It is a measure of the degree to which a given hypothesis will remain constant. Mathematically, it's defined as

$$Robustness = \frac{|C1| + |C2|}{|C1| + |C2| + |M1| + |E1| + |E2| + |I|}$$

- ***Rank***: It is a measure that a given hypothesis (a single physical fault along with multiple cyber faults) completely explains the system events observed. Mathematically, it is defined as $Rank = |C1| + |C2| - |M1| - |U|$
- ***Failure Mode Count***: is a measure of how many failure modes are listed by the hypothesis. The reasoner gives preference to hypotheses that explain the alarm events with a limited number of failure modes (i.e., it follows the parsimony principle). This metric plays an important role while pruning out **H0** from the final hypothesis report.

There are three types of events that invoke the reasoner to update the hypotheses. The first two are external physical events related to a change in the physical state of a monitored discrepancy and system mode. The third event is an internal timeout event that corresponds to the expectation of an alarm. A physical event is formally defined as a tuple e = (da, t), where da $\in D_0 \cup M$ is either an observable discrepancy or a system mode. The timeout event is described as a tuple $e = < h_f$, da, $t >$ which implies as per hypotheses h_f, any alarm related to discrepancy *da* should have been signaled by time t. Figures 8.5 and 8.6 give an overview of the underlying algorithm of reasoner response to three different type of events.

Timeout Event Whenever the observed state of a discrepancy does not change as expected by the reasoner, an internal timeout event, (h, da, t) is generated, where h denotes the set of hypotheses to be updated and da is the expected discrepancy and t is the current time. This event causes reasoner to update the expected sets of all hypotheses, h. If the expected sets, E1(E2), of any hypothesis in h, list da, then it is moved to missing sets M1(M2).

Mode Change Event If any actuator component in the protection layer changes its state, a mode change event is triggered by the corresponding observer. This event causes reasoner to update the expected sets of all hypotheses as the new actuator state might influence the operating modes and disable or enable failure propagation edges.

Discrepancy Mode Change Event This event is triggered if any observer detects appearance or disappearance of failure effects in both plants and protection devices. The event is denoted by (da, t), where da is a discrepancy that activated or deactivated at time t. If the observed state of this alarm is ON (activated), then reasoner iterates over all the hypotheses at time, t, to find hypotheses that explain this discrepancy (which lists da in expected sets). If found, expected and consist

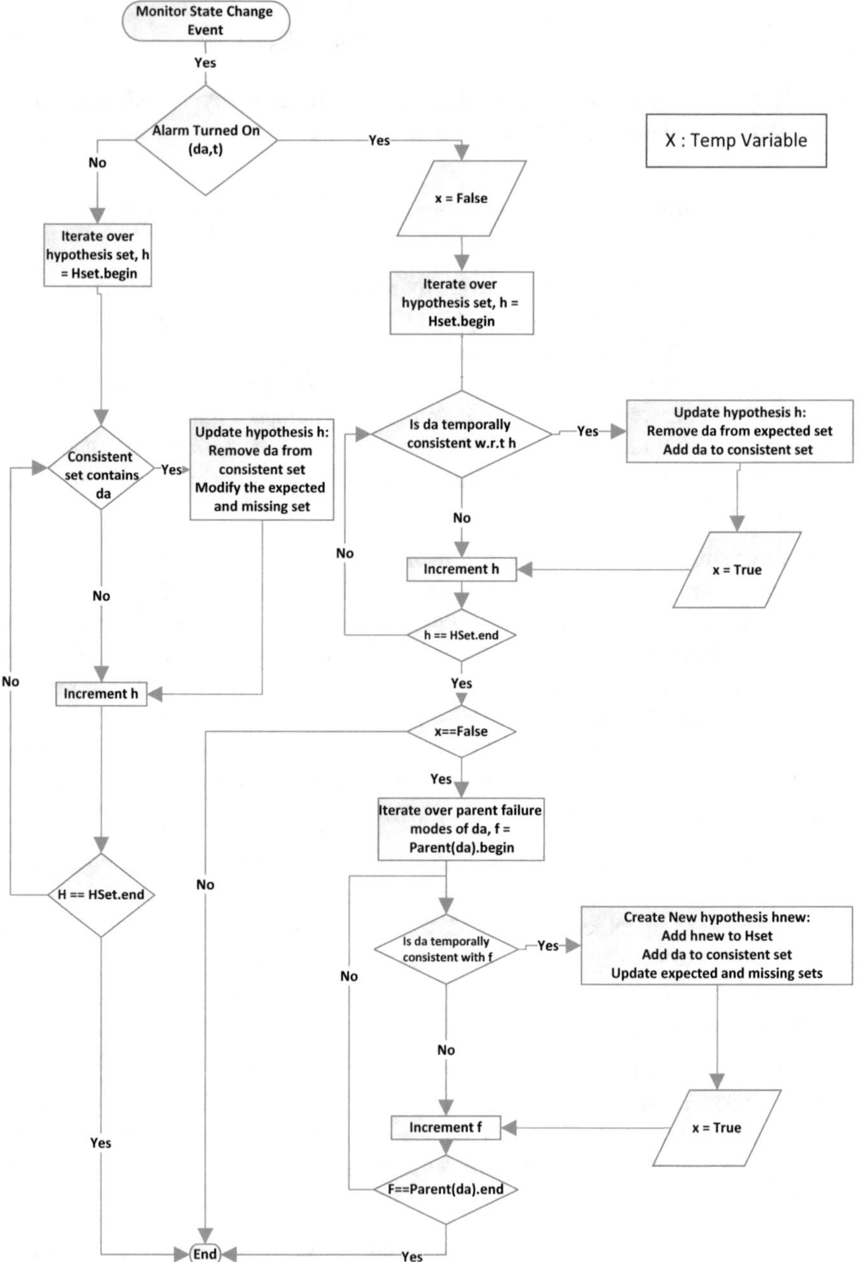

Fig. 8.5 Flowchart for handling Monitor or Discrepancy State Change Event

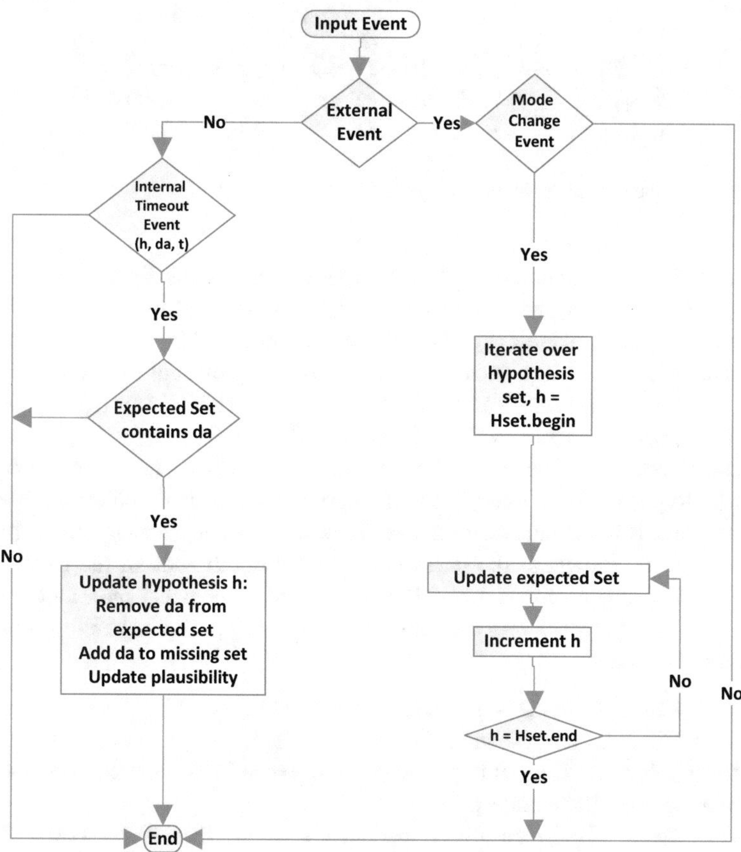

Fig. 8.6 Flowchart for handling Timeout and Mode Change Event

sets of those hypotheses are updated. In case, no hypothesis is discovered, a new hypothesis is generated and added to the hypothesis set. On the other hand, if the observed state of the discrepancy is OFF (deactivated), reasoner iterates over all hypotheses and update the consistent and expected sets of all hypotheses that list da in their consistent sets.

8.4 Example System: Electric Transmission Network

8.4.1 System Under Test

An electric power system can be considered as a tripartite graph with sources at one end and loads at the other with a complex transmission and distribution system in the middle. Figure 8.7 shows a segment of a transmission network where a load, L1

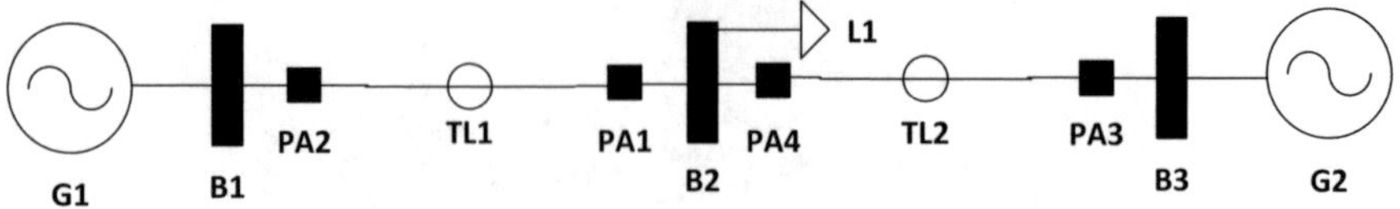

Fig. 8.7 A simple two transmission line system

is being fed by two generators G1, G2 through transmission lines TL1, TL2. The transmission lines are connected by buses B1, B2, B3. All these components are protected by specialized relays and breaker assemblies. In this work, we are focusing on transmission lines only, each transmission line is protected by a set of distance relays and breaker assemblies, installed at each end, collectively represented as a protection assembly labeled as PA1, PA2, PA3, PA4 in Fig. 8.7.

Distance relays are used for detecting two types of faults in transmission lines: (1) phase to phase faults, and (2) phase to ground faults. Both phase to phase and phase to ground faults cause an increase in current flowing through the conductor and decrease in voltage at the buses connected on both ends of the transmission line. This decrease in impedance (V/I) is detected as physical fault and typically categorized by the relay into the following three categories depending upon the calculated impedance:

- **Zone 1 Fault**: If the measured impedance is less than $(0.7 - 0.8) * Z_{TL}$ and the phase angle is between 0 and $\pi/2$, where Z_{TL} is the impedance of the line. The distance relay acts as a primary protection device and instructs the corresponding breaker to open immediately.
- **Zone 2 Fault**: If the measured impedance is greater than $(0.7 - 0.8) * Z_{TL}$ but less than $1.25 * Z_{TL}$ with phase angle being in first quadrant. After detecting a zone 2 fault, distance relay waits for 0.05–0.1 s before sending trip signal to the breaker. This wait time ensures the distance relay to act as a secondary or back-up protection element. If the fault is in any neighboring transmission line, then the wait time ensures the primary protection associated with that line to engage first. In case, the primary distance relays fail, then secondary protection kicks in after the waiting period expires.
- **Zone 3 Fault**: If the measured impedance is in the range $(1.25 - 2) * Z_{TL}$ with phase angle between 0 and $\pi/2$, then the fault considered as zone 3 fault. Similar to zone 2, the protection device acts as a back-up element in case primary device fails to engage. The wait time is of the order of $1 \ldots 1.5$ s.

The time to detect fault depends upon the sampling period of the relay and is of the range 16–30 ms.

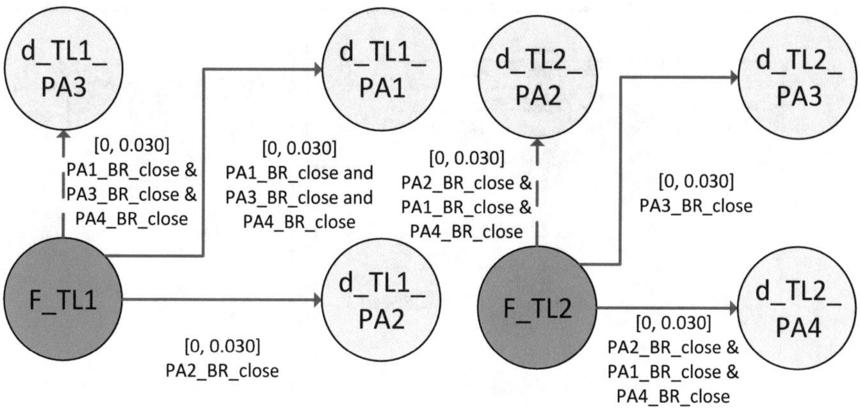

Fig. 8.8 Fault propagation graph for faults in two different transmission lines

Table 8.1 Discrepancy–Alarm Association Map

Discrepancy	Alarms
d_TL1_PA1	PA1_DR_Z1, PA1_DR_Z2
d_TL1_PA2	PA2_DR_Z1, PA2_DR_Z2
d_TL1_PA3	PA3_DR_Z2, PA3_DR_Z3
d_TL2_PA3	PA3_DR_Z1, PA3_DR_Z2
d_TL2_PA4	PA4_DR_Z1, PA4_DR_Z2
d_TL2_PA2	PA2_DR_Z2, PA2_DR_Z3

8.4.2 TCD: Fault Propagation Graph

The fault detection events are recorded by Sequence Event Recorders installed at substations. Using these events as alarms fault propagation graph can be created. Figure 8.8 shows such a graph for the segment of transmission network. The set of nodes labeled as F_TLn represents physical fault in transmission line, TLn. The discrepancy d_TLn_PAk represents the effect of failure F_TLn and the node represents the decrease in impedance as detected by relay in PAk. The edge between nodes represents the fault propagation and is constrained by the timing and operating conditions. The operating conditions are modeled in terms of the physical state of the breakers. The distance relay in PA4 will detect the failure mode F_TL1 as long as all the breakers in the path between G2 and TL1 are in close state. Table 8.1 lists the alarms that can signal discrepancies shown in Fig. 8.8, where the columns identify discrepancies, alarms, and the uncertainty associated to it. The failure edges that link failure source and discrepancy related to secondary protection relay are marked uncertain, i.e. ND(e) = false, depicted as dotted lines in Fig. 8.8.

A primary protection element will always signal Zone 1 or Zone 2 alarm for fault injected at any point in the transmission line. The secondary protection devices will always signal either Zone 2 or Zone 3 alarm depending upon the location of the fault.

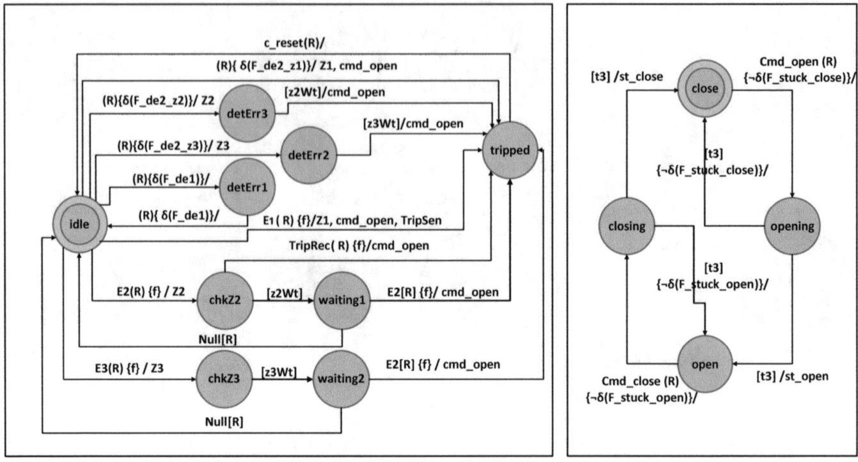

Fig. 8.9 Protection System Behavior Components (Left: Distance Relay; Right: Breaker), where f is a failure mode constraint defined as f:$\neg\delta$(F_de1) $\wedge\neg\delta$(F_de2_z1) $\wedge\neg\delta$(F_de2_z2) $\wedge\neg\delta$(F_de2_z3)

The failure graph captures the propagation of failures under different conditions (breaker states) but does not contain any information to diagnose faults related with the behaviors of breakers and relays. Figure 8.9 shows the TTA model of a protection assembly (distance relay and breaker).

8.4.3 TCD: Distance Relay Behavioral Model

Modern relays are reactive devices that monitor the health of the physical devices at a fixed rate, R secs. Figure 8.9 shows a time triggered model of a distance relay configured to detect Zone 1, 2, 3 faults. The time triggered automaton appropriately models the behavior of a relay under both faulty and non-faulty conditions. The model considers two types of faults, $F = f1 \cup f2$, where $f1 = \{Fde1\}$ is a set of missed detection faults and $f2 = \{Fde2z1, Fde2z2, Fde2z3\}$ is the collection of spurious detection faults related to three zones. As the name implies, a missed detection fault forces the relay to skip the detection of any fault conditions and a spurious detection fault, $Fde2zk$, ensues incorrect inference of zone k fault by the relay. Figure 8.9 lists five different failure mode constraints, namely, $\delta(Fde1)$, $\delta(Fde2z1)$, $\delta(Fde2z2)$, $\delta(Fde2z3)$, $\neg\delta(Fde1) \wedge \neg\delta(Fde2z1) \wedge \neg\delta(Fde2z2) \wedge \neg\delta(Fde2z3)$, where the first four imply the presence of a failure mode, i.e. its state is ON while the last means none of the failure modes in F are present.

There are a total of nine events used to model the behavior of the relay. Out of nine events, three are unobservable, labeled as E1, E2, and E3. These events represent the presence of zone 1, 2, 3 fault conditions. The state machine consists of nine locations, with idle being the initial location. In the idle location, automaton check for events—E1, E2, E3, and the status of failure modes every R seconds. If the distance relay detects zone 1 fault (modeled by the presence of the event E1), then the distance relay moves to the tripped location and issues a Z1 alarm and commands the breaker to open by emitting event, `cmd_open`. For zone 2 and zone 3 faults conditions (E2, E3), the protection relay does not issue an open command after moving to the `chkZ2` or `chkZ3` locations. The state machine waits for predefined time, $zn2wt, zn3wt \in \mathbb{R}_+$ and confirms again the presence of the fault conditions, once the time expires. If the fault is still present, the relay commands the breaker to open and transitions to tripped location, otherwise moves back to idle location. Additionally, distance relays may be configured with overreach trip transfer protocols. In this case, the primary relays associated with a transmission line send permissive trip signals to each other, `TripSen`, in order to avoid zone 2 wait time.

The deviation in the normal behavior of the relay is caused if any of the failure mode constraints evaluates to true. For instance, if the current location of the automaton is idle and failure mode $Fde1$ is present then automaton jumps to detErr1 location and stays there until the fault is persistent. Similarly if any of the spurious detection faults are present, then irrespective of the presence of E1, E2, and E3, the state machine jumps to `detErr2` or `detErr3` and finally transitions to tripped state. In this model, the faults (`F_de1`, `F_de2_z1`, `F_de2_z2`, `F_de2_z3`) are assumed to be mutually exclusive, i.e. one of the cyber faults can be present at a given time.

8.4.4 TCD: Breaker Behavioral Model

Figure 8.9 also shows TTA model of a breaker with two failure modes, F = *{F_stuck_close, F_stuck_open}*. The breaker automaton has four states labeled as `open`, `opening`, `close`, and `closing`, with `close` being the initial state. All the events used in the state machine are observable. The events cmd_open, cmd_close represent the commands received by the breaker assembly and st_open, st_close signify change in the physical state of the breaker. The transition from open to close and vice verse is not instantaneous. The lag is caused due to mechanical nature of the breaker and zero crossing detection, which is modeled by parameter t3. Automaton consists of two failure mode constraints, $\neg\delta(Fstuckclose)$, $\neg\delta(Fstuckopen)$, which evaluates to true when respective failure modes are not present.

The breaker is also modeled as reactive component which is periodically checking for commands. While in the `close` location, the automaton looks for event cmd_open and evaluates the failure constraint every R secs. If the event is present and F_stuck_close fault is absent, the state machine transitions to opening state. After t3 secs, the automaton moves to open state if failure mode constraint still evaluates to false. Similarly in open location, the presence of the event cmd_close and validity of failure constraint is checked.

8.4.5 TCD Diagnosis System: Observers

The TCD based diagnosis system employs a hierarchical framework as shown in Fig. 8.3. The lower layer includes observers that track the operation of cyber components (distance relays and circuit breakers) to detect and locally diagnose faults in physical and protection systems. The observers feed their results to the reasoning engine as explained in previous section. The TCD reasoning engine produces a set of hypotheses that explain the current system states as per the output of various observers by traversing the fault propagation graph. The traversal is constrained by the state of the protection system as predicted by observers tracking it. The following sections provide a detailed description of the model and operation of the observers related with power system protection devices.

8.4.5.1 Observer: Distance Relay

The TTA model of a distance relay observer can be seen in Fig. 8.10. The state machine has eight locations with `idle` being the initial state. The observer machine consumes the observable zone alarms (Z1, Z2, Z3), commands sent to breaker (cmd_open) and reset events and produce h_Z1, h_Z2, h_Z3 to indicate or confirm the presence of zone 1, 2, 3 faults. The observer also produces h_Z1′, h_Z2′ and h_Z3′ to indicate absence of zone 1, 2, 3 fault conditions. The observer remains in the `idle` position until zone fault conditions are reported by the corresponding distance relay. Once the distance relay fires a *Z1* event, the observer machine jumps to the `chkZ1` location while emitting h_Z1 event. The observer machine waits for *t2* seconds for open command (cmd_open event). If received, the observer moves

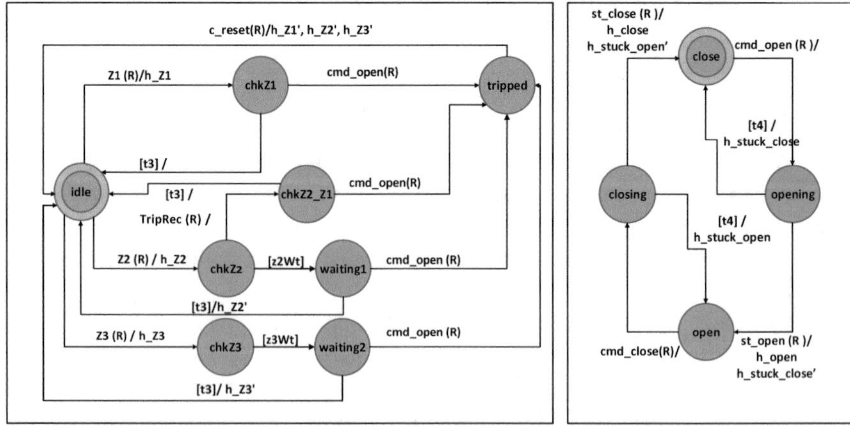

Fig. 8.10 Protection System Observer Models, Distance Observer Model (Left); Breaker Observer Model (Right)

to the `tripped` state, otherwise transitions back to `idle` state. *t2* is a parameter of the distance relay observer machine that models propagation delay and relay frequency. Please note that the transition from `chkZ1` state to the *idle* state implies a communication channel fault, but in this paper we are not considering such faults.

Similarly, the observer machine moves to the `chkZ2` state when the distance relay reports a Z2 event after detecting zone 2 fault conditions. Upon the confirmation of zone 2 fault, the observer waits *t3* seconds for the arrival of the *cmd_open* command. *t3* is a parameter which is equal to the sum of zone 2 wait time and t2. If the *cmd_open* event is not observed within *t3* seconds the automaton moves back to the `idle` state and concludes that the zone 2 fault condition has disappeared by generating h_Z2′ event. The observer machine moves from `chkZ2` state to `chkZ2_Z1` state if the event *TripRec* occurs and waits for the *cmd_open* event and concludes the presence of fault by producing h_Z2 event. In a similar fashion, the distance relay observer diagnoses zone 3 faults.

8.4.5.2 Observer: Circuit Breaker

The breaker observer model is shown in the right side of Fig. 8.10. It consists of four states labeled as `open`, `close`, `opening`, and `closing` and correlate directly to the four states of the breaker automaton. Initially the state machine is in the `close` state and jumps to the `opening` state after observing *cmd_open* event. The breaker observer transitions to the `open` state if it receives an *st_open* event from the breaker assembly within *t4* seconds. *t4* is a model parameter that is equal to the sum of propagation time and the maximum time required to open the breaker. If the event is observed in the time limit, the observer concludes the physical state of breaker is open and stuck close fault is not present by producing an event, h_stuck_close′. Otherwise it hypothesizes that the breaker has the stuck close fault. The fault is signaled by generating an event, *h_stuck_close*. Similarly, when the breaker is in the `open` state it has the same timed behavior and an *h_stuck_open* event is generated if an *st_close* event is not observed within *t4* seconds of receiving the *cmd_close* event.

8.4.6 Results

Figure 8.11 shows the sequence of events generated by protection devices, observers, and reasoning engine when a three phase to ground fault is injected in transmission line TL2 along with the presence of missed detection fault in PA4_DR and stuck close fault in PA2_BR. At $t = 0.501$, PA3_DR_OBS and PA2_DR_OBS report h_Z1 and h_Z3 alarms. These alarms produce two hypotheses H0, H1. H1 lists faults in line TL2 with two consistent discrepancies and expects an alarm from PA4_DR_OBS (h_Z1 or h_Z2). At $t = 0.531$, timeout forces the expected discrepancy to shift to the missing set. H1 and H0 both list two failure modes.

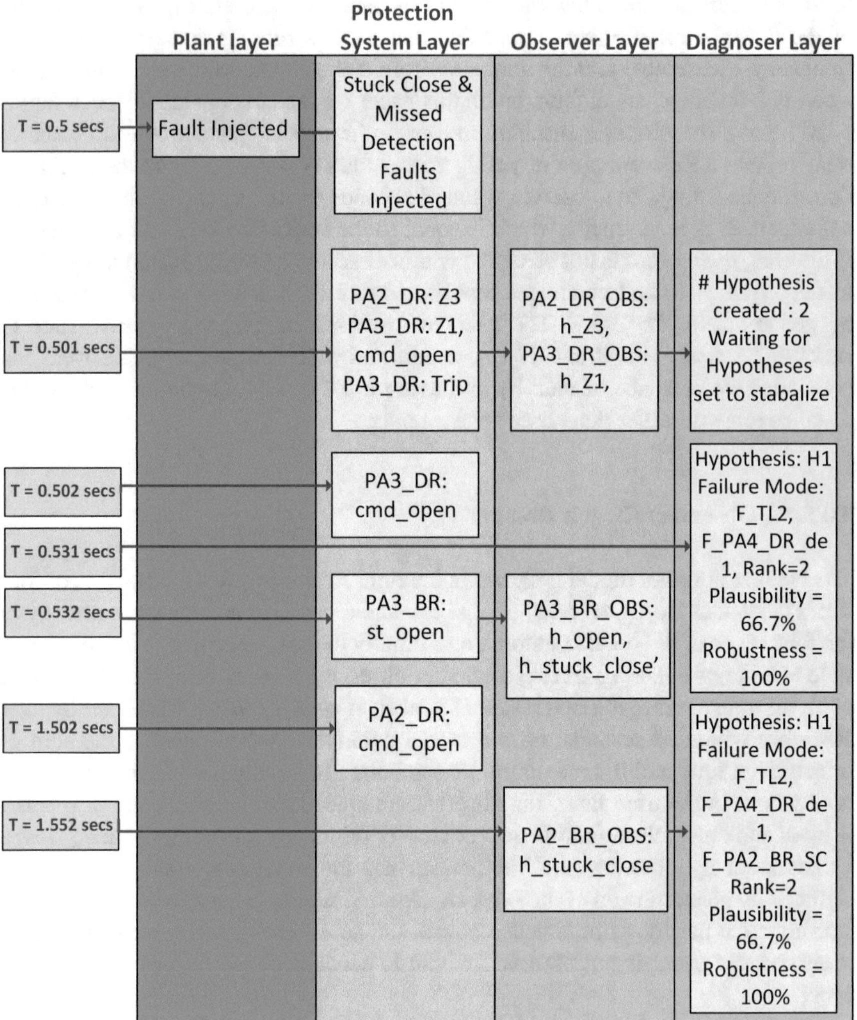

Fig. 8.11 Diagnosis results for scenario 4

H1 lists physical faults associated with line TL2 along with a missed detection fault in PA4_DR whereas H0 blames both the distance relays for having spurious detection faults. At $t = 1.552$, PA2_BR_OBS concludes a stuck fault in breaker PA2_BR after failing to receive a state change event (st_open). Both hypotheses are updated to reflect the breaker fault. The hypothesis H1 is given preference over H0 as the probability of two cyber faults is less than a physical and a cyber fault [62]. Figure 8.11 shows the events sequence and hypotheses evolution.

8.5 Conclusion

We have presented a new formalism: Temporal Causal Diagrams with the aim of diagnosing failures in cyber-physical systems that include local fast-acting protection devices. Specifically, we have demonstrated the capability of the TCD model to capture the discrete fault propagation and behavioral model of a segment of a power transmission system protected by distance relays and breakers. The paper also presented hierarchical TCD-based reasoner to diagnose faults in the physical system and its protection elements.

Acknowledgements This work is funded in part by the National Science Foundation under the award number CNS-1329803. Any opinions, findings, and conclusions or recommendations expressed in this material are those of the author(s) and do not necessarily reflect the views of NSF. The authors would like to thank Rishabh Jain, Srdjn Lukic, Saqib Hasan, Scott Eisele, and Amogh Kulkarni for their help and discussions related to the work presented here.

References

1. North American Electric Reliability Corporation, 2012 state of reliability, Tech. Rep. (2012). Available: http://www.nerc.com/files/2012_sor.pdf
2. S. Abdelwahed, G. Karsai, G. Biswas, A consistency-based robust diagnosis approach for temporal causal systems, in *The 16th International Workshop on Principles of Diagnosis* (2005), pp. 73–79
3. N. Mahadevan, A. Dubey, G. Karsai, Application of software health management techniques, in *Proceedings of the 6th International Symposium on Software Engineering for Adaptive and Self-managing Systems*, ser. SEAMS '11 (ACM, New York, 2011), pp. 1–10. Available: http://doi.acm.org/10.1145/1988008.1988010
4. P. Seifried, Fault detection and diagnosis in chemical and petrochemical processes, Bd. 8 der Serie „Chemical Engineering Monographs". Von D. M. Himmelblau, herausgegeben von S. W. Churchill, Elsevier Scientific Publishing Company, Amsterdam – New York 1978. 1. Aufl., X, 414 S., 137 Abb., 66 Tab., DM 145,–. Chem. Ing. Tech. **51**, 766 (1979). https://doi.org/10.1002/cite.330510726
5. N. Viswanadham, T.L. Johnson, Fault detection and diagnosis of automated manufacturing systems, in *27th IEEE Conference on Decision and Control* (1988)
6. R. Hessian, B. Salter, E. Goodwin, Fault-tree analysis for system design, development, modification, and verification. IEEE Trans. Reliab. **39**(1), 87–91 (1990)
7. Y. Ishida, N. Adachi, H. Tokumaru, Topological approach to failure diagnosis of large-scale systems. IEEE Trans. Syst. Man Cybern. **15**(3), 327–333 (1985)
8. S.V.N. Rao, N. Viswanadham, Fault diagnosis in dynamical systems: a graph theoretic approach. Int. J. Syst. Sci. **18**(4), 687–695 (1987)
9. S.V.N. Rao, N. Viswanadham, A methodology for knowledge acquisition and reasoning in failure analysis of systems. IEEE Trans. Syst. Man Cybern. **17**(2), 274–288 (1987)
10. J. Richman, K.R. Bowden, The modern fault dictionary, in *International Test Conference* (1985), pp. 696–702
11. W.T. Scherer, C.C. White, A survey of expert systems for equipment maintenance and diagnostics, in *Knowledge-Based System Diagnosis, Supervision and Control*, ed. by S.G. Tzafestas (Plenum, New York, 1989), pp. 285–300

12. S. Tzafestas, K. Watanabe, Modern approaches to system/sensor fault detection and diagnosis. J. A. IRCU Lab. **31**(4), 42–57 (1990)
13. P. Frank, Fault diagnosis in dynamic systems using analytical and knowledge-based redundancy – a survey and some new results. Automatica **26**(3), 459–474 (1990)
14. W. Hamscher, L. Console, J. de Kleer, *Readings in Model-Based Diagnosis* (Morgan Kaufmann Publishers Inc., San Francisco, 1992)
15. R. Patton, Robust model-based fault diagnosis: the state of the art, in *IFAC Fault Detection, Supervision and Safety for Technical Processes*, Espoo (1994), pp. 1–24
16. R. Patton, P. Frank, R. Clark, *Fault Diagnosis in Dynamic Systems: Theory and Application* (Prentice Hall International, Englewood Cliffs, 1989)
17. M. Sampath, R. Sengupta, S. Lafortune, K. Sinnamohideen, D. Teneketzis, Failure diagnosis using discrete event models. IEEE Trans. Control Syst. Technol. **4**, 105–124 (1996)
18. A.N. Srivastava, Discovering system health anomalies using data mining techniques, in *Proceedings of the Joint Army Navy NASA Air Force Conference on Propulsion* (2005)
19. R. Reiter, A theory of diagnosis from first principles. Artif. Intell. **32**(1), 57–95 (1987)
20. J. de Kleer, A. Mackworth, R. Reiter, Characterizing diagnoses and systems. Artif. Intell. **56**, 197–222 (1992)
21. A. Darwiche, Model-based diagnosis using structured system descriptions. J. Artif. Intell. Res. **8**, 165–222 (1998)
22. A. Darwiche, G. Provan, Exploiting system structure in model-based diagnosis of discrete-event systems, in *Proceedings of the Seventh International Workshop on Principles of Diagnosis* (1996), pp. 95–105
23. J. Gamper, A temporal reasoning and abstraction framework for model-based diagnosis systems. Ph.D. dissertation, RWTH, Aachen, 1996
24. L. Console, P. Torasso, On the co-operation between abductive and temporal reasoning in medical diagnosis. Artif. Intell. Med. **3**(6), 291–311 (1991)
25. A. Misra, Sensor-based diagnosis of dynamical systems. Ph.D. dissertation, Vanderbilt University, 1994
26. A. Misra, J. Sztipanovits, J. Carnes, Robust diagnostics: structural redundancy approach, in *SPIE's Symposium on Intelligent Systems* (1994)
27. S. Padalkar, J. Sztipanovits, G. Karsai, N. Miyasaka, K.C. Okuda, Real-time fault diagnostics. IEEE Expert **6**(3), 75–85 (1991)
28. G. Karsai, J. Sztipanovits, S. Padalkar, C. Biegl, Model based intelligent process control for cogenerator plants. J. Parallel Distrib. Syst. **15**, 90–103 (1992)
29. P.J. Mosterman, G. Biswas, Diagnosis of continuous valued systems in transient operating regions. IEEE Trans. Syst. Man Cybern. **29**(6), 554–565 (1999)
30. G. Karsai, G. Biswas, S. Abdelwahed, Towards fault-adaptive control of complex dynamic systems, in *Software-Enabled Control: Information Technology for Dynamical Systems*, ch. 17, ed. by T. Samad, G. Balas (IEEE Publication, Piscataway, 2003)
31. S. Abdelwahed, G. Karsai, G. Biswas, System diagnosis using hybrid failure propagation graphs, in *The 15th International Workshop on Principles of Diagnosis*, Carcassonne, 2004
32. V. Brusoni, L. Console, P. Terenziani, D.T. Dupre, A spectrum of definitions for temporal model-based diagnosis. Artif. Intell. **102**(1), 39–79 (1998)
33. Z. Yongli, Y.H. Yang, B.W. Hogg, W.Q. Zhang, S. Gao, An expert system for power systems fault analysis. IEEE Trans. Power Syst. **9**(1), 503–509 (1994)
34. Y.-C. Huang, Fault section estimation in power systems using a novel decision support system. IEEE Trans. Power Syst. **17**(2), 439–444 (2002)
35. G. Cardoso, J.G. Rolim, H.H. Zurn, Identifying the primary fault section after contingencies in bulk power systems. IEEE Trans. Power Deliv. **23**(3), 1335–1342 (2008)
36. J. Jung, C.-C. Liu, M. Hong, M. Gallanti, G. Tornielli, Multiple hypotheses and their credibility in on-line fault diagnosis. IEEE Trans. Power Deliv. **16**(2), 225–230 (2001)
37. G. Cardoso, J.G. Rolim, H.H. Zurn, Application of neural-network modules to electric power system fault section estimation. IEEE Trans. Power Delivery **19**(3), 1034–1041 (2004)

38. R.N. Mahanty, P.B.D. Gupta, Application of RBF neural network to fault classification and location in transmission lines. IEE Proc. Gener. Transm. Distrib. **151**(2), 201–212 (2004)
39. D. Thukaram, H.P. Khincha, H.P. Vijaynarasimha, Artificial neural network and support vector machine approach for locating faults in radial distribution systems. IEEE Trans. Power Delivery **20**(2), 710–721 (2005)
40. T. Bi, Z. Yan, F. Wen, Y. Ni, C. Shen, F.F. Wu, Q. Yang, On-line fault section estimation in power systems with radial basis function neural network. Int. J. Electr. Power Energy Syst. **24**(4), 321–328 (2002)
41. Y.-X. Wu, X.N. Lin, S.H. Miao, P. Liu, D.Q. Wang, D.B. Chen, Application of family eugenics based evolution algorithms to electric power system fault section estimation, in *Transmission and Distribution Conference and Exhibition: Asia and Pacific, 2005 IEEE/PES* (2005), pp. 1–5
42. F. Wen, C. Chang, Probabilistic approach for fault-section estimation in power systems based on a refined genetic algorithm. IEE Proc. Gener. Transm. Distrib. **144**(2), 160–168 (1997)
43. Z. He, H.-D. Chiang, C. Li, Q. Zeng, Fault-section estimation in power systems based on improved optimization model and binary particle swarm optimization, in *IEEE Power & Energy Society General Meeting, 2009. PES'09* (IEEE, Piscataway, 2009), pp. 1–8
44. W. Guo, F. Wen, G. Ledwich, Z. Liao, X. He, J. Liang, An analytic model for fault diagnosis in power systems considering malfunctions of protective relays and circuit breakers. IEEE Trans. Power Deliv. **25**(3), 1393–1401 (2010)
45. J. Sun, S.-Y. Qin, Y.-H. Song, Fault diagnosis of electric power systems based on fuzzy petri nets, IEEE Trans. Power Syst. **19**(4), 2053–2059 (2004)
46. W.-H. Chen, C.-W. Liu, M.-S. Tsai, Fast fault section estimation in distribution substations using matrix-based cause-effect networks. IEEE Trans. Power Deliv. **16**(4), 522–527 (2001)
47. W.H. Chen, S.H. Tsai, H.I. Lin, Fault section estimation for power networks using logic cause-effect models. IEEE Trans. Power Deliv. **26**(2), 963–971 (2011)
48. W. Guo, L. Wei, F. Wen, Z. Liao, J. Liang, C.L. Tseng, An on-line intelligent alarm analyzer for power systems based on temporal constraint network, in *International Conference on Sustainable Power Generation and Supply, 2009. SUPERGEN '09* (2009), pp. 1–7
49. W.H. Chen, Online fault diagnosis for power transmission networks using fuzzy digraph models. IEEE Trans. Power Deliv. **27**(2), 688–698 (2012)
50. Z. Yongli, H. Limin, L. Jinling, Bayesian networks-based approach for power systems fault diagnosis. IEEE Trans. Power Deliv. **21**(2), 634–639 (2006)
51. Y. Sekine, Y. Akimoto, M. Kunugi, C. Fukui, S. Fukui, Fault diagnosis of power systems. Proc. IEEE **80**(5), 673–683 (1992)
52. 1962. Available: http://www2.ee.washington.edu/research/pstca/pf14/pg_tca14bus.htm
53. 1962. Available: http://icseg.iti.illinois.edu/ieee-14-bus-system/
54. 2016. Available: http://www2.ee.washington.edu/research/pstca/formats/cdf.txt
55. R. Dugan, *OpenDSS Manual.* Electrical Power Research Institute, 2016. Available: http://sourceforge.net/apps/mediawiki/electricdss/index.php
56. S. Padalkar, G. Karsai, C. Biegl, J. Sztipanovits, K. Okuda, N. Miyasaka, Real-time fault diagnostics. IEEE Expert **6**(3), 75–85 (1991)
57. S. Abdelwahed, G. Karsai, Notions of diagnosability for timed failure propagation graphs, in *2006 IEEE Autotestcon*, Sept 2006, pp. 643–648
58. A. Dubey, G. Karsai, N. Mahadevan, Model-based software health management for real-time systems, in *2011 IEEE Aerospace Conference* (IEEE, Piscataway, 2011), pp. 1–18
59. P. Krčál, L. Mokrushin, P. Thiagarajan, W. Yi, Timed vs. time-triggered automata, in *CONCUR 2004-Concurrency Theory* (Springer, Berlin, 2004), pp. 340–354
60. M. Sampath, R. Sengupta, S. Lafortune, K. Sinnamohideen, D. Teneketzis, Diagnosability of discrete-event systems. IEEE Trans. Autom. Control **40**(9), 1555–1575 (1995)
61. S. Tripakis, Fault diagnosis for timed automata, in *International Symposium on Formal Techniques in Real-Time and Fault-Tolerant Systems* (Springer, Berlin, 2002), pp. 205–221
62. E. Schweitzer, B. Fleming, T.J. Lee, P.M. Anderson et al., Reliability analysis of transmission protection using fault tree methods, in *Proceedings of the 24th Annual Western Protective Relay Conference* (1997), pp. 1–17

Chapter 9
Passive Diagnosis of Hidden-Mode Switched Affine Models with Detection Guarantees via Model Invalidation

Farshad Harirchi, Sze Zheng Yong, and Necmiye Ozay

9.1 Introduction

Sensor-rich networked cyber-physical systems, which integrate physical processes and embedded computers, shape the basis of our future smart systems. Such systems, that include critical infrastructures such as traffic, power and water networks, as well as autonomous vehicles, aircrafts, home appliances, and manufacturing processes, are becoming increasingly common and will affect many aspects of our daily lives. As such, the reliability and security of these cyber-physical systems is paramount for their successful implementation and operation. However, some major incidents involving these critical infrastructure systems as a result of cyber-attacks and system failures have taken place in recent years and are a big source of concern. Scalable and reliable fault and attack diagnosis monitors play a crucial role in enhancing the robustness of these systems to failures and adversarial attacks. In addition, a thorough understanding of the vulnerability of system components against such events can be incorporated in future design processes to better design such systems. Hybrid systems provide a convenient means to model many cyber-physical systems. In this chapter, we consider hidden-mode switched affine models with parametric uncertainty subject to process and measurement noise and present a fault/attack detection and isolation framework for such systems.

F. Harirchi (✉) · N. Ozay
University of Michigan, Ann Arbor, MI, USA
e-mail: harirchi@umich.edu; necmiye@umich.edu

S. Z. Yong
Arizona State University, Tempe, AZ, USA
e-mail: szyong@asu.edu

M. Sayed-Mouchaweh (ed.), *Diagnosability, Security and Safety of Hybrid Dynamic and Cyber-Physical Systems*, https://doi.org/10.1007/978-3-319-74962-4_9

9.1.1 Literature Review

The study of fault detection began with the introduction of the first failure detection filter by Beard in 1971 [1]. Since then, fault diagnosis has attracted a great deal of attention and has become an integral part of most, if not all system designs. The problem of fault diagnosis has been approached by researchers from a wide variety of perspectives including signal processing and control theory. The most popular methods in the literature employ either data-driven techniques or model-based approaches. In this paper, we consider a model-based fault diagnosis approach.

Model-based fault detection and isolation schemes in the literature can be categorized into two classes, i.e., approaches that are based on residual generation and on set-membership. The former approach is more common in the fault diagnosis literature, and in this approach, the difference between the measurements and the estimates is defined as a residual or a symptom [2]. Two major trends in the residual generation techniques are the observer-based [3–5] and the parameter estimation based [6, 7] methods. Even though the residual generation based approaches are efficient and are thus widely used in the industry, their performance is highly dependent on the preciseness of the observers or the parameter estimates and also the employed residual evaluation approach. In addition, these methods do not provide any guarantees for the detection of faults. Residual-based methods are also employed for fault detection and isolation in non-linear and hybrid models [8–11]. In particular, an observer-based method is proposed for fault diagnosis of hybrid systems in [12], in which an extended Kalman filter is used to track the continuous behavior of the system, and a mode estimator to estimate the discrete state.

On the other hand, set-membership based fault detection and isolation techniques are proposed with the goal of providing guarantees for the detection of some specific faults. Most of these methods operate by discarding models that are not compatible with observed data, rather than identifying the most likely model. There is an extensive literature on set-membership based methods for active fault diagnosis of linear models [13–15]. These active fault diagnosis methods can only handle systems with linear models, and even so, they are still computationally demanding. Recently, we proposed set-membership based guaranteed passive fault diagnosis approaches for the class of switched affine models with parametric uncertainty and subject to process and measurement noise [16, 17] and for the class of polynomial state space models [18]. These approaches are developed by leveraging ideas from model invalidation [19–21] and taking advantage of recent advances in optimization.

In this chapter, we address three problems related to switched affine models: (1) model invalidation; (2) fault[1] detection, and (3) fault isolation. In the model invalidation problem, one starts with a family of models (i.e., a priori or admissible

[1]For convenience, we will use the term "fault" to refer to any fault, attack or anomaly throughout this chapter. Note that our proposed approach is primarily concerned with the detection and isolation of changes in dynamical system behavior and is indifferent to the nature of the observed changes, i.e., whether they are accidental faults or strategic attacks, either cyber or physical.

model set) and experimental input/output data collected from a system (i.e., a finite execution trace) and tries to determine whether or not the experimental data can be generated by one of the models in the initial model family. It was originally proposed as a way to build trust in models obtained through a system identification step by discarding/improving these models before using them in robust control design [19], but we employ it as a tool for detection and isolation of faults.

In addition, we present some conditions under which model invalidation can be efficiently applied in an online receding horizon manner for the purpose of fault detection and isolation. In order to check these conditions, we introduce a property for model pairs—T-distinguishability. When one model is the nominal system model and the other is one of the fault models, this concept is also known as T-detectability [16, 17, 22], while when both models are fault models, this corresponds to I-isolability defined in [17]. If this property holds for a given set of models, it allows us to detect and isolate faults in a receding horizon manner with time horizon of size T or I without compromising detection or isolation guarantees. The concept of T-distinguishability is closely related to the input-distinguishability of linear systems [23, 24] and mode discernibility in hybrid systems [25]. Even though some of these conditions may appear rather strong, we show that they are necessary and sufficient for any guaranteed passive fault detection and isolation scheme that only uses data from a finite horizon.

9.2 Preliminaries

In this section, the notation used throughout the chapter and the modeling framework we consider are described.

9.2.1 Notation

Let $\mathbf{x} \in \mathbb{R}^n$ and $\mathbf{M} \in \mathbb{R}^{n \times m}$ denote a vector and a matrix, respectively. The infinity norm of a vector $\mathbf{x}$ is denoted by $\|\mathbf{x}\| \doteq \max_i |\mathbf{x}^i|$, where $\mathbf{x}^i$ denotes the ith element of vector $\mathbf{x}$. The set of positive integers up to n is denoted by $\mathbb{Z}_n^+$, and the set of non-negative integers up to n is denoted by $\mathbb{Z}_n^0$. Moreover, we denote with $\mathbf{x}_{0:N} = \begin{bmatrix}\mathbf{x}_0 \ \mathbf{x}_1 \ \ldots \ \mathbf{x}_N\end{bmatrix}$ the concatenation of vectors $\mathbf{x}_j$ for all $j \in \mathbb{Z}_N^0$. We also make use of integral constraints known as Special Ordered Set of degree 1 (SOS-1) constraints in our optimization solution, defined as follows:

Definition 1 (SOS-1 Constraints [26]) A special ordered set of degree 1 (SOS-1) constraint is a set of integer and/or real scalar variables for which at most one variable in the set may take a value other than zero, denoted as SOS-1: $\{v_1, \ldots, v_N\}$. For instance, if $v_i \neq 0$, then this constraint imposes that $v_1 = \ldots = v_{i-1} = v_{i+1} = \ldots = v_N = 0$.

9.2.2 Modeling Framework

We consider systems that can be represented by discrete-time switched affine (SWA) models.

Definition 2 (SWA Model) A switched affine model is a tuple:

$$\mathcal{G} = (\mathcal{X}, \mathcal{E}, \mathcal{U}, \{G_i\}_{i=1}^m), \tag{9.1}$$

where $\mathcal{X} \subset \mathbb{R}^n$ is the set of states, $\mathcal{E} \subset \mathbb{R}^{n_y+n_p}$ is the set of measurement and process noise signals, $\mathcal{U} \subset \mathbb{R}^{n_u}$ is the set of inputs, and $\{G_i\}_{i=1}^m$ is a collection of m modes. Each mode $i \in \mathbb{Z}_m^+$ is an affine model G_i:

$$G_i = \{\mathbf{A}_i, \mathbf{B}_i, \mathbf{C}_i, \mathbf{D}_i, \mathbf{f}_i, \mathbf{g}_i\}, \tag{9.2}$$

with system matrices $\mathbf{A}_i$, $\mathbf{B}_i$, $\mathbf{C}_i$, and $\mathbf{D}_i$, and (affine) vectors $\mathbf{f}_i$ and $\mathbf{g}_i$.

The evolution of $\mathcal{G}$ is governed by:

$$\begin{aligned} \mathbf{x}_{t+1} &= \mathbf{A}_{\sigma_t}\mathbf{x}_t + \mathbf{B}_{\sigma_t}\mathbf{u}_t + \mathbf{f}_{\sigma_t} + \boldsymbol{\nu}_t, \\ \mathbf{y}_t &= \mathbf{C}_{\sigma_t}\mathbf{x}_t + \mathbf{D}_{\sigma_t}\mathbf{u}_t + \mathbf{g}_{\sigma_t} + \boldsymbol{\eta}_t, \end{aligned} \tag{9.3}$$

where $\boldsymbol{\nu} \in \mathbb{R}^{n_p}$ and $\boldsymbol{\eta} \in \mathbb{R}^{n_y}$ denote the bounded process and measurement noise signals, and the mode signal $\sigma_t \in \mathbb{Z}_m^+$ indicates the active mode at time t.

Remark 1 We assume $\mathcal{X}, \mathcal{E}, \mathcal{U}$ are convex and compact sets. In particular, we consider the following form for the admissible sets:

$$\begin{aligned} &\mathcal{X} = \{\mathbf{x} \mid P\mathbf{x} \le p\},\ \mathcal{E} = \{[\boldsymbol{\eta}^\intercal \ \ \boldsymbol{\nu}^\intercal]^\intercal \mid \|\boldsymbol{\eta}\| \le \varepsilon_\eta, \|\boldsymbol{\nu}\| \le \varepsilon_\nu\}, \\ &\mathcal{U} = \{\mathbf{u} \mid \|\mathbf{u}\| \le U\}, \end{aligned} \tag{9.4}$$

where $P \in \mathbb{R}^{n_p \times n}$ and $p \in \mathbb{R}^{n_p}$. Note that our analysis holds true for any $\mathcal{X}, \mathcal{E}, \mathcal{U}$ that are convex sets, but for the sake of simplicity in notation, we use the above-mentioned admissible sets.

We define the fault model as follows:

Definition 3 (Fault Model) A *fault model* for a (nominal) switched affine model $\mathcal{G} = (\mathcal{X}, \mathcal{E}, \mathcal{U}, \{G_i\}_{i=1}^m)$ is another switched affine model $\bar{\mathcal{G}} = (\bar{\mathcal{X}}, \bar{\mathcal{E}}, \bar{\mathcal{U}}, \{\bar{G}_i\}_{i=1}^{\bar{m}})$ with the same number of states, inputs, and outputs.

Further, to describe our framework of model invalidation and T-distinguishability (will be defined in Definition 5) for fault detection and isolation in the next section, we define system behavior as the following:

Definition 4 (Length-N Behavior) The *length-N behavior* associated with an SWA system $\mathcal{G}$ is the set of all length-N input–output trajectories compatible with

$\mathcal{G}$, given by the following set:

$$\mathcal{B}^N_{swa}(\mathcal{G}) := \big\{\{\mathbf{u}_t, \mathbf{y}_t\}_{t=0}^{N-1} \mid \mathbf{u}_t \in \mathcal{U} \text{ and } \exists \mathbf{x}_t \in \mathcal{X}, \sigma_t \in \mathbb{Z}_m^+, [\boldsymbol{\eta}_t^\top \;\; \boldsymbol{\nu}_t^\top]^\top \in \mathcal{E},$$
$$\text{for } t \in \mathbb{Z}_{N-1}^0 \text{ s.t. (9.3) holds}\big\}.$$

Moreover, with a slight abuse of terminology, we call $\mathcal{B}^N_{swa}(\mathcal{G})$ the *behavior* of the system $\mathcal{G}$ for conciseness when N is clear from the context.

9.3 Model Invalidation

First, we present the model invalidation problem for switched affine models and formulate a tractable feasibility problem to *(in)validate* models. This model invalidation framework is our main tool for fault detection and isolation in Sect. 9.5.

Given an input–output data sequence and a switched affine model, the model invalidation problem is to determine whether or not the data is compatible with the model. More formally, the model invalidation problem is as follows:

Problem 1 (Model Invalidation) Given an SWA model $\mathcal{G}$ and an input–output sequence $\{\mathbf{u}_t, \mathbf{y}_t\}_{t=0}^{N-1}$, determine whether or not the input–output sequence is contained in the behavior of $\mathcal{G}$, i.e., whether or not the following is true:

$$\{\mathbf{u}_t, \mathbf{y}_t\}_{t=0}^{N-1} \in \mathcal{B}^N_{swa}(\mathcal{G}). \tag{9.5}$$

With this definition, it is clear that if the model is invalidated by data, i.e., (9.5) does not hold, and the model is reliable; one can conclude that the data represents a fault in the system generating it. Hence, model invalidation can be used for fault detection and isolation.

Using Definition 4, model invalidation problem can be recast as a feasibility problem as follows:

$$\begin{aligned}
&\text{Find } \mathbf{x}_t, \boldsymbol{\eta}_t, \boldsymbol{\nu}_t, \sigma_t, \forall t \in \mathbb{Z}_{N-1}^0\\
&\text{s.t. } \sigma_t \in \mathbb{Z}_m^+,\\
&\quad \mathbf{x}_{t+1} = \mathbf{A}_{\sigma_t}\mathbf{x}_t + \mathbf{B}_{\sigma_t}\mathbf{u}_t + \mathbf{f}_{\sigma_t} + \boldsymbol{\nu}_t,\\
&\quad \mathbf{y}_t = \mathbf{C}_{\sigma_t}\mathbf{x}_t + \mathbf{D}_{\sigma_t}\mathbf{u}_t + \mathbf{g}_{\sigma_t} + \boldsymbol{\eta}_t,\\
&\quad P\mathbf{x}_t \le p, \; \|\boldsymbol{\nu}_t\| \le \varepsilon_\nu, \; \|\boldsymbol{\eta}_t\| \le \varepsilon_\eta.
\end{aligned} \tag{9.6}$$

This feasibility problem has a solution, if at every time the input–output sequence satisfies the dynamics of at least one mode. However, the optimization problem (9.6) is not stated in a form that can be readily solved due to system matrices' dependence on the mode signal σ, another variable. Next, we show that this dependence can be eliminated and we pose the model invalidation problem as a Mixed-Integer Linear Programming (MILP) problem:

Proposition 1 *Given an SWA model $\mathcal{G}$ and an input–output sequence $\{\mathbf{u}_t, \mathbf{y}_t\}_{t=0}^{N-1}$, the model is invalidated if and only if the following problem is infeasible.*

$$
\begin{aligned}
&\text{Find } \mathbf{x}_t, \boldsymbol{\eta}_t, \boldsymbol{\nu}_t, a_{i,t}, \mathbf{s}_{i,t}, \mathbf{r}_{i,t} \text{ for } \forall t \in \mathbb{Z}_{N-1}^0, \forall i \in \mathbb{Z}_m^+\\
&\text{s.t.} \forall j \in \mathbb{Z}_n^+, \forall k \in \mathbb{Z}_{n_y}^+, \forall t \in \mathbb{Z}_{N-1}^0, \text{ we have :}\\
&\mathbf{x}_{t+1} = \mathbf{A}_i \mathbf{x}_t + \mathbf{B}_i \mathbf{u}_t + \mathbf{f}_i + \boldsymbol{\nu}_t + \mathbf{s}_{i,t},\\
&\mathbf{y}_t = \mathbf{C}_i \mathbf{x}_t + \mathbf{D}_i \mathbf{u}_t + \mathbf{g}_i + \boldsymbol{\eta}_t + \mathbf{r}_{i,t},\\
&P\mathbf{x}_t \le p,\ a_{i,t} \in \{0,1\},\ \textstyle\sum_{i \in \mathbb{Z}_m^+} a_{i,t} = 1,\ \|\boldsymbol{\nu}_t\| \le \varepsilon_\nu,\\
&\|\boldsymbol{\eta}_t\| \le \varepsilon_\eta,\ (a_{i,t}, \mathbf{s}_{i,t}^j) : SOS-1,\ (a_{i,t}, \mathbf{r}_{i,t}^k) : SOS-1,
\end{aligned}
\tag{P$_{MI}$}
$$

where $\mathbf{s}_{i,t}$ and $\mathbf{r}_{i,t}$ are slack variables. We refer to this problem as $Feas(\{\mathbf{u}_t, \mathbf{y}_t\}_{t=0}^{N-1}, \mathcal{G})$.

Proof In order to prove the result, it suffices to show the equivalence of (9.6) and (P$_{MI}$), by illustrating that a feasible point of (9.6) is indeed a feasible point of (P$_{MI}$), and vice versa.

A Feasible Point of (9.6) **Is Feasible in** (P$_{MI}$) **($\Rightarrow$)**

Let $(\mathbf{x}_{0:N}^*, \boldsymbol{\eta}_{0:N-1}^*, \boldsymbol{\nu}_{0:N-1}^*, \sigma_{0:N-1}^*)$ be a feasible point of (9.6). As the admissible set for states and process and measurement noise are identical in (9.6) and (P$_{MI}$), we only focus on the rest of the constraints. Suppose that $a_{\sigma_t^*,t} = 1$ for some $t \in \mathbb{Z}_{N-1}^0$, then in order to satisfy $\sum_{i \in \mathbb{Z}_m^+} a_{i,t} = 1$, we have: $a_{i,t} = 0$ for all $i \neq \sigma_t^*$. Then, because of the SOS-1 constraints, this means that the variables $\mathbf{s}_{i,t}, \mathbf{r}_{i,t}$ are unconstrained/free for all $i \neq \sigma_t^*$. Since this holds for any $t \in \mathbb{Z}_{N-1}^0$, the state and output equation constraints in (P$_{MI}$) are trivially satisfied for $(\mathbf{x}_{0:N}^*, \boldsymbol{\eta}_{0:N-1}^*, \boldsymbol{\nu}_{0:N-1}^*, \mathbf{s}_{i,t}, \mathbf{r}_{i,t})$ for all $i \neq \sigma_t^*$ and for all $t \in \mathbb{Z}_{N-1}^0$. It remains to check if the state and output constraints in (P$_{MI}$) are feasible for σ_t^* for all $t \in \mathbb{Z}_{N-1}^0$. Clearly, from the feasibility of $(\mathbf{x}_{0:N}^*, \boldsymbol{\eta}_{0:N-1}^*, \boldsymbol{\nu}_{0:N-1}^*, \sigma_{0:N-1}^*)$, the state and output equations in (P$_{MI}$) are satisfied with $\mathbf{s}_{\sigma_t^*,t} = 0$, $\mathbf{r}_{\sigma_t^*,t} = 0$, which is enforced by the SOS-1 constraints for σ_t^*. This proves the forward direction of the equivalence.

A Feasible Point of (P$_{MI}$) **is Feasible in** (9.6) **($\Leftarrow$)**

Now, let $(\mathbf{x}_{0:N}^*, \boldsymbol{\eta}_{0:N-1}^*, \boldsymbol{\nu}_{0:N-1}^*, a_{1:m,0:N-1}^*, \mathbf{s}_{1:m,0:N-1}^*, \mathbf{r}_{1:m,0:N-1}^*)$ be a feasible point of (P$_{MI}$). As before, since the admissible sets for states and process and measurement noise are identical, we place our attention on the rest of the constraints. As a result of the feasibility of $a_{1:m,0:N-1}^*$, there exists a sequence $\sigma_t^*, t \in \mathbb{Z}_{N-1}^0$ such that $a_{\sigma_t^*,t}^* = 1$. For such a sequence, $\mathbf{s}_{\sigma_t^*,t}^{*j} = 0,\ \forall j \in \mathbb{Z}_n^+,\ \mathbf{r}_{\sigma_t^*,t}^{*k} = 0,\ \forall k \in \mathbb{Z}_{n_y}^+$, which results in the satisfaction of state and output constraints in (9.6) for the sequence $\sigma_t^*, t \in \mathbb{Z}_{N-1}^0$. Thus, we showed that there is a switching sequence corresponding to the feasible solution of (P$_{MI}$) that satisfies the state and output equations in (9.6) and therefore, the feasibility of (9.6).

Since we have shown that the feasibility of each problem implies the feasibility of the other, the proof is complete. □

Intuitively, the infeasibility of (P_{MI}) indicates that there are no state, input, and noise values that can generate input–output sequence from the model, and hence it is impossible that the data is generated by the model. Proposition 1 enables us to solve the model invalidation problem by checking the feasibility of (P_{MI}), which is a MILP with SOS-1 constraints that can be efficiently solved with many off-the-shelf optimization softwares, e.g., [26, 27].

9.4 *T*-distinguishability

Next, we introduce a property for a pair of models (system and/or fault models) called T-distinguishability,[2] which imposes that the trajectory generated from the two models cannot be identical for a time horizon of length T for any initial state and any noise signals. This notion is very similar to the concept of input-distinguishability, which is defined for linear time-invariant models in [23, 24]. T-distinguishability for a pair switched affine models is formally defined as follows:

Definition 5 (T-distinguishability) A pair of switched affine models $\mathcal{G}$ and $\bar{\mathcal{G}}$ is called *T-distinguishable* if $\mathcal{B}^T_{swa}(\mathcal{G}) \cap \mathcal{B}^T_{swa}(\bar{\mathcal{G}}) = \emptyset$, where T is a positive integer.

Thus, given two SWA models and an integer T, *the T-distinguishability problem* is to check whether the two models are T-distinguishable or not. This problem can be addressed using a Satisfiability Modulo Theory approach[16], or a MILP feasibility check [17, 22].

Problem 2 (T-distinguishability Problem) Given a pair of SWA models and an integer T, the T-distinguishability problem checks if the two models are T-distinguishable or not. More precisely, whether or not the following problem is feasible:

$$
\begin{aligned}
&\text{Find } \mathbf{x}, \bar{\mathbf{x}}, \mathbf{u}, \boldsymbol{\eta}, \bar{\boldsymbol{\eta}}, \boldsymbol{\nu}, \bar{\boldsymbol{\nu}} \\
&\text{s.t. } \forall t \in \mathbb{Z}^0_{T-1} : \exists i \in \mathbb{Z}^+_m, \exists j \in \mathbb{Z}^+_{\bar{m}} \text{ such that} \\
&\qquad \mathbf{x}_{t+1} = \mathbf{A}_i \mathbf{x}_t + \mathbf{B}_i \mathbf{u}_t + \mathbf{f}_i + \boldsymbol{\nu}_t, \\
&\qquad \bar{\mathbf{x}}_{t+1} = \bar{\mathbf{A}}_j \bar{\mathbf{x}}_t + \bar{\mathbf{B}}_j \mathbf{u}_t + \bar{\mathbf{f}}_j + \bar{\boldsymbol{\nu}}_t, \\
&\qquad \mathbf{P}\mathbf{x}_t \le \mathbf{p},\ \bar{\mathbf{P}}\bar{\mathbf{x}}_t \le \bar{\mathbf{p}}, \\
&\qquad \mathbf{C}_i \mathbf{x}_t + \mathbf{D}_i \mathbf{u}_t + \mathbf{g}_i + \boldsymbol{\eta}_t = \bar{\mathbf{C}}_j \bar{\mathbf{x}}_t + \bar{\mathbf{D}}_j \mathbf{u}_t + \bar{\mathbf{g}}_j + \bar{\boldsymbol{\eta}}_t, \\
&\qquad \|\boldsymbol{\eta}_t\| \le \varepsilon_\eta,\ \|\bar{\boldsymbol{\eta}}_t\| \le \varepsilon_{\bar{\eta}},\ \|\boldsymbol{\nu}_t\| \le \varepsilon_\nu,\ \|\bar{\boldsymbol{\nu}}_t\| \le \varepsilon_{\bar{\nu}},\ \|\mathbf{u}_t\| \le U.
\end{aligned}
\tag{9.7}
$$

[2] When the pair of models consists of the nominal system model and the fault model, this is also known as T-detectability [16, 17, 22], whereas when both models are fault models, this is also referred to as I-isolability [17].

If problem (9.7) is infeasible, the two models are T-distinguishable. Otherwise they are not T-distinguishable.

As we show next, for a given T, T-distinguishability can be verified by solving a MILP feasibility problem. Note that in the following T-distinguishability test, we have added a decision variable δ that will be used later to quantify the level of distinguishability, which can be computed with little additional computational cost.

Theorem 1 *A pair of switched affine models $\mathcal{G}$ and $\bar{\mathcal{G}}$ is T-distinguishable, if and only if the following problem is infeasible.*

$$
\begin{aligned}
\bar{\delta} = & \min_{\mathbf{x},\bar{\mathbf{x}},\mathbf{u},\boldsymbol{\eta},\bar{\boldsymbol{\eta}},\boldsymbol{\nu},\bar{\boldsymbol{\nu}},\mathbf{s},\bar{\mathbf{s}},\mathbf{r},a,\delta} \quad \delta \\
& s.t. \ \forall t \in \mathbb{Z}_{T-1}^0, \forall i \in \mathbb{Z}_m^+, \ \forall j \in \mathbb{Z}_{\bar{m}}^+, \forall k \in \mathbb{Z}_n^+, \forall l \in \mathbb{Z}_{n_y}^+, \\
& \quad \forall h \in \mathbb{Z}_{n_p}^+, \bar{h} \in \mathbb{Z}_{n_{\bar{p}}}^+, \\
& \quad \mathbf{x}_{t+1} = \mathbf{A}_i \mathbf{x}_t + \mathbf{B}_i \mathbf{u}_t + \mathbf{f}_i + \boldsymbol{\nu}_t + \mathbf{s}_{i,t}, \\
& \quad \bar{\mathbf{x}}_{t+1} = \bar{\mathbf{A}}_j \bar{\mathbf{x}}_t + \bar{\mathbf{B}}_j \mathbf{u}_t + \bar{\mathbf{f}}_j + \bar{\boldsymbol{\nu}}_t + \bar{\mathbf{s}}_{j,t}, \\
& \quad \mathbf{P}\mathbf{x}_t \le \mathbf{p}, \ \bar{\mathbf{P}}\bar{\mathbf{x}}_t \le \bar{\mathbf{p}}, \qquad\qquad (\mathrm{P}_T) \\
& \quad \mathbf{C}_i \mathbf{x}_t + \mathbf{D}_i \mathbf{u}_t + \mathbf{g}_i + \boldsymbol{\eta}_t = \bar{\mathbf{C}}_j \bar{\mathbf{x}}_t + \bar{\mathbf{D}}_j \mathbf{u}_t + \bar{\mathbf{g}}_j + \bar{\boldsymbol{\eta}}_t + \mathbf{r}_{i,j,t}, \\
& \quad a_{i,j,t} \in \{0,1\}, \ \textstyle\sum_{i\in\mathbb{Z}_m^+} \sum_{j\in\mathbb{Z}_{\bar{m}}^+} a_{i,j,t} = 1, \\
& \quad \|\boldsymbol{\eta}_t\| \le \varepsilon_\eta, \ \|\bar{\boldsymbol{\eta}}_t\| \le \varepsilon_{\bar{\eta}}, \ \|\boldsymbol{\nu}_t\| \le \varepsilon_\nu, \ \|\bar{\boldsymbol{\nu}}_t\| \le \varepsilon_{\bar{\nu}}, \ \|\mathbf{u}_t\| \le U, \\
& \quad (a_{i,j,t}, \mathbf{s}_{i,t}^k) : \mathrm{SOS}-1, \ (a_{i,j,t}, \bar{\mathbf{s}}_{j,t}^k) : SOS-1, \ (a_{i,j,t}, \mathbf{r}_{i,j,t}^l) : SOS-1, \\
& \quad \left\| \begin{bmatrix} \boldsymbol{\eta}_t \\ \boldsymbol{\nu}_t \end{bmatrix} - \begin{bmatrix} \bar{\boldsymbol{\eta}}_t \\ \bar{\boldsymbol{\nu}}_t \end{bmatrix} \right\| \le \delta.
\end{aligned}
$$

We refer to the above-mentioned problem as $Feas_T(\mathcal{G}, \bar{\mathcal{G}})$.

Proof The proof follows essentially the same reasoning as the proof of Proposition 1, i.e., by showing the feasibility of Problem (9.7) is equivalent to Problem (P_T), and is omitted for brevity. Note that the last constraint does not appear in (9.7). However, this constraint clearly does not change the feasible set, therefore the feasibility of (P_T) is necessary and sufficient for T-distinguishability.

The optimization formulation (P_T) in Theorem 1 enables us to solve the T-distinguishability problem, i.e., to determine if the pair of models are sufficiently different based on their length-T behaviors. If the pair of models is T-distinguishable, then one of the two models is guaranteed to be invalidated by the model invalidation approach, discussed in the previous section, by using only data from the most recent T time steps.

9.5 Fault Detection and Isolation

In this section, we propose a tractable fault detection and isolation scheme for the diagnosis of faults that satisfy the T-distinguishability property. The proposed fault diagnosis scheme can handle multiple fault scenarios, and can be implemented in real-time via model invalidation for a large class of applications.

9.5.1 *Fault Detection*

We are interested in developing a model-based tool for guaranteed fault detection, i.e., one that can conclusively decide if a fault has occurred or not, given the nominal system model and a set of potential fault models of interest for the system. To this end, we show that the model invalidation framework we introduced in Sect. 9.3 with the right assumptions can naturally serve this purpose.

Our proposed fault detection approach is based on determining if the measured input–output data over a horizon is compatible with the behavior of the nominal model $\mathcal{G}$, i.e., if the nominal model is valid or more precisely, not invalidated. Thus, our fault detection approach consists simply of checking for the feasibility of the model invalidation problem with the nominal model $\mathcal{G}$, and equivalently, by checking the feasibility of (P_{MI}) with the nominal model $\mathcal{G}$ (by Proposition 1).

If the nominal model is invalidated, i.e., (P_{MI}) is infeasible, then we know for certain that a fault has occurred. However, the feasibility of (P_{MI}) for the nominal system does not imply that a fault did not occur. This is because a fault model $\bar{\mathcal{G}}$ may also have a similar behavior of some given length as the nominal model $\mathcal{G}$. Thus, in order to achieve guaranteed fault detection even in this case, we take advantage of the assumed knowledge of the set of fault models.

With a given set of fault models, in order to conclusively establish if the fault has occurred or not, we need to find a long enough time horizon T such that the input–output trajectory generated from the fault models $\bar{\mathcal{G}}$ cannot be contained within the length-T behavior of the nominal system model $\mathcal{G}$. This coincides with the T-distinguishability property that we defined in the previous section. In other words, if we have a sufficiently long time horizon that guarantees T-distinguishability for the system model and all the faults, we can implement model invalidation in a receding horizon manner with this time horizon of size T with guarantees of fault detection or that no fault has occurred. This is formalized in the next section, where the multiple faults scenario is considered.

9.5.1.1 Multiple Faults Scenario

We consider the scenario when there are multiple possible faults, each described by a different fault model. It is easy to verify that in order to conclusively determine if any fault has occurred or not for this scenario, all pairs of nominal system and fault models need to be T-distinguishable. Thus, for each of the N_f fault models $\bar{\mathcal{G}}_j$, $j \in \mathbb{Z}^+_{N_f}$, we assume the following:

Assumption 1 (Detectability Assumption) *We assume that for all $j \in \mathbb{Z}^+_{N_f}$, there exists a finite T_j such that the pair of nominal system $\mathcal{G}$ and the fault model $\bar{\mathcal{G}}_j$ is T_j-distinguishable. In addition, we assume that the faults are persistent, i.e., once they occur, the system continues to evolve according to the fault dynamics.*

Then, for guaranteed detection of all possible fault models, the following condition is necessary and sufficient:

Proposition 2 (T-detectability for Multiple Faults) *Consider N_f fault models that satisfy Assumption 1. Then, the existence of a fault can be detected in at most $T = \max_j T_j$ steps after the occurrence of a fault. We refer to such a set of faults as T-detectable.*

Proof Under Assumption 1, and since $T \geq T_j$, all pairs of nominal system model and faults for all $j \in \mathbb{Z}^+_{N_f}$ are T-distinguishable. This means that if any of the faults occurs persistently, it will be detected by observing at most T samples from the time of occurrence.

Now, if all pairs of system and fault models are T-distinguishable, then the previously discussed model invalidation based fault detection approach guarantees the determination of the occurrence or non-occurrence of any fault in this set. In brief, we check for the feasibility of the nominal system model, i.e., if the problem (P_{MI}) with the nominal model is infeasible, then a fault is detected. Otherwise, if (P_{MI}) is feasible, then we know for certain that no fault has occurred. Note, however, that the detection of faults is not sufficient for uniquely determining which fault has occurred, which is the subject of the next section.

9.5.2 Fault Isolation

In addition to fault detection, i.e., to determine if a fault has occurred, it is also important and of interest in many applications to uniquely determine which specific fault has occurred, i.e., to isolate the source of faults. The ability to do this can in turn save a significant amount of effort in accommodating the isolated fault. Hence, we develop a model-based fault isolation approach in this section.

In particular, given the nominal system model and a set of potential fault models of interest for the system, we wish to determine which fault model is validated or rather, not invalidated, based on the measured input–output data. Thus, similar to fault detection, our proposed fault isolation approach consists simply of checking for the feasibility of the model invalidation problem with each fault model $\bar{\mathcal{G}}$, and equivalently, by checking the feasibility of (P_{MI}) with each of the fault models $\bar{\mathcal{G}}$ (by Proposition 1).

If all but one fault model is invalidated (i.e., (P_{MI}) is infeasible for all fault models except for one), then we definitively know that particular fault has occurred. Therefore, in order to guarantee the isolation of a fault after it has occurred, we need to find a sufficiently long time horizon T such that the input–output trajectory generated from all the fault models cannot be contained within the length-T behavior of each other. With this sufficiently long time horizon T, we can then implement model invalidation in a receding horizon manner with this time horizon of size T for all fault models (can be executed in parallel) with guarantees of fault isolation. Note that this fault isolation approach is similar to our receding horizon fault detection approach but with the fault models in place of the nominal system model. Next, we formalize the notion of sufficiently long time horizon based on the previously introduced property of T-distinguishability for the multiple faults scenario.

9.5.2.1 Multiple Faults Scenario

Let us consider N_f fault models $\bar{\mathcal{G}}_j$, $j \in \mathbb{Z}^+_{N_f}$ that may occur for system model $\mathcal{G}$. In order to isolate a fault, e.g., identify which of the faults has occurred, it is straightforward to verify that all pairs of fault models need to be T-distinguishable, i.e., the following assumption is necessary:

Assumption 2 (Isolability Assumption) *We assume that for all $m, n \in \mathbb{Z}^+_{N_f}$, $m \neq n$, there exists a finite $I_{m,n}$ such that $\bar{\mathcal{G}}_m$ and $\bar{\mathcal{G}}_n$ are $I_{m,n}$-distinguishable.*

Remark 2 This isolability assumption and the detectability assumption in Assumption 1 are indeed relatively strong assumptions. However, they are necessary and sufficient for providing guarantees for passive fault detection and isolation approaches with a receding horizon. In fact, these "strong" assumptions are the reason that we have also considered active fault diagnosis methods [28, 29], which, at the cost of perturbing the desired input to the system, make fault diagnosis possible for a wider class of faults. Fortunately, as we show with examples in Sect. 9.7, these assumptions hold for many parametric fault scenarios in real-world applications.

Next, we provide a necessary and sufficient condition for guaranteed fault isolation when there are multiple faults:

Proposition 3 (I-isolability for Multiple Faults) *Consider N_f fault models that satisfy Assumption 2. If a fault occurs, it can be isolated in at most $I = \max_{m,n,\, m\neq n} I_{m,n}$ steps after the occurrence. We refer to such a set of faults as I-isolable.*

Proof Under Assumption 2, and because $I \geq I_{m,n}$ for all possible pairs of fault models, all pairs of faults are I-distinguishable. Therefore, if any of the faults occur persistently, by observing at most I samples, it will be isolated. This is because the length-I behavior of the occurred fault does not have any intersection with the length-I behavior of any of the other faults.

9.5.3 FDI Scheme

In this section, we combine the results from previous subsections to obtain a fault detection and isolation (FDI) scheme, which consists of two steps:

1. *Off-line step*: In the off-line step, under Assumptions 1 and 2, we calculate the following quantities:

$$\text{Isolability index}: I = \max_{m,n} I_{m,n},\ m,n \in \mathbb{Z}^+_{N_f},\ m \neq n;$$

$$\text{Isolability index for fault i}: \tilde{I}_i = \max_{j \in \mathbb{Z}^+_{N_f},\, j \neq i} I_{i,j};$$

$$\text{Detectability index}: T = \max_{j \in \mathbb{Z}^+_{N_f}} T_j;$$

$$\text{Length of memory}: K = \max\{T, I\}.$$

2. *On-line step*: In this step, we leverage $N_f + 1$ *parallel* monitors corresponding to system and fault models. The monitors are labeled as $\{\mathcal{M}_0, \mathcal{M}_1, \ldots, \mathcal{M}_{N_f}\}$, where $\mathcal{M}_0$ corresponds to the system model and $\mathcal{M}_i$ corresponds to the i-th fault model. First, only $\mathcal{M}_0$ is active for fault detection. The rest of the monitors stay "off" until a fault is detected by $\mathcal{M}_0$. The inputs to each monitor at time t are the input–output sequence of length $K_i = \max\{\tilde{I}_i, T_i\}$, $\{\mathbf{u}_k, \mathbf{y}_k\}_{k=t-K_i+1}^t$, and the corresponding model $\bar{\mathcal{G}}_i$. For instance, $\mathcal{M}_0$ knows $\mathcal{G}$, and at each time step, it solves the model invalidation problem, *Feas*$(\{\mathbf{u}_k, \mathbf{y}_k\}_{k=t-T+1}^t, \mathcal{G})$. If the problem is feasible, the monitor outputs 0, otherwise it outputs 1. In the latter case, the bank of fault monitors is activated and *parallelly* solves the model invalidation problems for all fault models, i.e., to check if $\mathcal{M}_j$ solves *Feas*$(\{\mathbf{u}_k, \mathbf{y}_k\}_{k=t-K_j+1}^t, \bar{\mathcal{G}}_j)$ for each $j \in \mathbb{Z}^+_{N_f}$. By Assumptions 1 and 2, it is guaranteed that in this case, the problem of at most one monitor is feasible. The output block receives the signal from all the monitors and shows two elements. The first element is 1 or 0, which indicates that a fault has occurred or not, respectively. The second element is $k_f \in \mathbb{Z}^+_{N_f}$ if the fault matches k_f-th fault model or 0 if the fault does not match any of the fault models.

Such an FDI scheme is illustrated in Fig. 9.1. As we can see, at every time step t, this FDI scheme acts as a function:

$$[\mathcal{H}, \mathcal{F}] = \psi(\{\mathbf{u}_k, \mathbf{y}_k\}_{k=t-K+1}^t, \mathcal{G}, \{\bar{\mathcal{G}}_j\}_{j=1}^{N_f}), \tag{9.8}$$

where $\mathcal{H}$ is either 0 or 1 to indicate healthy or faulty behaviors, and $\mathcal{F}$ either indicates the fault model that is active or claims that none of the fault models matches the faulty behavior.

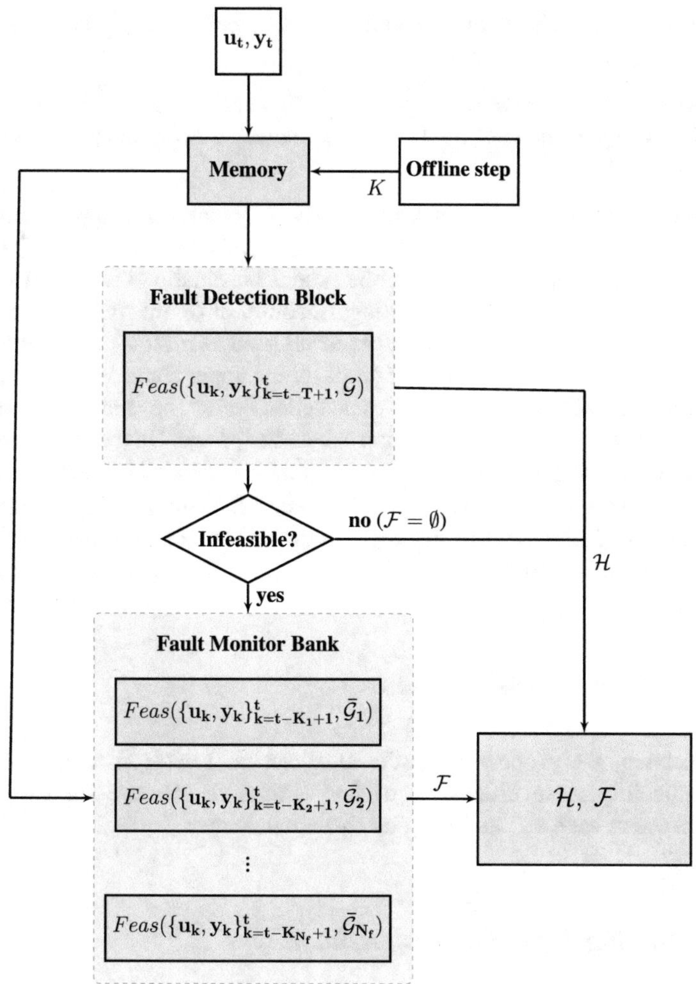

Fig. 9.1 Flowchart of the proposed FDI scheme

9.5.4 Detection and Isolation Delays

Next, we describe the notion of delays in fault detection and isolation, and provide theoretical bounds on these delays using the detectability and isolability indices.

Definition 6 (Detection/Isolation Delay) Detection/isolation delay is the number of time samples it takes from the occurrence of the fault to its detection/isolation. We denote detection and isolation delays with τ_T and τ_I, respectively.

Using the above definition and given a T_i-distinguishable pair of system and fault model $(\mathcal{G}, \bar{\mathcal{G}}_i)$, the detection delay of the proposed fault detection scheme is then bounded by T_i. In addition, the isolation delay of a pair of $I_{i,j}$-distinguishable fault models $(\bar{\mathcal{G}}_i, \bar{\mathcal{G}}_j)$ is bounded by $I_{i,j}$.

Theorem 2 *The detection delay for fault $\bar{\mathcal{G}}_i$ using FDI scheme proposed in Sect. 9.5.3 is upper-bounded by T_i, and the isolation delay is upper-bounded by $K_i = \max\{\tilde{I}_i, T_i\}$.*

Proof Suppose fault i occurs at time t^*. The FDI approach implements model invalidation with a time horizon size of $T \geq T_i$. At the time $t^* + T_i - 1$, the input–output trajectory that is fed to the model invalidation contains a length T_i trajectory that is in $\mathcal{B}^{T_i}(\bar{\mathcal{G}}_i)$. By T_i-distinguishability of $\bar{\mathcal{G}}_i$, this trajectory cannot be generated by $\mathcal{G}$. Therefore, the model is invalidated at most by observing T_i data points from fault i. This concludes the proof for the upper bound on detection delay. For isolation, the FDI approach requires detection before the isolation monitors are activated, and in the worst case detection occurs in T_i steps. On the other hand, if we observe any trajectory from t^* to $t^* + \tilde{I}_i - 1$ that is generated by fault i, it is not in $\mathcal{B}^{\tilde{I}_i}(\bar{\mathcal{G}}_j), j \neq i$. This is because $\tilde{I}_i \geq I_{i,j}, j \neq i$. Hence, the fault is isolated with at most $\tilde{I}_i$ observations of the fault. Considering that the fault needs to be detected first, the isolation delay is upper-bounded by $K_i = \max\{\tilde{I}_i, T_i\}$. This concludes the proof.

9.6 Practical Considerations

In this section, we propose heuristics that can be leveraged to find T for T-distinguishability more efficiently, to find a reliable measure for existence or non-existence of such a T, and to reduce isolation delays.

9.6.1 Finding T for T-distinguishability

The following issues are important in practice: (1) If there is a finite T, how can one search for the smallest such T? (2) What happens if the systems are not T-distinguishable? How can the non-existence of a finite T be verified and what can be done in terms of fault detection in this case?

Consider two switched affine models. If the two models are not T-distinguishable for a given T, i.e., the solution to (P_T) is feasible, the optimization formulation additionally outputs the value $\bar{\delta}$, which we argue is a good indication and measure for the separability of two models. In essence, $\bar{\delta}$ can be interpreted as the noise effort that is required to make the trajectories of the two models identical. A larger value for $\bar{\delta}$ indicates a larger separation between the two models that the noise has to compensate for. Hence, we refer to the normalized version of $\bar{\delta}$ as the *distinguishability index*, given by

$$\delta^* = \frac{\bar{\delta}}{\delta^{\max}}, \tag{9.9}$$

where $\delta^{\max} \doteq \min\{\max\{\varepsilon_\eta + \varepsilon_{\bar{\eta}}, \varepsilon_\nu + \varepsilon_{\bar{\nu}}\}, \max\{\varepsilon_\eta, \varepsilon_\nu\} + \max\{\varepsilon_{\bar{\eta}}, \varepsilon_{\bar{\nu}}\}\}$ is an upper bound on $\bar{\delta}$; hence, $0 \leq \delta^* \leq 1$.

First, it is noteworthy that if the pair of models is T-distinguishable, then the pair is also necessarily T^+-distinguishable for any $T^+ \geq T$. So once the problem (P_T) is infeasible for some T, it will remain infeasible for larger values of T. This suggests that a smallest time horizon T may exist, which is very useful in terms of computational complexity because the number of variables and constraints for the formulation in (P_{MI}) increases with the size of the time horizon. Hence, the complexity of the solutions to the model invalidation problem grows with the length of the input–output data sequence. To find the smallest T for which we have T-distinguishability, one could use binary search starting from $T = 1$ until the smallest T is obtained that makes the T-distinguishability problem in Theorem 1 infeasible. The upper increments of the line search can also be guided by the value of the *distinguishability index* δ^*, to make larger increments in T when the problem is feasible and δ^* is small for the current T.

On the other hand, not all pairs of models are T-distinguishable. Clearly, if the models are identical, then no finite T exists, hence a finite T for T-distinguishability can only be obtained for certain pairs of models. System theoretic conditions under which a T exists are the subject of current research [22]. Nonetheless, we would like to suggest that the trend of δ^* with increasing T can be used as a heuristic to determine when the iterations with increasing T can be terminated with some confidence that a finite T does not exist. In particular, we propose to terminate the iterations when the trend of δ^* with increasing T reaches a plateau. This is demonstrated to be effective in a simulation example in Sect. 9.7.3. In addition, when this index does reach a (non-zero) plateau and the problem remains not T-distinguishable, then this $\delta^*_{plateau}$ is also a useful parameter for the pair of models, which can be interpreted as the maximum allowed uncertainty beyond which the behaviors of the pair of models are not distinguishable. This may suggest possible design remedies involving the choice of sensors with better precision or the employment of noise isolation platforms to reduce the amount of noise, in order to facilitate fault detection and isolation.

In some practical examples, detectability and isolability of all faults cannot be achieved by design, i.e., Assumptions 1 and 2 are not satisfied. In these cases, one can still pick an arbitrary T for which online computation is tractable and apply the FDI approach. Any infeasibility of the model invalidation problem for the nominal model is still a certificate that the system has deviated from the nominal behavior and a fault occurred, however some of the fault can remain undetected and unisolated among the given fault models. Moreover, the FDI approach can be simply modified such that $\mathcal{F}$ outputs either the set of faults that matches the data (because some fault models may not be isolable) along with their corresponding "likelihoods" in terms of their distinguishability indices or the empty set if none of the models matches the data. Another aspect that is worth mentioning is the case of non-persistent, and in particular, cascading faults. The proposed modeling framework can easily handle such cascading faults if one defines a switched system model for the fault cascade with possibly additional constraints on the switching signal as detailed in [22].

9.6.2 Adaptive Fault Isolation

The bound on isolation delays represents the worst case scenario, where the data created by a fault model falls within the behavior of some other models up until the very last time step. However, the worst case scenario is rarely encountered in practice, where the faults can be isolated much prior to this bound. Here, in this section, we propose an adaptive fault isolation scheme that may reduce isolation delay, which is based on the idea of validation of only one of the fault models. Since the data prior to the time of detection is likely to invalidate all the fault models (in fact, this is guaranteed before the occurrence of a fault), we propose to reduce isolation delays by using an adaptive receding horizon that considers only the data starting from the detection time (fixed horizon lower bound) with increasing horizon until only one fault model matches or validates the data. In practice, we can achieve this by considering model invalidation for each of the fault models with the adaptive receding horizon until only one fault model remains that matches the data.

Since we assumed that the fault is among the predefined set of models and is persistent, it is guaranteed that the fault will be isolated with this approach. Such an approach has the potential to significantly reduce isolation delays, as we have observed in simulation in Sect. 9.7.2 (cf. Fig. 9.4 (bottom row)).

9.7 Illustrative Examples

First, we demonstrate in Sect. 9.7.1 that our new formulations for model invalidation and T-distinguishability in Proposition 1 and Theorem 1, respectively, are computationally superior to the previous formulation in [16, 22]. Then, we illustrate the performance of the proposed FDI scheme using a numerical model for the Heating, Ventilating, and Air Conditioning (HVAC) system that is proposed in [30] in Sect. 9.7.2. Moreover, we provide a numerical example in Sect. 9.7.3 to illustrate the practical merits of the distinguishability index that was introduced in Sect. 9.4. All the simulations in this section are implemented on a 3.5 GHz machine with 32 GB of memory that runs Ubuntu. For the implementation of the MILP feasibility check problems, we utilized YALMIP [31] and Gurobi [26]. All the approaches and examples are implemented in MATLAB.

9.7.1 Run-Time Comparison

In this section, we compare the run-time for the formulations proposed in this paper with the one in [22]. Consider a hidden-mode switched affine model, $\mathcal{G}$, with admissible sets $\mathcal{X} = \{\mathbf{x} \mid \|\mathbf{x}\| \leq 11\}$, $\mathcal{U} = \{\mathbf{u} \mid \|\mathbf{u}\| \leq 1000\}$ and $\mathcal{E} = \{\eta \mid \|\eta\| \leq 0.1\}$. We assume there is no process noise. We also assume $B = [1 \;\; 0 \;\; 1]^\intercal$ and $C = [1 \;\; 1 \;\; 1]$ for all modes. The system matrices of the modes are:

$$\mathbf{A}_1 = \begin{bmatrix} 0.5 & 0.5 & 0.5 \\ 0.1 & -0.2 & 0.5 \\ -0.4 & 0.6 & 0.2 \end{bmatrix},\ \mathbf{f}_1 = \begin{bmatrix} 1 \\ 0 \\ 0 \end{bmatrix},\ \mathbf{A}_2 = \begin{bmatrix} 0.5 & 0.5 & 0.5 \\ -0.3 & -0.2 & 0.3 \\ 0.1 & -0.3 & -0.5 \end{bmatrix},\ \mathbf{f}_2 = \begin{bmatrix} 0 \\ 1 \\ 0 \end{bmatrix},$$

$$\mathbf{A}_3 = \begin{bmatrix} 0.5 & 0.2 & 0.6 \\ 0.2 & -0.2 & 0.2 \\ -0.9 & 0.7 & 0.1 \end{bmatrix},\ \mathbf{f}_3 = \begin{bmatrix} 0 \\ 0 \\ 1 \end{bmatrix}.$$

In addition, consider a fault model, $\mathcal{G}^f$, with:

$$\mathbf{A}^f = \begin{bmatrix} 0.8 & 0.7 & 0.6 \\ 0.1 & -0.2 & 0.3 \\ -0.4 & 0.3 & -0.2 \end{bmatrix},\ \mathbf{B}^f = \begin{bmatrix} 1 \\ 0 \\ 0 \end{bmatrix},\ \mathbf{f}^f = \begin{bmatrix} 1 \\ 1 \\ 1 \end{bmatrix}.$$

The implementation of the T-distinguishability approach proves that the system and fault model pairs is 12-distinguishable. We first randomly generate input–output trajectories (5 for each time horizon length) from $\mathcal{G}^f$. We then compare the model invalidation approaches that use the proposed formulation in Proposition 1 and the one in [16, 22]. The average run-time for each time horizon length as well as the standard deviation of run-times for both formulations are illustrated in Fig. 9.2. Clearly, the results indicate the superiority of the proposed formulation to the one in [16, 22]. Similar improvements were also observed for the proposed T-distinguishability formulation in Theorem 1 when compared to [16, 22] (plots are omitted for brevity).

9.7.2 Fault Diagnosis in HVAC Systems

In [30], a single-zone HVAC system in cooling mode (cf. schematic in Fig. 9.3) is considered. This HVAC system is represented by a non-linear model as follows:

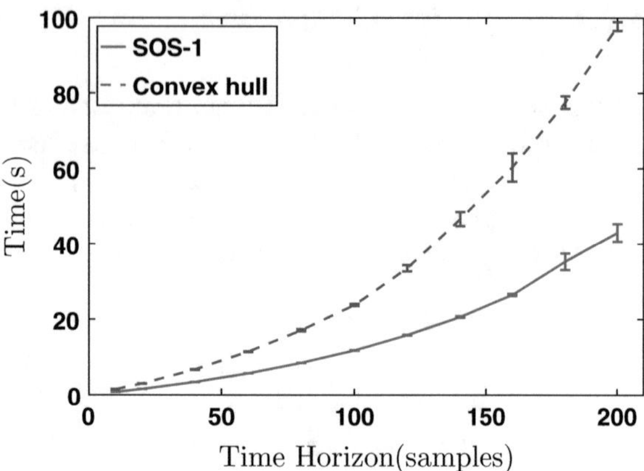

Fig. 9.2 Average execution time (with standard deviations) for invalidating data generated by $\mathcal{G}^f$ with various time horizons

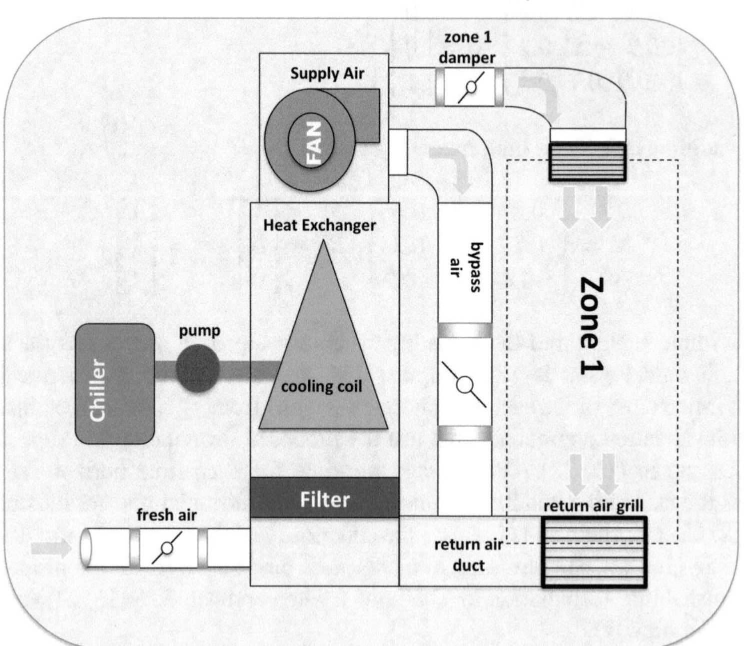

Fig. 9.3 Schematic of a single-zone HVAC system

$$
\begin{pmatrix} \dot{T}_{TS} \\ \dot{W}_{TS} \\ \dot{T}_{SA} \end{pmatrix} = \begin{pmatrix} -\frac{f}{V_s} & \frac{h_{fg}f}{C_pV_s} & \frac{f}{V_s} \\ 0 & -\frac{f}{V_s} & 0 \\ 0.75\frac{f}{V_{he}} & -0.75\frac{fh_w}{C_pV_{he}} & -\frac{f}{V_{he}} \end{pmatrix} \begin{pmatrix} T_{TS} \\ W_{TS} \\ T_{SA} \end{pmatrix}
+ \begin{pmatrix} -\frac{h_{fg}f}{C_pV_s}W_s + \frac{4}{C_pV_s}(Q_o - h_{fg}M_o) \\ \frac{f}{V_s}W_s + \frac{M_o}{\rho V_s} \\ \frac{f}{4V_{he}}(T_o - \frac{h_w}{C_p}W_o) + \frac{fh_w}{C_pV_{he}}W_s - 6000\frac{gpm}{\rho C_pV_{he}} \end{pmatrix}, \tag{9.10}
$$

Table 9.1 Parameters of the model

Parameter	Description	Value
h_w	Enthalpy of liquid water	180 (Btu/lb)
h_{fg}	Enthalpy of water vapor	1078.25 (Btu/lb)
W_o	Humidity ratio of outdoor air	0.018 (lb/lb)
W_s	Humidity ratio of supply air	0.007 (lb/lb)
W_{TS}	Humidity ratio of thermal space	State variable
C_p	Specific heat of air	0.24 (Btu/lb.°F)
T_o	Temperature of outdoor air	85 (°F)
T_{SA}	Temperature of supply air	State variable (°F)
T_{TS}	Temperature of thermal space	State variable (°F)
V_s	Volume of thermal space	58,464 (ft^3)
V_{he}	Volume of heat exchange space	60.75 (ft^3)
M_o	Moisture load	[150 180] (lb/h)
Q_o	Sensible heat load	[289, 800 289, 950] (Btu/h)
ρ	Air mass density	0.074 (lb/ft^3)
f	Volumetric flow rate of air	17,000 (ft^3/min)
gpm	Flow rate of chilled water	$\{0, 58\}$ (gal/min)

where f, gpm, M_o and Q_o are time varying parameters. The parameters of the model are defined in Table 9.1.

We leverage an augmented state-space model with additional states Q_0 and M_0 that is obtained in [30]. To further simplify the model, we assume that the fan is always turned on and the flow rate is fixed at 17,000 ft^3/min and the chiller pump is either "off" or "on" with a fixed flow rate of 58 gal/min. These assumptions along with a discretization with a sampling time of 5 min convert the nonlinear system (9.10) to a switched affine model parameterized by

$$A_1 = A_2 = \begin{pmatrix} 0.98 & 229.63 & 0.001 & 0 & -0.0035 \\ 0 & 0.94 & 0 & 0 & 0 \\ 0.74 & -360.61 & 0.0008 & 0 & -0.0030 \\ 0 & 0 & 0 & 1 & 0 \\ 0 & 0 & 0 & 0 & 1 \end{pmatrix}, \; f_2 = \begin{pmatrix} 0.3886 \\ 0.0001 \\ -22.576 \\ 0 \\ 0 \end{pmatrix},$$

$$C_1 = C_2 = \begin{pmatrix} 1 & 0 & 0 & 0 & 0 \\ 0 & 1 & 0 & 0 & 0 \end{pmatrix}, \; f_1 = \mathbf{0},$$

where the system evolves according to the following continuous dynamics in mode i:

$$\begin{aligned} \mathbf{x}_{t+1} &= A_i \mathbf{x}_t + f_i + \mathbf{v}_t, \\ \mathbf{y}_t &= C_i(\mathbf{x}_t + \mathbf{x}_e) + \boldsymbol{\eta}_t, \; \mathbf{x}_e = \begin{bmatrix} 71 \; 0.0092 \; 55 \; 289,897.52 \; 166.06 \end{bmatrix}^\mathsf{T}. \end{aligned} \tag{9.11}$$

Table 9.2 Detectability and isolability indices

$T_1 = 4$	$T_2 = 16$	$T_3 = 8$	$I_{1,2} = 4$	$I_{1,3} = 4$	$I_{2,3} = 16$

The states in the SWA model represent the deviation of T_{TS}, W_{TS}, T_{SA}, Q_0, and M_0 from their equilibria, $\mathbf{x}_e$. In addition, the HVAC model is represented by $\mathcal{G}_H = (\mathcal{X}, \mathcal{E}, \mathcal{U}, \{G_i\}_{i=1}^2)$, where $\mathcal{X} = \{\mathbf{x} \mid [-100 \;\; -0.05 \;\; -50 \;\; -75 \;\; -15]^\intercal \leq \mathbf{x} \leq [100 \;\; 0.05 \;\; 50 \;\; 75 \;\; 15]^\intercal\}$, $\mathcal{E} = \{\boldsymbol{\eta} \mid |\boldsymbol{\eta}| \leq [0.2 \;\; 0.002]^\intercal\}$ and $\mathcal{U} = \emptyset$. The last two bounds on the states are for the augmented states, which are assumed to stay within a small range of their equilibria. The first mode corresponds to chiller being "on" and the second mode represents the model when it is "off." The controller keeps the temperature in the comfort zone of 65–75°F by turning the chiller on and off. Control signals are not observed by the FDI scheme.

We consider three fault models[3]:

1. Faulty fan: The fan rotates at half of its nominal speed.
2. Faulty chiller water pump: The pump is stuck and spins at half of its nominal speed.
3. Faulty humidity sensor: The humidity measurements are biased by an amount of +0.005.

By implementing the proposed approach on these fault models, we can calculate the T-distinguishability and I-isolability indices, which are given in Table 9.2.

Next, we consider three fault scenarios, where for each scenario i ($i \in \{1, 2, 3\}$), we generate data from the nominal system for 4 h and from fault i afterwards. The times at which the faults occur and their detection times, as well as the upper bounds on isolation delays are indicated in Fig. 9.4 (top and middle rows), which show the output trajectories for each scenario. Furthermore, we plot in Fig. 9.4 (bottom row) the detection and isolation signals for all three faults to show that only the occurred fault is isolated in all scenarios before their upper bounds are exceeded, and that the proposed adaptive isolation scheme reduces the isolation delay, as desired.

Moreover, to illustrate the practical use of the distinguishability index, δ^*, in Fig. 9.5, we plotted the growth trend of the distinguishability index δ^* as the time horizon increases for T-distinguishability of fault 3 and I-isolability of faults 2 and 3. The plot shows that the distinguishability index we introduced does indeed deliver a nice measure of how far two models are from detectability or isolability, and at the same time, it allows us to estimate the size of time horizon, T or I, to achieve T-distinguishability or I-Isolability.

[3]These faults can also be consequences of cyber or physical attacks. For instance, the bias in the humidity sensor can be a result of a false data injection attack (a common form of cyberattack).

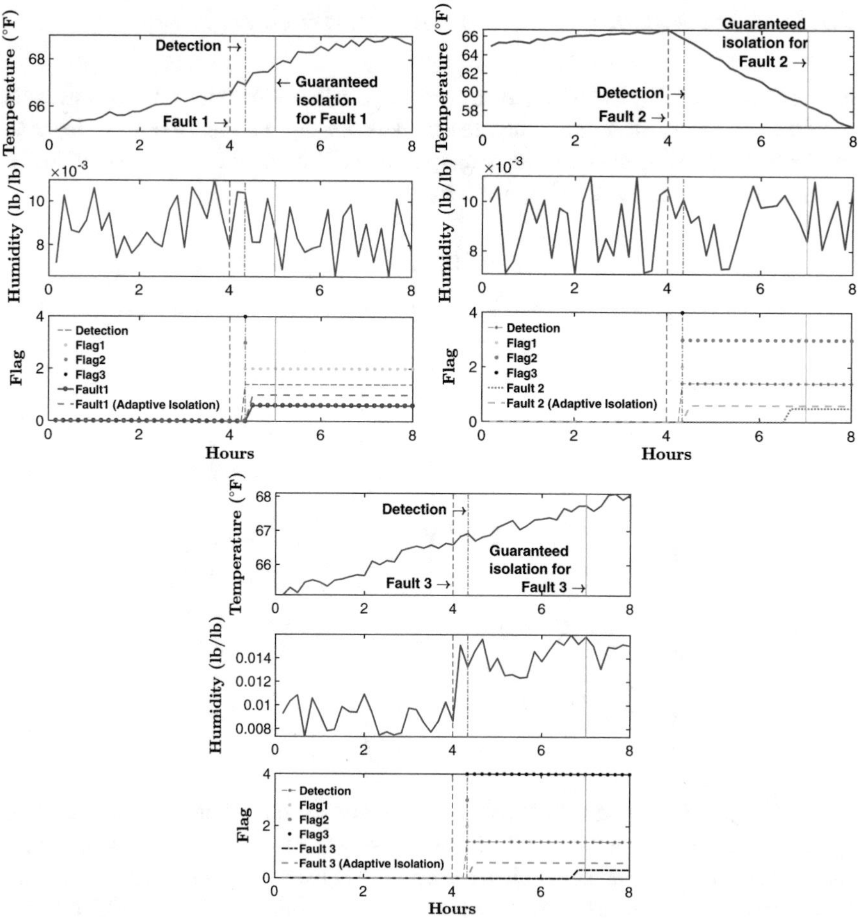

Fig. 9.4 The outputs (top two rows) of three fault scenarios; Detection, isolation, and adaptive isolation signals for all faults (bottom row). Flag i is non-zero when the model invalidation problem associated with fault i using the adaptive horizon length is validated. Adaptive isolation occurs when only one Flag is non-zero

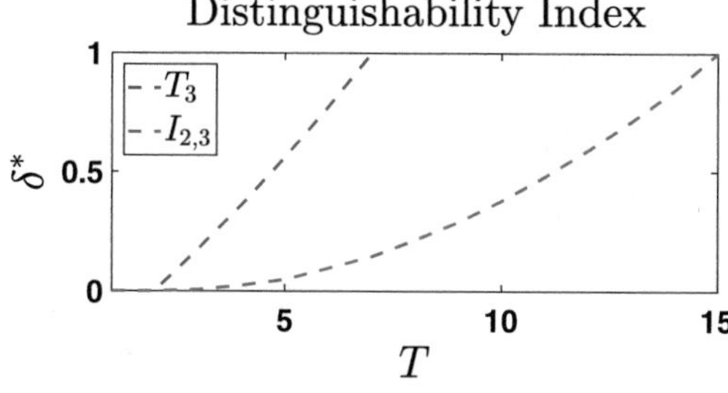

Fig. 9.5 Increase in the detectability index for fault 3, T_3, and the isolability index for faults 2 and 3, $I_{2,3}$, in the HVAC example

9.7.3 Distinguishability Index and System Uncertainty

The distinguishability index, δ^*, does not always achieve the final value of 1 as in the previous example, especially when the models are not distinguishable or isolable. To demonstrate this, we consider two synthetic SWA models $\mathcal{G}$ and $\bar{\mathcal{G}}$ subject to measurement and process noise, given by

$$\mathcal{G}:\begin{cases} A_1 = \begin{pmatrix} 0.1 & 0 & 0.1 \\ 0 & 0.1 & 0.2 \\ 0.2 & 0.12 & 0 \end{pmatrix}, \; A_2 = \begin{pmatrix} 0 & 0 & 0.15 \\ 0.1 & 0 & 0 \\ 0.1 & 0.12 & 0.1 \end{pmatrix}, \\ C_1 = C_2 = I, \, f_1 = \begin{pmatrix} 0.5 \\ 0.2 \\ 1 \end{pmatrix}, \, f_2 = \begin{pmatrix} 1 \\ 0 \\ 0.5 \end{pmatrix}, \end{cases}$$
$$\bar{\mathcal{G}}:\begin{cases} \bar{A}_1 = \begin{pmatrix} 0.1 & 0 & 0.1 \\ 0 & 0.1 & 0.2 \\ 0.2 & 0.1 & 0 \end{pmatrix}, \; \bar{A}_2 = \begin{pmatrix} 0 & 0 & 0.1 \\ 0.1 & 0 & 0 \\ 0.1 & 0.1 & 0.1 \end{pmatrix}, \\ \bar{C}_1 = \bar{C}_2 = I, \, \bar{f}_1 = \begin{pmatrix} 0.3 \\ 0 \\ 0.9 \end{pmatrix}, \, \bar{f}_2 = \begin{pmatrix} 0.8 \\ 0.2 \\ 0.3 \end{pmatrix}, \end{cases} \tag{9.12}$$

where the rest of the parameters are zero. The bounds on the process and measurement noise are set to be 0.2 and 0.25, respectively. Figure 9.6 depicts the change of the distinguishability index with increasing T. We observe that the distinguishability index increases nonlinearly and reaches a plateau at a value of $\delta^*_{plateau} = 0.7581 < 1$. In this case, the distinguishability index δ^* provides a practical indication that these two models are very unlikely to be isolable for any finite I.

In addition, for the process and measurement noise with bounds 0.2 and 0.25, we can compute $\delta^{\max} = 0.5$; hence, we can correspondingly calculate $\bar{\delta}_{plateau} = 0.379$

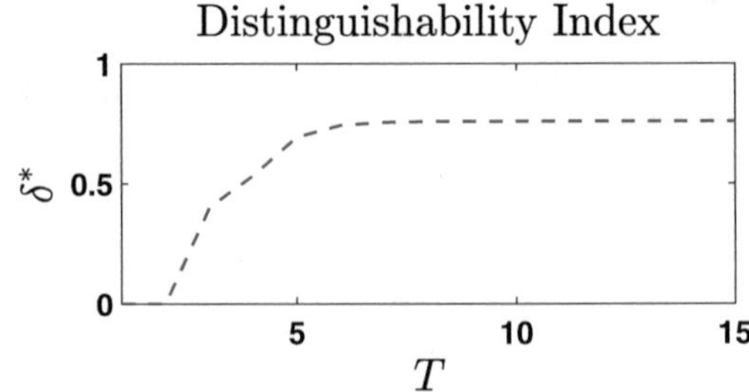

Fig. 9.6 Nonlinear increase of the distinguishability index with a plateau at around $T = 5$, for the numerical example described by (9.12)

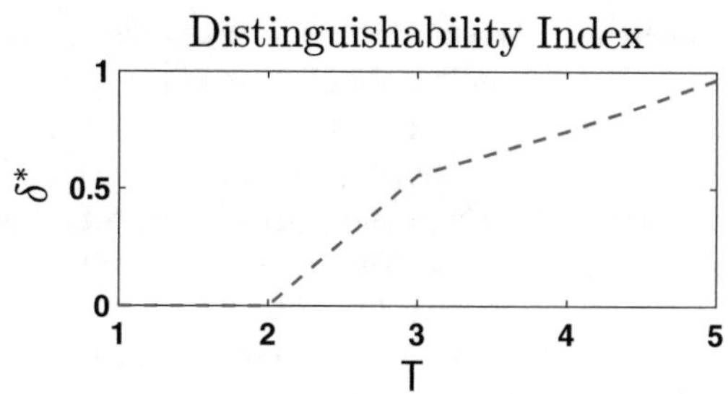

Fig. 9.7 Distinguishability index with redesigned system (9.12) with noise bounds at 0.18

using $\delta^*_{plateau} = 0.7581$ and (9.9). Note that $\bar{\delta}_{plateau}$ represents the uncertainty effort to make the outputs of the two models identical. If Problem (P_T) is feasible with a minimum of $\bar{\delta}_{plateau}$ for some given process and measurement noise bounds, then the distinguishability index plot will have a plateau as shown in Fig. 9.6. In fact, any bigger noise bounds will yield such as plateau. On the other hand, any smaller noise bounds will make Problem (P_T) infeasible, and the two models are then guaranteed to be T-distinguishable for some finite T, and the distinguishability index plot will reach its maximum at time $T-1$. Therefore, if we can redesign the system such that with the new noise bounds we have:

$$\bar{\delta} < \bar{\delta}_{plateau} = 0.379,$$

then we can be sure that these faults are isolable. For instance, if we can reduce the process and measurement noise bounds to 0.18, the minimum of Problem (P_T) will be obtained at $\bar{\delta} = 0.36$. In this case, since $\bar{\delta} < \bar{\delta}_{plateau}$, the two models in (9.12) are found to be 6-distinguishable. The growth trend of the corresponding distinguishability index with these new noise bounds is plotted in Fig. 9.7.

Thus, this example illustrates that the distinguishability index can also be exploited to derive the maximum allowed uncertainty for a system such that certain faults are guaranteed to be detectable or isolable. In turn, this suggests possible measures for ensuring fault detection and isolation through the reduction of noise levels, either with a better choice of sensors or with the use of noise isolation platforms.

9.8 Conclusion

In this paper, we considered the FDI problem for switched affine models using a model invalidation approach. First, we proposed new model invalidation and T-distinguishability formulations using SOS-1 constraints, that are demonstrated to be computationally more efficient and do not require a complicated change of variables. Further, we introduced a *distinguishability index* as a measure of separation between models and showed that this index is also a practical tool for finding the smallest receding time horizon that is needed for fault detection and isolation, as well as for recommending system design changes for ensuring fault detection and isolation.

Moreover, we introduced a fault detection and isolation scheme for switched affine models, which guarantees the detection and isolation of faults when certain conditions are met. The scheme is built upon an optimization-based method, which formulates the fault-detection and isolation as MILP feasibility check and optimization problems. The detection and isolation monitors can be implemented independently on several processing units, hence it can be efficiently implemented for a large number of faults. Moreover, we introduced adaptive time horizons in order to isolate faults faster. Finally, we illustrated the efficiency of the proposed approaches with several examples, including with an HVAC system model that is equipped with our FDI scheme.

As future work, we are interested to find system theoretic upper bounds on the time horizon T or I such that the incremental search for the smallest T or I can be efficiently terminated with some formal guarantees.

Acknowledgements This work is supported in part by DARPA grant N66001-14-1-4045 and an Early Career Faculty grant from NASA's Space Technology Research Grants Program.

References

1. R. Beard, Failure accommodation in linear systems through self-reorganization. PhD thesis, MIT, 1971
2. S. Simani, C. Fantuzzi, R.J. Patton, *Model-Based Fault Diagnosis in Dynamic Systems Using Identification Techniques* (Springer, London, 2003)
3. P. Frank, Advances in observer-based fault diagnosis, in *International Conference on Fault Diagnosis: TOOLDIAG* (1993)
4. P. Frank, X. Ding, Survey of robust residual generation and evaluation methods in observer-based fault detection systems. J. Process Control **7**(6), 403–424 (1997)
5. H. Sneider, P.M. Frank, Observer-based supervision and fault detection in robots using nonlinear and fuzzy logic residual evaluation. IEEE Trans. Control Syst. Technol. **4**(3), 274–282 (1996)
6. R. Isermann, Fault diagnosis of machines via parameter estimation and knowledge processing–tutorial paper. Automatica **29**(4), 815–835 (1993)
7. X. Liu, H. Zhang, J. Liu, J. Yang, Fault detection and diagnosis of permanent-magnet DC motor based on parameter estimation and neural network. IEEE Trans. Ind. Electron. **47**(5), 1021–1030 (2000)
8. H. Hammouri, M. Kinnaert, E.H. El Yaagoubi, Observer-based approach to fault detection and isolation for nonlinear systems. IEEE Trans. Autom. Control **44**(10), 1879–1884 (1999)
9. S. Paoletti, A. Garulli, J. Roll, A. Vicino, A necessary and sufficient condition for input-output realization of switched affine state space models, in *47th IEEE Conference on Decision and Control*, Dec 2008, pp. 935–940
10. A. Abdo, S.X. Ding, J. Saijai, W. Damlakhi, Fault detection for switched systems based on a deterministic method, in *IEEE Conference on Decision and Control (CDC)* (2012), pp. 568–573

11. W. Pan, Y. Yuan, H. Sandberg, J. Gonçalves, G. Stan, Online fault diagnosis for nonlinear power systems. Automatica **55**, 27–36 (2015)
12. S. Narasimhan, G. Biswas, Model-based diagnosis of hybrid systems. IEEE Trans. Syst. Man Cybern. Part A **37**(3), 348–361 (2007)
13. S. Campbell, R. Nikoukhah, *Auxiliary Signal Design for Failure Detection* (Princeton University Press, Princeton, 2004)
14. J.K. Scott, R. Findeisen, R.D. Braatz, D.M. Raimondo, Input design for guaranteed fault diagnosis using zonotopes. Automatica **50**(6), 1580–1589 (2014)
15. P. Rosa, C. Silvestre, J.S. Shamma, M. Athans, Fault detection and isolation of LTV systems using set-valued observers, in *IEEE Conference on Decision and Control (CDC)* (2010), pp. 768–773
16. F. Harirchi, N. Ozay, Model invalidation for switched affine systems with applications to fault and anomaly detection. IFAC ADHS Conf. **48**(27), 260–266 (2015)
17. F. Harirchi, S.Z. Yong, N. Ozay, Guaranteed fault detection and isolation for switched affine models, in *IEEE Conference on Decision and Control* (2017)
18. F. Harirchi, Z. Luo, N. Ozay, Model (in)validation and fault detection for systems with polynomial state-space models, in *American Control Conference (ACC)*, July 2016, pp. 1017–1023
19. R.S. Smith, J.C. Doyle, Model validation: a connection between robust control and identification. IEEE Trans. Autom. Control **37**(7), 942–952 (1992)
20. J. Anderson, A. Papachristodoulou, On validation and invalidation of biological models. BMC Bioinf. **10**(1), 1 (2009)
21. N. Ozay, M. Sznaier, C. Lagoa, Convex certificates for model (in)validation of switched affine systems with unknown switches. IEEE Trans. Autom. Control **59**(11), 2921–2932 (2014)
22. F. Harirchi, N. Ozay, Guaranteed model-based fault detection in cyber-physical systems: a model invalidation approach (2016). arXiv:1609.05921 [math.OC]
23. H. Lou, P. Si, The distinguishability of linear control systems. Nonlinear Anal. Hybrid Syst. **3**(1), 21–38 (2009)
24. P. Rosa, C. Silvestre, On the distinguishability of discrete linear time-invariant dynamic systems, in *IEEE CDC-ECC* (2011), pp. 3356–3361
25. M. Babaali, M. Egerstedt, Observability of switched linear systems, in *International Workshop on Hybrid Systems: Computation and Control* (Springer, Berlin, 2004), pp. 48–63
26. Gurobi Optimization, Inc., *Gurobi Optimizer Reference Manual* (2015)
27. CPLEX, IBM ILOG, V12. 1: User's manual for CPLEX. Int. Bus. Mach. Corp. **46**(53), 157 (2009)
28. F. Harirchi, S.Z. Yong, E. Jacobsen, N. Ozay, Active model discrimination with applications to fraud detection in smart buildings, in *IFAC World Congress*, Toulouse (2017)
29. Y. Ding, F. Harirchi, S.Z. Yong, E. Jacobsen, N. Ozay, Optimal input design for affine model discrimination with applications in intention-aware vehicles, in *International Conference on Cyber-Physical Systems (ICCPS)*, (Porto, 2018)
30. B. Argüello-Serrano, M. Vélez-Reyes, Nonlinear control of a heating, ventilating, and air conditioning system with thermal load estimation. IEEE Trans. Control Syst. Technol. **7**(1), 56–63 (1999)
31. J. Löfberg, YALMIP: a toolbox for modeling and optimization in MATLAB, in *CACSD Conference*, Taipei (2004)

Chapter 10
Diagnosability of Discrete Faults with Uncertain Observations

Alban Grastien and Marina Zanella

10.1 Introduction

A *Hybrid System* (HS) [11] is a dynamic system whose behavior can be described by means of an automaton consisting in a set of states, where state transitions are triggered by a finite number of discrete events, each of which is either observable or unobservable. Each state represents an operating mode described by (possibly abnormal) continuous dynamics. Reasoning about HSs is a difficult task because it requires handling the discrete aspects (which involve search) as well as the continuous aspects (which lead to infinite spaces). A popular method to diagnose such systems is to discretize the model and reason only about the underlying *Discrete-Event System* (DES). Traditionally the discretization was performed manually. Indeed the examples of DESs used in the literature are often relevant to cyber-physical systems that one would naturally model as HSs: HVAC systems [21], power networks [1, 2], telecommunication networks [19], water distribution networks [7], etc. In more recent works [3, 10, 29] this discretization is performed automatically through the use of analytical redundancy relations [24] or possible conflicts [20]. Diagnosis of DESs is a research field that has developed a number of techniques to scale to large systems [1, 8, 9, 15, 17, 19, 22, 26]. For this reason, we assume that the model of the HS is a DES.

A DES [4] is a conceptual model of a dynamical system where the system behavior is described by transitions over a finite set of states and each transition

A. Grastien (✉)
Data61, The Australian National University, Canberra, ACT, Australia
e-mail: alban.grastien@data61.csiro.au

M. Zanella
University of Brescia, Department of Information Engineering, Brescia, Italy
e-mail: marina.zanella@unibs.it

M. Sayed-Mouchaweh (ed.), *Diagnosability, Security and Safety of Hybrid Dynamic and Cyber-Physical Systems*, https://doi.org/10.1007/978-3-319-74962-4_10

is associated with an event out of a finite set of events. Model-based diagnosis of DESs is a task that takes as input the DES model of a (natural or man-made) system along with a relevant observation and produces as output a *diagnosis*, i.e. some pieces of information explaining whether what has been observed is consistent either with a normal behavior or an abnormal one. There are several notions of diagnosis of DESs in the literature featuring different levels of abstraction. According to a common notion, the diagnosis of a DES is a set of *candidates*, each candidate being a set of *faults*, where a fault is an undesired state transition. The definition of a candidate requires that the faults included in a candidate are consistent with both the DES model and the given observation. However, distinct candidates may bring conflicting information. This is the case, for instance, when according to a candidate the system is free of faults while according to another it is affected by faults. A DES that is repeatedly diagnosed while it is being monitored (that is, a new set of candidates is produced every time a new observable event or a chunk of observable events is processed) is *diagnosable* if such ambiguity can be removed once a bounded sequence of observable events have taken place.

DES diagnosability was introduced by the diagnoser approach [21], where a necessary and sufficient condition is proposed to check diagnosability based on the construction of a so-called *diagnoser*. The problem of deciding diagnosability was then proved to be polynomial by using the *twin plant* method [13]. Similar approaches to diagnosability checking can be found in [5, 28].

Existing works are focused on how to verify the intrinsic diagnosability of a DES and assume that candidates are computed by an *exact* diagnostic algorithm that takes as input a completely *certain* observation. Exceptions include diagnosability under imperfect conditions for modular structures [6], decentralized analysis [23], and approximate diagnosers [25]. As remarked in this latest paper, the diagnosability property can be exhibited even when some *incomplete* or *approximate* diagnostic algorithms are used, i.e. algorithms that do not perform a complete search of the behavioral space of the DES. However, this work still relies on a completely certain observation while in the real world the observation may be uncertain, as remarked by some contributions on diagnosis of DESs [8, 14, 18]. In a broader perspective, one can see that the ability to remove ambiguities in candidates depends not only on the DES and the diagnostic algorithm at hand but also on the available observations.

This chapter, which extends a recent work by the authors [27], investigates whether the ability to disambiguate DES candidates, i.e. the diagnosability property, holds for a diagnosable system with uncertain observations. The uncertainty is measured by a parameter, which allows one to study the level of noise that can affect the observation without impacting the performance of diagnosis. Several kinds of uncertainties in the observations can be envisaged and several parameters can be adopted to measure the same type of uncertainty.

The remaining of this chapter is organized as follows. The next section presents the relevant background information on diagnosability, including an introduction to the twin plant approach to check diagnosability. Section 10.3 discusses the notion of uncertain observations, and presents three types of uncertainty: temporal, logical, and combined. Section 10.4 extends the original definition of DES diagnosability to the case when an uncertain observation is considered, and shows how the twin plant approach can be applied with uncertain observations. Section 10.5 describes an HS, and performs its diagnosability analysis. Finally, Sect. 10.6 draws conclusions and hints at directions for future research, some of which are specific to the diagnosability analysis of HSs.

10.2 Background

A DES diagnosis problem consists in a DES D and a (finite) observation O, the latter representing what has been observed while D was running during a time interval of interest.

10.2.1 Discrete-Event Systems

A (partially observable) DES D is a 4-tuple (Σ, L, obs, flt) where Σ is the finite set of events that can take place in the system; $L \subseteq \Sigma^*$ is the *behavior space*, which is a prefix-closed and live, i.e. deadlock-free, language that models all (and only) the possible sequences of events, or *traces*, that can take place in the system. Function obs associates each trace τ with an observation $obs(\tau) \in \Sigma_o^*$ and is defined as the projection of τ on the subset $\Sigma_o \subseteq \Sigma$ of observable events, i.e. $obs(\tau)$ is a copy of τ where all non-observable events have been removed. The length of the sequence of events in $obs(\tau)$ is denoted $|obs(\tau)|$. The prefix-closed and live observable language relevant to L is denoted as $obs(L)$.[1] The set of unobservable *faulty events*, or *faults*, is denoted as Σ_f where $\Sigma_f \subseteq \Sigma \setminus \Sigma_o$. Function flt associates each trace τ with the set $flt(\tau) \in \Sigma_f^*$ of faulty events that appear in the trace itself.

Language L of DES $D = (\Sigma, L, obs, flt)$ can be represented by a finite automaton (FA) $G = (X, \Sigma, \delta, x_0)$, called the *behavioral model*, where X is the set of states and $\delta \subseteq X \times \Sigma \times X$ is the set of state transitions. Each $x \in X$ represents a state that D can be in and each triple $(x, \sigma, x') \in \delta$ represents a possible state change. State $x_0 \in X$ is the *initial* one, i.e. the state of the system at the moment when we have started to observe its evolution. A *path* in automaton G is a sequence of transitions starting from the initial state, concisely represented as $x_0 \xrightarrow{\sigma_1} x_1 \xrightarrow{\sigma_2} \cdots \xrightarrow{\sigma_n} x_n$ where $n \geq 1$. A trace is a projection of a path on Σ, e.g. $\sigma_1 \cdot \cdots \cdot \sigma_n$. Figure 10.1 displays the behavioral model G of a DES D that will be used as a running example throughout this chapter. Such a model encompasses one faulty event f, another unobservable

[1]Notice that the fact that L is assumed to be live does not imply that $obs(L)$ is live. However, following the diagnoser approach [21], we also assume that $obs(L)$ is live.

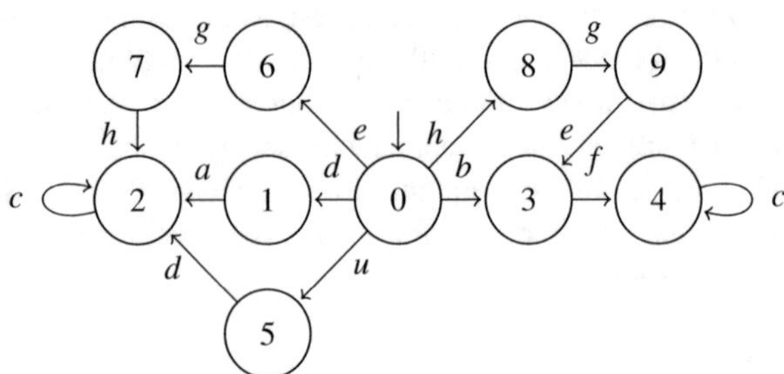

Fig. 10.1 DES used to illustrate this chapter

event u, and seven observable events (a–e, g, and h). A possible path is $0 \xrightarrow{h} 8 \xrightarrow{g} 9 \xrightarrow{e} 3 \xrightarrow{f} 4 \xrightarrow{c} 4$, corresponding to the trace $h.\,g.\,e.f.\,c$, where . is the concatenation operator.

Given a diagnosis problem (D, O), a diagnosis *candidate* is a pair $(x, \varphi) \in X \times 2^{\Sigma_f}$ where x represents the state that system D has reached by a path generating O and φ represents the set of faults of this path. The *diagnosis* is the set of all the candidates relevant to the diagnosis problem (D, O). The diagnosis relevant to our sample system D in Fig. 10.1 and observation $O = h.\,g.\,e.\,c$ is $\{(4, \{f\})\}$. Such a diagnosis consists of just one candidate, meaning that, once observation O has been perceived, the state of D is certainly 4 and fault f has necessarily occurred.

10.2.2 Diagnosability

Following [21], a DES D exhibits the diagnosability property as far as a fault $f \in \Sigma_f$ is concerned if the occurrence of such a fault can always be detected and isolated without any ambiguity once a finite sequence of observable events has been recorded. Given an observation, an exact diagnostic algorithm is able to draw all the sets of faults relevant to all the traces consistent with such an observation. If, for whichever path that has preceded the occurrence of the fault, and for whichever sequence of transitions (generating k observable events) that has followed it, all the traces that are consistent with such an observation include the fault, then such a fault is certain (and it is said to be diagnosable) as it belongs to the intersection of all the candidate sets of faults. The system is said to be diagnosable if all its faults are diagnosable. We denote $L_f = (\Sigma^* f \Sigma^*) \cap L$ the set of traces that include fault f and $\bar{L_f} = (\Sigma^* f) \cap L$ the set of traces that end with fault f.

Definition 1 (Diagnosability [21]) Given a DES $D = (\Sigma, L, obs, flt)$ whose set of faults is $\Sigma_f \subseteq \Sigma$, a fault $f \in \Sigma_f$ is *diagnosable* if

$$\forall \tau_1 \in \bar{L_f}, \exists k \in \mathbf{N}, \forall \tau_2 : \tau_1.\,\tau_2 \in L, |obs(\tau_2)| \geq k$$
$$\Rightarrow (\forall \tau \in L), (obs(\tau) = obs(\tau_1.\,\tau_2) \Rightarrow (\tau \in L_f)).$$

System D is *diagnosable* if all its faults are diagnosable.

DES D of our example in Fig. 10.1 is diagnosable with $k = 1$ since the occurrence of fault f is precisely detected once the occurrence of observable event c has been perceived immediately after having perceived either b or $h.g.e$. One can appreciate that such a notion of diagnosability relies on function obs, which provides the sequence of observable events that have occurred in the system during its evolution, where such a sequence reflects the chronological order of the occurrence of events within a trace. The above definition of diagnosability implicitly assumes that, if a DES follows a trace u, the observation O processed by the diagnostic engine equals $obs(u)$. An observation like this is *certain*.

10.2.3 Twin Plant Method

The most popular approach to DES diagnosability analysis is the so-called *twin plant* method [13]. The name refers to the product of a system behavioral model (or of some FA drawn from it) by itself.

Given a (nondeterministic) FA G, representing the behavioral model of a DES, the twin plant method draws from G a completely observable (nondeterministic) FA G_o, whose set of events is Σ_o, that is, it consists of all and only the observable events of G. Each state of G_o is a pair (x, ϕ), where x is either the initial state x_0 of G or a state in G that is the target of an observable transition, and ϕ is a set of faults. If $x = x_0$, then $\phi = \emptyset$, that is, it is assumed that G is initially free of faults, the same as in the diagnoser approach [21]. Each transition from a pair (x, ϕ) to a pair (x_1, ϕ_1) in G_o represents a path in G from state x to state x_1, where the only observable transition in such a path is the last one. Set ϕ_1 is the set-theoretic union of ϕ with all the faults corresponding to the transitions on the path from x to x_1 in G. Thus the constraint holds that $\phi_1 \supseteq \phi$.

Intuitively G_o is a (nondeterministic) FA generating the observable language of G, hence each state (x, ϕ) in G_o includes the set ϕ of all the faults that manifest themselves along a path (at least) in G that produce the same sequence of observable events as a path in G_o from the initial state $(x_0, \emptyset)$ to state (x, ϕ).

Once G_o is available, the product [4] of G_o with itself ($G_o \otimes G_o$) is computed, and denoted G_d. Thus each state in G_d is a pair of pairs, $((x_1, \phi_1); (x_2, \phi_2))$.

Finally, an algorithm checks whether in G_d there exists a cycle that includes an *ambiguous* state $((x_1, \phi_1); (x_2, \phi_2))$, that is, a state such that ϕ_1 does not equal ϕ_2: if this condition holds, G is not diagnosable. This check can be performed by first identifying all the states in G_d for which ϕ_1 does not equal ϕ_2, then deleting all the other states and their incoming and exiting transitions, and finally checking whether the remaining automaton contains a cycle. Notice that, in this last step, the observable events associated with transitions are irrelevant, therefore G_d can be considered as a graph.

Now we try and explain the rationale behind this method. Each pair of states of G contained in the same state of G_d can be reached by producing the same sequence of observable events. If the sets of faults associated with such states are different, for instance $\phi_1 = \{f_1\}$ and $\phi_2 = \{f_1,f_2\}$, then, based on the sequence of events the observer has received so far, we cannot find out whether only fault f_1 has occurred, or both faults f_1 and f_2. The system is diagnosable only if we can isolate the occurred failure within a finite delay following the sequence of events observed up to this point.

As already remarked, given a transition $(x, \phi) \to (x_1, \phi_1)$ in G_o, the constraint holds that $\phi_1 \supseteq \phi$. Hence, in G_o the set of failure types is the same for all the states belonging to the same cycle. Consequently, in G_d, if a cycle includes a state $((x_1, \phi_1); (x_2, \phi_2))$, all the other states in the same cycle are $((-, \phi_1); (-, \phi_2))$, that is, they include the same sets of faults. Thus, if in a state in a cycle in G_d such sets are different from each other, the system is not diagnosable since it may indefinitely produce the observable events relevant to the cycle, in which case we cannot decide within a finite delay which faults have occurred.

The number of states in G_o is at most $|X|2^{|\Sigma_f|}$, where (as before) X is the set of states in G, and Σ_f is the domain of faults in G (this upper bound is obtained by assuming that, for each state in G, there is a state in G_o including it and each subset of the set of failure types). The number of transitions in G_o is at most $|X|^2 2^{2|\Sigma_f|}|\Sigma_o|$, where, as above, Σ_o is the set of observable events in G (this upper bound is obtained by assuming that, for each observable event, there is a transition exiting from each state and directed to each state, where the number of considered states is the maximum one). The number of states in G_d is at most $|X|^2 2^{2|\Sigma_f|}$ (this upper bound is obtained by assuming that in G_d there is a state for each distinct pair of states of G_o), and the number of transitions in G_d is at most $|X|^4 2^{4|\Sigma_f|}|\Sigma_o|$ (this upper bound is obtained by assuming the complete connectivity of the states in G_d).

The complexity of G_o construction is thus $O(|X|^2 2^{2|\Sigma_f|}|\Sigma_o|)$, while the complexity of G_d construction is $O(|X|^4 2^{4|\Sigma_f|}|\Sigma_o|)$ (in both cases, the complexity is given by the maximum number of transitions, since the growth of such a number is asymptotically higher than that of the number of states). The complexity relevant to detecting the presence of a certain "offending" cycle in an appropriately pruned subgraph of G_d is linear in the number of states and transitions of the subgraph.

Thus the complexity of the whole method is $O(|X|^4 2^{4|\Sigma_f|}|\Sigma_o|)$, which is exponential in the number of faults. However, it can be reduced to polynomial in the number of faults by noticing that a system is diagnosable with respect to all the faults if and only if it is diagnosable with respect to each individual fault. In other words, one can apply the algorithm iteratively (a number of times that equals the number of distinct failures), for testing the diagnosability with respect to each singleton set of faults. In case an individual fault $f \in \Sigma_f$ is considered, the set of faults within each pair (x, ϕ) of G_o is either $\phi = \{f\}$, which can conveniently be denoted as F, or $\phi = \emptyset$, which is denoted as N. Automaton G_o can be referred to as the *verifier* of fault f. A state of the twin plant G_d is *ambiguous* if it matches the pattern $((x, N); (x', F))$ or $((x, F); (x', N))$. As already remarked, if a state in a loop is ambiguous, then all the states in the same loop are ambiguous.

Finally the complexity of the test relevant to each individual fault is $O(|X|^4 2^{4|1|}|\Sigma_o|) = O(|X|^4|\Sigma_o|)$. So, the overall complexity of the method for testing diagnosability is $O(|X|^4|\Sigma_o||\Sigma_f|)$, which is polynomial.

10.3 Temporal and Logical Uncertainty of Observations

In the literature, prior to [27], diagnosability had been confined to certain observations. However, observations are uncertain in many applications. We present two types of uncertainties, i.e. temporal and logical, and for each of them, a measure of how uncertain the observation is. These measures are by no means the only ones possible, as it will be remarked in Sect. 10.3.2.

10.3.1 Temporally Uncertain Observations

In observation $O = h.g.e.c$ used in Sect. 10.2.1, the occurrence order of the observable events is known. This observation is depicted in the top graph of Fig. 10.2, where the order is represented by the arrows between observed events. Implicit arrows, e.g. from h to e, are not displayed. We say that the observation is *temporally certain*. However, the temporal order of the observable events that have occurred within the DES is not always known, in particular when they occur in a short time span. The bottom graph of Fig. 10.2 shows a *temporally uncertain* observation O' where the order between observable events g and e is unknown. Since we do not know which sequence, i.e. either $h.g.e.c$ or $h.e.g.c$, actually occurred, an exact diagnostic algorithm has to consider both of them. The pair of observable events e and g can altogether be considered as a *temporally compound event* $e//g$, which cumulatively represents both sequences $e.g$ and $g.e$. We can describe the uncertain observation as a sequence $O' = h.e//g.c$.

Definition 2 (Temporally Compound Observable Event) A *temporally compound observable event* of level ℓ (with $\ell \geq 1$) is a multiset of ℓ reciprocally temporally unrelated instances of observable events. When $\ell > 1$, not all the ℓ instances are identical. A temporally compound event of level 1 is a *single* observable event.

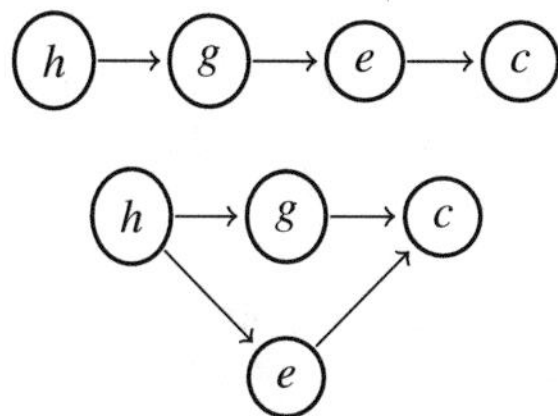

Fig. 10.2 Certain (top) and temporally uncertain (bottom) observations

We use $\left(\!\binom{\Sigma_o}{\ell}\!\right)$ and $\left(\!\binom{\Sigma_o}{\leq \ell}\!\right)$ to denote the collection of multisets of Σ_o of cardinality ℓ and of cardinality ℓ or less, respectively. Notice that $\left(\!\binom{\Sigma_o}{\leq \ell}\!\right) = \bigcup_{i \leq \ell} \left(\!\binom{\Sigma_o}{i}\!\right)$. Although a temporally compound event is univocally identified by writing the values of all the instances that it includes, independently of their order, using // as a separator, we put such values in alphabetical order in this chapter.

Definition 3 (Temporal Uncertainty Level) A *temporally uncertain observation* is a sequence of temporally compound observable events. The *temporal uncertainty level* of a temporally uncertain observation O is the maximum level of the compound observable events that O includes.

The lowest temporal uncertainty level of an observation is 1, corresponding to a certain observation. The temporal uncertainty level of the observation in the bottom graph of Fig. 10.2 instead is 2, since events e and g are reciprocally temporally unrelated. Notice that the temporally uncertain observations defined above do not encompass all the temporally uncertain observations as defined in [14]. However, the class of temporally uncertain observation we are addressing is meaningful. If an exact diagnostic algorithm is adopted to diagnose a DES in a monitoring context, the diagnosis output is *monotonic* [18] for whichever temporally uncertain observation that is a sequence of temporally compound events, provided that a new set of candidates is output only after all the observable events in a temporally compound event have been processed.

10.3.2 Logically Uncertain Observations

A temporally uncertain observation is a certain observation where some temporal constraints have been relaxed. Similarly, a logically uncertain observation is a *logical* relaxation of a certain one. The so-called logical content [14, 16] of an observed event is its (discrete) value. If such a value is not known with certainty, the relevant observation is logically uncertain.

Figure 10.3 shows a logically uncertain observation where the logical uncertainty comes from the fact that (1) the first observed event is not known with certainty, i.e. it could be either e or h, and (2) whether the third observed event (e) actually occurred is not certain, which is represented by ε. Since we do not know which sequence (either $h.g.e.c$ or $h.g.c$ or $e.g.e.c$ or $e.g.c$) actually occurred, an exact diagnostic algorithm has to consider all of them.

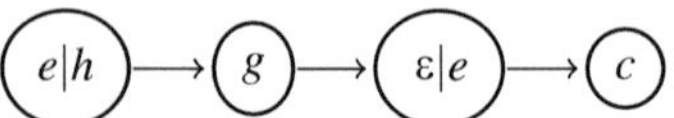

Fig. 10.3 Logically uncertain observation

Because logically uncertain observations may include ε, we use the notation $\Sigma_{o+} = \Sigma_o \cup \{\varepsilon\}$. We define a logically compound observable event as a set of events belonging to Σ_{o+}. The *degree* of this compound event, tantamount to the level of a temporally compound observable event, is here defined as the maximal *distance* according to a specified distance matrix (increased by 1 for normalization) between any pair of events included in the compound event.

Definition 4 (Distance Matrix and Logically Compound Observable Event) The *distance matrix* M is a map that associates any pair in $\Sigma_o \times \Sigma_{o+}$ with a (possibly infinite) non-negative integer while respecting these constraints, i.e. $\forall(e_1, e_2) \in \Sigma_o \times \Sigma_{o+}$, $M(e_1, e_2) = 0$ if $e_1 = e_2$, $M(e_1, e_2) > 0$, if $e_1 \neq e_2$, $M(e_2, e_1) = M(e_1, e_2)$, if $e_2 \neq \varepsilon$. A *logically compound observable event* o of degree d (with $d \geq 1$) is a non-empty subset of elements from Σ_{o+}, that is not the singleton $\{\varepsilon\}$, such that $d = \max\{M(e_1, e_2) \mid \{e_1, e_2\} \subseteq o)\} + 1$. A logically compound event of degree 1 is a *single* observable event.

Given the system D in Fig. 10.1, we assume that observable events b and d are hard to distinguish. We also assume that a, e, and h are less difficult to distinguish, and sometimes it is difficult to find out whether what has been perceived is either pure noise (that is, no observable event has occurred in the DES) or observable event e. This can be modeled by $M(b, d) = 1$; $M(a, e) = M(a, h) = M(e, h) = 2$; $M(e, \varepsilon) = 3$; and $M(\cdot) = \infty$ for any other pair of distinct elements.

We use $2^{\Sigma_{o+}, d}$ and $2^{\Sigma_{o+}, \leq d}$ to denote the collection of subsets of Σ_{o+} where each subset represents a logically compound observable event of degree d, and d or less, respectively. Notice that $2^{\Sigma_{o+}, \leq d} = \bigcup_{i \leq d} 2^{\Sigma_{o+}, i}$. We use the symbol # as a separator to represent the logically compound observable event, e.g. $e\#h$. Although a logically compound event is univocally identified by writing all the values that are included, independently of their order, we put such values in an order in this chapter so that ε precedes any other event and all the other events are in alphabetical order.

Definition 5 (Logical Uncertainty Degree) A *logically uncertain observation* is a sequence of logically compound observable events. The *logical uncertainty degree* of a logically uncertain observation is the maximum degree of the logically compound events that are included.

The logical uncertainty degree of observation $e\#h.\, g.\, \varepsilon\#e.\, c$ depicted in Fig. 10.3 is 4 because the distance between ε and e is 3.

We have already claimed that several distinct parameters can be adopted to measure the "noise" in an uncertain observation. Focusing on logically uncertain observations, here and in the following, we will only use as a parameter the logical uncertainty degree as defined above. However, in order to provide an instance of another parameter, we mention also the maximum number of observable events that can be included in a logically compound observable event.

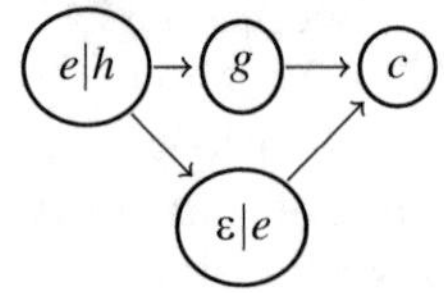

Fig. 10.4 Observation that is both temporally and logically uncertain

10.3.3 Observations with Combined Uncertainty

An observation with a combined uncertainty is a temporally uncertain observation where events are logically uncertain.

Figure 10.4 shows an observation exhibiting a combined uncertainty: the logical content of two events ($e\#h$ and $\varepsilon\#e$) is uncertain and the relative emission order of a pair of events (g and $\varepsilon\#e$) is unknown.

Definition 6 (Combined Compound Observable Event) A *combined compound observable event* of level ℓ (with $\ell \geq 1$) and degree d (with $d \geq 1$) is a multiset of ℓ reciprocally temporally unrelated instances of logically compound observable events whose maximum degree is d. When $\ell > 1$, not all the ℓ instances are identical. A combined compound observable event whose level and degree are both 1 is a *single* observable event.

We use $\left(\!\!\binom{2^{\Sigma_{o+},\leq d}}{\ell}\!\!\right)$ and $\left(\!\!\binom{2^{\Sigma_{o+},\leq d}}{\leq \ell}\!\!\right)$ to denote the collection of multisets of $2^{\Sigma_{o+},\leq d}$ of cardinality ℓ and of cardinality ℓ or less, respectively. Notice that $\left(\!\!\binom{2^{\Sigma_{o+},\leq d}}{\leq \ell}\!\!\right) = \bigcup_{i \leq \ell} \left(\!\!\binom{2^{\Sigma_{o+},\leq d}}{i}\!\!\right)$.

Definition 7 (Combined Uncertainty Index) An *observation with combined (temporal and logical) uncertainty* is a sequence of combined compound observable events. The *combined uncertainty index* of an observation O with combined uncertainty is the pair given by the maximum level ℓ and the maximum degree d of the combined compound observable events included in O. Such an index is denoted d^{ℓ}.

The lowest combined uncertainty index of an observation is 1^1, corresponding to a certain observation. If the combined uncertainty index of an observation is 1^{ℓ}, then such an observation is affected only by temporal uncertainty (of level ℓ). Conversely, if the combined index of an observation is d^1, then such an observation is affected only by logical uncertainty (of degree d).

The combined uncertainty index of observation $(e\#h).(\varepsilon\#e)//g.c$ shown in Fig. 10.4 is 4^2 since its maximum level is 2 (event g is temporally unrelated with respect to $\varepsilon\#e$) and its maximum degree is 4 (since the distance between ε and e is 3).

10.3.4 Unifying Uncertainty Representations

To make the next definitions independent of the specific type of the considered uncertainty and of its measure (since several measures can be envisaged for the same uncertainty type), we introduce the *extension* of an observation. Such a notion encapsulates both the specific kind of uncertainty and its value according to a specific measure.

Definition 8 (Extension of a Certain Observation) Given a type of uncertainty $\|$, the value m of a specific measure of this uncertainty, and a certain observation O, the *extension* $||O||^{\|m}$ of the observation is the set of certain and uncertain observations that O could produce according to the given uncertainty type and up to the given uncertainty measure, where $O \in ||O||^{\|m}$.

We now talk about $\|^m$ as the uncertainty that can affect the observation produced by the system. For instance, we denote the extension up to temporal uncertainty of level ℓ as $||O||^{//\ell}$, the extension up to logical uncertainty of degree d as $||O||^{\#d}$, and the extension up to combined uncertainty of index d^ℓ as $||O||^{d^\ell}$. In our example, given the trace τ whose certain observation is $obs(\tau) = h.g.e.c$, the extension of such an observation to the second temporal uncertainty level is $||obs(\tau)||^{//2} = \{h.g.e.c, h.g.c//e, h.e//g.c, g//h.e.c, g//h.c//e\}$. The extension of $obs(\tau)$ to the third logical uncertainty degree based on the distance matrix provided in Sect. 10.3.2 is $||obs(\tau)||^{\#3} = \{x.g.y.c\}$, which is a set including 16 sequences where $x \in \{h, a\#h, e\#h, a\#e\#h\}$ and $y \in \{e, a\#e, e\#h, a\#e\#h\}$. The extension of $obs(\tau)$ to the combined index 3^2 is $||obs(\tau)||^{3^2} = \{x.g.y.c, x.g.c//y, x.g//y.c, g//x.y.c, g//x.c//y\}$, which includes 80 sequences.

Notice that $||obs(\tau)||^{//1} = ||obs(\tau)||^{\#1} = ||obs(\tau)||^{1^1} = \{h.g.e.c\}$, i.e. these extensions give a certain observation. This is a general property, i.e. for any trace τ, $||obs(\tau)||^{//1} = ||obs(\tau)||^{\#1} = ||obs(\tau)||^{1^1} = \{obs(\tau)\}$ since $||^{//1}$, $||^{\#1}$, and $||^{1^1}$ comprise no uncertainty. Given an uncertain observation O_u, the fact $O_u \in ||O_1||^{\|m} \cap ||O_2||^{\|m}$ for certain observations O_1 and O_2 means that observable behavior O_1 can be mistaken for observable behavior O_2 if the observation is affected by uncertainty $\|^m$.

10.4 Diagnosability with Uncertain Observations

This section proposes a definition of diagnosability with an observation affected by a (generic) uncertainty $\|^m$. According to this generalized definition, a faulty behavior should always eventually produce an observation that cannot be mistaken for an observation produced by a nominal behavior.

Definition 9 (Diagnosability Under Uncertainty) Given a DES $D = (\Sigma, L, obs, flt)$ with a set of faults $\Sigma_f \subseteq \Sigma$ and an uncertainty $\|^m$, a fault $f \in \Sigma_f$ is $\|^m$*-diagnosable* if

$$\forall \tau_1 \in \bar{L_f}, \exists k \in \mathbf{N}, \forall \tau_2 : \tau_1 . \tau_2 \in L, |obs(\tau_2)| \geq k$$
$$\Rightarrow (\forall \tau \in L), \left(||obs(\tau_1 . \tau_2)||^{\|m} \cap ||obs(\tau)||^{\|m} \neq \emptyset \Rightarrow (\tau \in L_f) \right).$$

System D is $\|^m$*-diagnosable* if every fault $f \in \Sigma_f$ is $\|^m$-diagnosable.

Comparing Definition 1 with Definition 9, it is easy to see that the latter one is a generalization of the former since $||O||^{\|m}$ is a singleton when $\|^m$ comprises no uncertainty. System D in Fig. 10.1 is $||^{//2}$-diagnosable. Indeed, fault f is identified by observing either $b.c^*$ or $e.c^*$; changing the order of two consecutive observed events does not eliminate the fact that b will be observed; a temporally uncertain observation with level $\ell \leq 2$ will not modify the order between h and e. However, the system is not $||^{//3}$-diagnosable since observation $e//g//h.c^*$ cannot be precisely diagnosed as it is relevant both to a normal and a faulty trace.

Given the distance matrix provided in Sect. 10.3.2, let us now consider the logical uncertainty degree as 2. Hence, only events whose distance value between them is up to 1 need to be taken into account, i.e. the only uncertainty lies in event b, which may be confused with d (and vice versa). If the logically uncertain observation $b\#d.c^*$ is perceived, the diagnosis task cannot find out whether fault f has occurred or not, which proves that D is not $||^{\#2}$-diagnosable.

Notice how the definition of diagnosability is well-behaved w.r.t. increasing uncertainty. If uncertainty $||''$ is more permissive than $||'$, i.e. $||O||'' \supseteq ||O||'$ for any certain observation O, then $||''$-diagnosability implies $||'$-diagnosability. Given a sequence $||^1, ||^2, ||^3, \ldots$ of increasingly more permissive uncertainties, the maximum value i such that the system is $||^i$-diagnosable (or 0, if the DES is not diagnosable even for certain observations, or $+\infty$, if there does not exist any upper bound for i) defines the robustness of the system w.r.t. uncertainty. Since temporal uncertainty is increasingly more permissive for increasing values of the uncertainty level and logical uncertainty is increasingly more permissive for increasing values of the uncertainty degree (for any given distance matrix), we can conclude that the sample DES D is not $||^{//\ell}$-diagnosable for any $\ell > 2$ and it is not $||^{\#d}$-diagnosable for any $d > 1$. As to the permissiveness of the combined (temporal and logical) uncertainty, $||O||^{\hat{d}^{\hat{\ell}}} \supseteq ||O||^{d^\ell}$ holds if and only if $\hat{d} \geq d$ and (at the same time) $\hat{\ell} \geq \ell$. DES D, since it is not $||^{\#d}$-diagnosable for any $d > 1$, cannot be $||^{d^\ell}$-diagnosable for any $d > 1$; hence, its $||^{d^\ell}$-diagnosability collapses to $||^{//\ell}$-diagnosability.

10.4.1 Diagnosability and Temporal Uncertainty

We now adapt the twin plant method to DESs with temporally uncertain observations. First, we present the $||^{//\ell}$-verifier, which is the verifier that incorporates temporal uncertainty. This is nothing but the classical twin plant verifier where the alphabet of observable events, instead of being Σ_o, is $\left(\!\!\binom{\Sigma_o}{\leq \ell}\!\!\right)$, i.e. the alphabet of all temporally compound observable events up to a given level ℓ.

Definition 10 (**$||^{//\ell}$-Verifier**) Let $D = (\Sigma, L, obs, flt)$ be a DES where $\Sigma_o \subseteq \Sigma$ and $\Sigma_f \subseteq \Sigma$ are the sets of observable and faulty events, respectively. Let $G = (X, \Sigma, \delta, x_0)$ be an FA generating L. The $||^{//\ell}$-*verifier* relevant to a fault $f \in \Sigma_f$ is an FA $G^{//\ell} = (X^{//\ell}, \Sigma^{//\ell}, \delta^{//\ell}, x_0^{//\ell})$ defined as follows:

- $X^{//\ell} = X \times \{N, F\}$ and $x_0^{//\ell} = (x_0, N)$;
- $\Sigma^{//\ell} = \left(\!\!\binom{\Sigma_o}{\leq \ell}\!\!\right)$; and
- $\delta^{//\ell} = \{((x, \phi), w, (x', \phi')) \in X^{//\ell} \times \Sigma^{//\ell} \times X^{//\ell} \mid \exists$ a path in $G: x \xrightarrow{\sigma_1} \ldots \xrightarrow{\sigma_n} x'$, $n \geq 1$, $\sigma_n \in \Sigma_o$, $w \in ||obs(\sigma_1 \cdot \cdots \cdot \sigma_n)||^{//\ell}$ and $\phi' = N \Leftrightarrow (\phi = N \wedge f \notin \{\sigma_1, \ldots, \sigma_n\})\}$.

Based on the complexity analysis performed in Sect. 10.2.3, the computational complexity of the construction the $||^{//\ell}$-verifier relevant to a fault is $O(|X|^2|\Sigma^{//\ell}|) = O(|X|^2|\Sigma_o|^\ell)$.

Once $G^{//\ell}$ has been built, it has to be synchronized with itself, which results in the twin plant. Diagnosability holds if no loop in the twin plant includes ambiguous states, as stated in Theorem 1.

Theorem 1 *Given a DES D whose behavior is represented by FA* $G = (X, \Sigma, \delta, x_0)$ *and the* $||^{//\ell}$-*verifier* $G^{//\ell}$ *of fault f drawn from it, fault f is* $||^{//\ell}$-*diagnosable iff* $G^{//\ell} \otimes G^{//\ell}$ *contains no loop of ambiguous states.*

Proof Outline The proof is similar to the corresponding one in the classical twin plant approach [13]. A loop of ambiguous states proves that there is an infinite ambiguous path in the twin plant. By construction, an infinite ambiguous path in the twin plant betrays the existence of two infinite behaviors of the DES, an N one (which does not include any occurrence of fault f) and an F one (in which fault f has certainly occurred), that can indefinitely generate the same temporally uncertain observation. This shows that the finite delay k in Definition 9 after which the fault can be diagnosed does not exist. □

The complexity of the whole method to check the $||^{//\ell}$-diagnosability of a fault (see Sect. 10.2.3) is $O(|X|^4|\Sigma_o|^\ell)$.

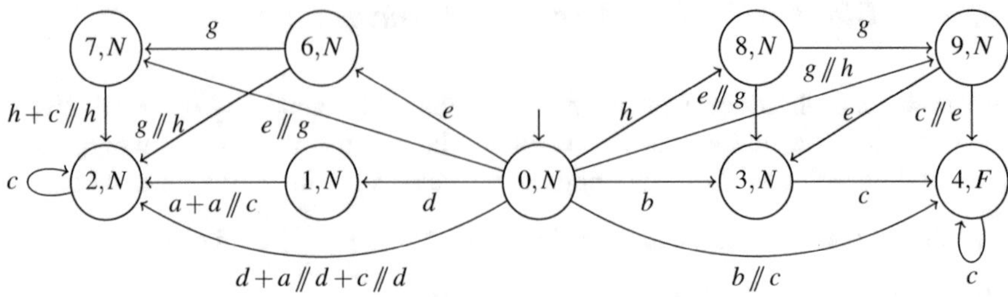

Fig. 10.5 $||^{//2}$-Verifier for the DES in Fig. 10.1

The $||^{//2}$-verifier for our sample DES D in Fig. 10.1 is depicted in Fig. 10.5. Instead of displaying all the transitions having the same source and target nodes, just one is shown, which is labeled by all the events triggering these transitions, where $+$ is a separator.

10.4.2 Diagnosability and Logical Uncertainty

We now present the $||^{\#d}$-verifier, which incorporates logical uncertainty. This is nothing but the classical twin plant verifier where the alphabet of observable events, instead of being Σ_o, is $2^{\Sigma_o+,\leq d}$, i.e. the alphabet of all logically compound observable events up to a given degree d (according to the considered distance matrix M).

Definition 11 (**$||^{\#d}$-Verifier**) Let $D = (\Sigma, L, obs, flt)$ be a DES where $\Sigma_o \subseteq \Sigma$ and $\Sigma_f \subseteq \Sigma$ are the sets of observable and faulty events, respectively. Let $G = (X, \Sigma, \delta, x_0)$ be an FA generating L. Let M be the distance matrix. The $||^{\#d}$-*verifier* relevant to a fault $f \in \Sigma_f$ is an FA $G^{\#d} = (X^{\#d}, \Sigma^{\#d}, \delta^{\#d}, x_0^{\#d})$ defined as follows:

- $X^{\#d} = X \times \{N, F\}$ and $x_0^{\#d} = (x_0, N)$;
- $\Sigma^{\#d} = 2^{\Sigma_o+,\leq d}$; and
- $\delta^{\#d} = \{((x, \phi), w, (x', \phi')) \in X^{\#d} \times \Sigma^{\#d} \times X^{\#d} \mid \exists$ a path in $G: x \xrightarrow{\sigma_1} \ldots \xrightarrow{\sigma_n} x'$, $n \geq 1$, $obs(\sigma_1 . \cdots . \sigma_n) = \sigma_n$, $w \in ||\sigma_n||^{\#d}$ and $\phi' = N \Leftrightarrow (\phi = N \wedge f \notin \{\sigma_1, \ldots, \sigma_n\})\}$.

Basically, each path from a source state to a target state in the behavioral model of the DES, where such a path terminates with an observable transition (marked with an event e), this being the only observable transition in the path, is mirrored by a transition from the same source to the same target for each of the possible logically compound events relevant to e (including $\varepsilon\#e$, if proper according to matrix M and the value of degree d).

The semantics of a transition $((x, \phi), w, (x', \phi'))$ of the $||^{\#d}$-verifier, where w is a logically compound event that contains ε (say, an ε-*event*), such as $\varepsilon\#e$, is that,

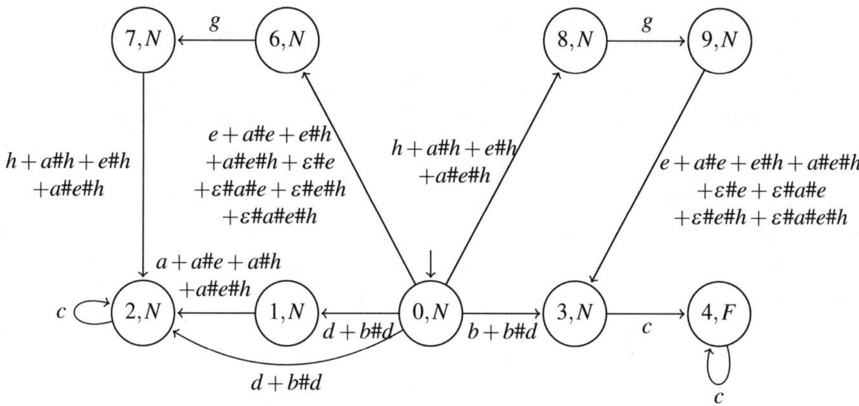

Fig. 10.6 $||^{\#4}$-Verifier for the DES in Fig. 10.1

when the DES is in state x and event $\varepsilon\#e$ is perceived, this can actually denote the state change to x', since such an event is not necessarily noise. If, instead, in the $||^{\#d}$-verifier there is no transition exiting from (x,ϕ) that is marked with $\varepsilon\#e$, this means that, if event $\varepsilon\#e$ is perceived when the state of the DES is x, then $\varepsilon\#e$ is necessarily noise.

The computational complexity of the construction of the $||^{\#d}$-verifier relevant to a fault is $O(|X|^2 2^{|\Sigma_{o+}|})$. This estimate is quite general and very pessimistic: a more accurate estimate should take into account the value of degree d and the specific distance matrix M.

Figure 10.6 depicts the $||^{\#4}$-verifier relevant to the sample DES in Fig. 10.1. Building this verifier is just an exercise since we already know that the sample DES is not diagnosable for any degree of logical uncertainty greater than 1.

A transition $((x,\phi), w, (x',\phi'))$ of the $||^{\#d}$-verifier of a fault f marked with a logically compound event including ε, such as $\varepsilon\#e$, denotes that, if such an event is perceived when the current candidate diagnosis (as far as fault f is concerned) is (x,ϕ), then if we assume that what we have observed is not noise, the next candidate diagnosis is (x',ϕ'). However, the diagnosis task, when processing $\varepsilon\#e$, considers also the chance that no observable event has been generated by the system. Since this is not explicitly represented by any transition in the $||^{\#d}$-verifier, it has to be taken into account in the twin plant.

Definition 12 (Logical Product) Let $G^{\#d} = (X^{\#d}, \Sigma^{\#d}, \delta^{\#d}, x_0^{\#d})$ be a $||^{\#d}$-verifier. The *logical product* of $G^{\#d}$ by itself, denoted as $G^{\#d} \otimes_L G^{\#d}$, is a nondeterministic FA $(X^L, \Sigma^L, \delta^L, x_0^L)$ defined as follows:

- $X^L \subseteq X^{\#d} \times X^{\#d}$;
- $x_0^L = (x_0^{\#d}, x_0^{\#d})$;
- $\Sigma^L = \Sigma^{\#d}$; and
- $\delta^L = \delta_1^L \cup \delta_2^L \cup \delta_3^L$ where

– $\delta_1^L = \{((x_1, x_2), w, (x_1', x_2')) \in X^L \times \Sigma^L \times X^L \mid (x_1, w, x_1') \in \delta^{\#d}, (x_2, w, x_2') \in \delta^{\#d}\}$;
– $\delta_2^L = \{((x_1, x_2), w, (x_1', x_2)) \in X^L \times \Sigma^L \times X^L \mid (x_1, w, x_1') \in \delta^{\#d}, \varepsilon \in w\}$;
– $\delta_3^L = \{((x_1, x_2), w, (x_1, x_2')) \in X^L \times \Sigma^L \times X^L \mid (x_2, w, x_2') \in \delta^{\#d}, \varepsilon \in w\}$.

If δ^L were equal to δ_1^L, the logical product would just be the product of $G^{\#d}$ by itself. However, this would not account for the fact that any perceived ε-event may be pure noise. This is the reason for δ_2^L and δ_3^L are needed. The former encompasses the cases when, given a state of the twin plant, this representing a pair of candidate diagnoses (x_1, x_2), an ε-event is perceived and the diagnosis task assumes (if it is possible, that is, if in the $||^{\#d}$-verifier there is a transition exiting from x_1 marked with such an event) that it is not noise for x_1 while it is noise for x_2. The dual cases are encompassed by δ_3^L.

Theorem 2 *Given a DES D whose behavior is represented by FA $G = (X, \Sigma, \delta, x_0)$, a distance matrix M, and the $||^{\#d}$-verifier $G^{\#d}$ of fault f, fault f is $||^{\#d}$-diagnosable iff $G^{\#d} \otimes_L G^{\#d}$ contains no loop of ambiguous states.*

Proof Outline The proof amounts to showing that the logical product builds a twin plant such that (i) each path represents a pair of evolutions of the considered DES (and, consequently, of the relevant candidate diagnoses) that are compatible with the same (logically uncertain) observation, and (ii) any pair of distinct evolutions compatible with the same (logically uncertain) observation is represented by a path in the twin plant. Point (i) can be understood by observing that (*a*) the initial state of the twin plant is a pair of candidate diagnoses according to which the DES is in its initial state and it is free of faults, and (*b*) each transition in the twin plant brings to a new pair of candidate diagnoses, relevant to a pair of evolutions that are driven by the same perception of a new (logically compound) observable event. Such an evolution is compliant with the $||^{\#d}$-verifier, which in turn is compliant with the behavioral model of the DES.

As to the completeness claimed by point (ii), let us assume that the observation is a logical relaxation of the sequence of observable events that take place in the system, possibly intermixed by ε-events that are actually pure noise. Let us consider a state (x_1, x_2) of the twin plant. Any new perceived (logically compound) observable event falls into one of the following categories:

- it is not an ε-event, in which case in the twin plant there is a δ_1^L transition exiting from (x_1, x_2) marked with such an event;
- it is an ε-event that can actually correspond to an observable event perceivable both when the DES is in state x_1 and when it is in state x_2. In this case, there are three transitions in the twin plant exiting from (x_1, x_2) and marked with such an event: a δ_1^L transition (which corresponds to assuming that the event is noise neither starting from x_1 nor starting from x_2), a δ_2^L transition (which corresponds to assuming that the event is noise starting from x_2), and a δ_3^L transition (which corresponds to assuming that the event is noise starting from x_1). Also the case when the event is noise for both the evolutions is encompassed by the twin plant, since this is represented by the current state (x_1, x_2);

- it is an ε-event that can actually correspond to an observable event perceivable only when the DES is either in state x_1 or in state x_2, in which case in the twin plant there is a transition exiting from (x_1, x_2) marked with such an event. Such a transition is a δ_2^L transition, if the event is observable starting from x_1, or a δ_3^L transition, if the event is observable starting from x_2. As above, also the case when the event is noise for both the evolutions is encompassed by the twin plant, since this is represented by the current state (x_1, x_2);
- it is an ε-event that cannot correspond to any observable event perceivable when the DES is in state x_1 or in state x_2, that is, if such event is perceived when the system is in state x_1 or in state x_2, such event necessarily corresponds to noise. This is correctly registered in the twin plant since there is no transition marked with such an event exiting from (x_1, x_2), which implies that, according to Theorem 2, an infinite sequence of ε-events that are necessarily noise do not concur to identify a lack of $||^{\#d}$-diagnosability (as in fact such an infinite sequence is not represented by any loop in the twin plant).

An infinite ambiguous path in the twin plant betrays the existence of two infinite behaviors of the DES, an N one (which does not include any occurrence of fault f) and an F one (in which fault f has certainly occurred), that can indefinitely generate the same logically uncertain observation, possibly intermixed by ε-events that are actually pure noise. □

The complexity of the logical product and of whole method to check the $||^{\#d}$-diagnosability of a fault is $O(|X|^4 2^{|\Sigma_{o+}|})$.

10.4.3 Diagnosability and Combined Uncertainty

We now present the $||^{d^\ell}$-verifier, which is the verifier that incorporates the combined (temporal and logical) uncertainty. Its construction basically consists in building the $||^{//\ell}$-verifier of the $||^{\#d}$-verifier.

Definition 13 ($||^{d^\ell}$-Verifier) Let $D = (\Sigma, L, obs, flt)$ be a DES where $\Sigma_o \subseteq \Sigma$ and $\Sigma_f \subseteq \Sigma$ are the sets of observable and faulty events, respectively. Let $G = (X, \Sigma, \delta, x_0)$ be an FA generating L. Let M be the distance matrix. Let $G^{\#d} = (X^{\#d}, \Sigma^{\#d}, \delta^{\#d}, x_0^{\#d})$ be the $||^{\#d}$-verifier of a fault $f \in \Sigma_f$. The $||^{d^\ell}$*-verifier* relevant to the same fault is an FA $G^{d^\ell} = (X^{d^\ell}, \Sigma^{d^\ell}, \delta^{d^\ell}, x_0^{d^\ell})$ defined as follows:

- $X^{d^\ell} = X^{\#d}$ and $x_0^{d^\ell} = x_0^{\#d}$;
- $\Sigma^{d^\ell} = \left(\!\!\binom{\Sigma^{\#d}}{\leq \ell}\!\!\right)$; and
- $\delta^{d^\ell} = \{((x, \phi), w, (x', \phi')) \in X^{d^\ell} \times \Sigma^{d^\ell} \times X^{d^\ell} \mid \exists$ a path in $G^{\#d}$: $(x, \phi) \xrightarrow{\sigma_1} \ldots \xrightarrow{\sigma_n} (x', \phi'), 1 \leq n \leq \ell, w \in ||\sigma_1. \cdots . \sigma_n||^{//\ell}\}$.

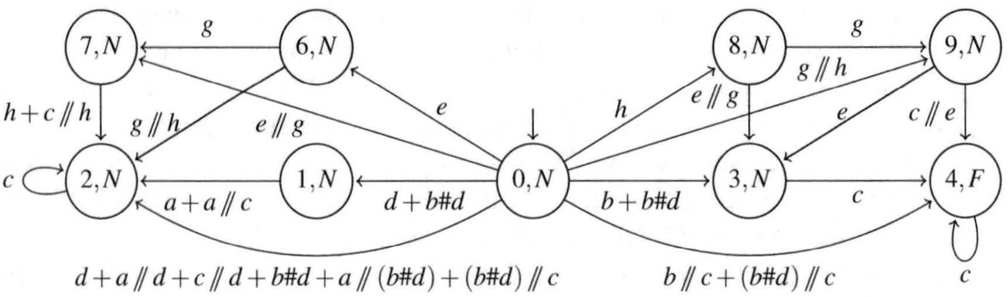

Fig. 10.7 $||^{2^2}$-Verifier for the DES in Fig. 10.1

Notice that, in Definition 13, $||\sigma_1.\cdots.\sigma_n||^{//\ell}$ applies temporal uncertainty to a sequence of logically compound events.

The complexity of the construction of the $||^{d^\ell}$-verifier relevant to a fault is $O(|X^{d^\ell}|^2|\Sigma^{d^\ell}|) = O(|X^{\#d}|^2|\Sigma^{\#d}|^\ell) = O(|X|^2|2^{\Sigma_o+,\leq d}|^\ell) = O(|X|^2 2^{\ell|\Sigma_o+|})$.

Figure 10.7 depicts the $||^{2^2}$-verifier for the sample DES. No ε-event can be found since value 2 of the logical uncertainty degree causes just observable events b and d to possibly be confused with each other.

The perception of a logically compound event that includes ε (that is, an ε-event) corresponds either to an observable event that took place in the DES or to no observable event at all. While the $||^{d^\ell}$-verifier (the same as the $||^{\#d}$-verifier) explicitly represents just the former option, the twin plant has to take into account both of them.

Given a combined compound event w, let us denote as $Sub(w)$ the set of all the submultisets of w such that each submultiset is obtained by casually removing from w one or more instances of ε-events.

Definition 14 (Combined Product) Let $G^{d^\ell} = (X^{d^\ell}, \Sigma^{d^\ell}, \delta^{d^\ell}, x_0^{d^\ell})$ be a $||^{d^\ell}$-verifier. The *combined product* of G^{d^ℓ} by itself, denoted as $G^{d^\ell} \otimes_C G^{d^\ell}$, is a nondeterministic FA $(X^C, \Sigma^C, \delta^C, x_0^C)$ defined as follows:

- $X^C \subseteq X^{d^\ell} \times X^{d^\ell}$;
- $x_0^C = (x_0^{d^\ell}, x_0^{d^\ell})$;
- $\Sigma^C = \Sigma^{d^\ell}$; and
- $\delta^C = \delta_1^C \cup \delta_2^C \cup \delta_3^C$ where
  - $\delta_1^C = \{((x_1, x_2), w, (x_1', x_2')) \in X^C \times \Sigma^C \times X^C \mid (x_1, w, x_1') \in \delta^{d^\ell}, (x_2, w, x_2') \in \delta^{d^\ell}\}$;
  - $\delta_2^C = \{(x_1, x_2), w, (x_1', x_2') \in X^C \times \Sigma^C \times X^C \mid (x_1, w, x_1') \in \delta^{d^\ell}, w' \in Sub(w), (x_2, w', x_2') \in \delta^{d^\ell}$ if w' is not empty, $x_2' = x_2$ otherwise$\}$;
  - $\delta_3^C = \{(x_1, x_2), w, (x_1', x_2') \in X^C \times \Sigma^C \times X^C \mid (x_2, w, x_2') \in \delta^{d^\ell}, w' \in Sub(w), (x_1, w', x_1') \in \delta^{d^\ell}$ if w' is not empty, $x_1' = x_1$ otherwise$\}$.

In the combined product definition, δ_1^C includes the transitions that result from strict synchronization (i.e., the synchronization of two transitions in the $||^{d^\ell}$-verifier that are marked with the same combined compound event). Instead, δ_2^C includes the transitions that result from the synchronization of a transition (relevant to x_1) in the $||^{d^\ell}$-verifier that is marked with a combined compound event w with a transition (relevant to x_2) in the $||^{d^\ell}$-verifier that is marked with a (possibly empty) submultiset of w. In other words, a multiset of logically compound events can be synchronized with any of its submultisets obtained by removing one or more ε-events (which means that the removed events are assumed to be just noise). Finally, δ_3^C is the dual of δ_2^C, as w is relevant to x_2 while w' is relevant to x_1.

Theorem 3 *Given a DES D whose behavior is represented by FA $G = (X, \Sigma, \delta, x_0)$, a distance matrix M, and the $||^{d^\ell}$-verifier G^{d^ℓ} of a fault f, fault f is $||^{d^\ell}$-diagnosable iff $G^{d^\ell} \otimes_C G^{d^\ell}$ contains no loop of ambiguous states.*

Proof Outline The proof is similar to that of Theorem 2, but while in the logical product each transition was marked with a logically compound event, here each transition is marked with a combined compound event, this being a multiset of logically compound events. The construction of the twin plant for the combined uncertainty, i.e. the combined product, is similar to that inherent to the logical product, where δ_1^C, δ_2^C, and δ_3^C subsume δ_1^L, δ_2^L, and δ_3^L, respectively (as δ_i^C equals δ_i^L when the considered combined compound event w is indeed a multiset containing a single logically compound event). This construction guarantees that (1) each path in the combined product represents a pair of evolutions of the considered DES (and, consequently, of the relevant candidate diagnoses) that are compatible with the same combined uncertain observation, and (2) any pair of distinct evolutions compatible with the same combined uncertain observation is represented by a path in the twin plant. An infinite ambiguous path in the twin plant betrays the existence of two infinite behaviors of the DES, a normal and a faulty one, that can indefinitely generate the same combined uncertain observation, where any ε-event contained in a combined compound event can actually be pure noise. □

A very pessimistic estimate of the complexity of the combined product and of the whole method to check the $||^{d^\ell}$-diagnosability of a fault is $O(|X|^4 2^{\ell|\Sigma_o + |})$.

10.5 Example

We illustrate the results of this chapter, in particular those relevant to diagnosability of HSs with logically uncertain observations, with an example inspired from the monitoring of power distribution systems. The example is very much trimmed down to make it manually manageable.

In a power distribution network, components can be connected in a tree. The very simple network we consider here contains three components (one root and two

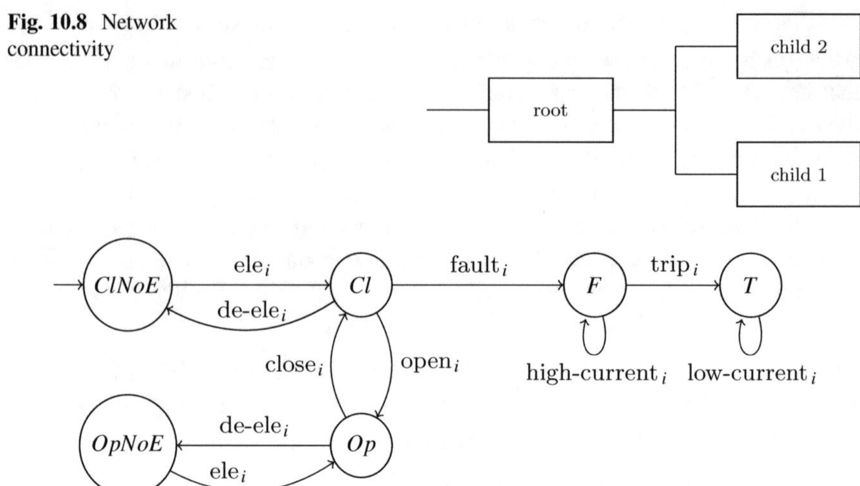

Fig. 10.8 Network connectivity

Fig. 10.9 Component model

children), as depicted in Fig. 10.8. The power comes from a source (on the left of the root in the figure), and it has to be distributed from the root to its children.

10.5.1 Discrete Model

Each component in the network has six operating modes that can be represented as the states of a DES (cf. Fig. 10.9): closed with no electricity (*ClNoE*), closed (*Cl*), open with no electricity (*OpNoE*), open (*Op*), faulty (*F*), and tripped (*T*). In case the component is not fed, no electricity is circulating, hence the component is either in its initial mode, *ClNoE*, or in mode *OpNoE*. When the component is fed, it lets the power flow when it is closed (mode *Cl*), whereas, when it is open (mode *Op*) or tripped (mode *T*), it behaves as an open circuit; finally, when it is faulty (mode *F*), the component behaves as a short circuit. We assume that the component cannot become faulty when not fed or when open. Analogously, the component can trip only if it is faulty. We also make a single fault assumption, meaning that no two components can be faulty.

While the DES model is shared by all components, the events in the model are specific to each component, as underlined by the parametric subscript *i*. In the following of this example, as far as events are concerned, subscripts 0, 1, and 2 refer to the root, child 1, and child 2, respectively.

The set of observable events of the overall network is $\Sigma_o = \{$ele$_0$, de-ele$_0$, open$_0$, close$_0$, high-current$_0$, low-current$_0$, open$_1$, close$_1$, high-current$_1$, low-current$_1$, open$_2$, close$_2$, high-current$_2$, low-current$_2\}$. The set of faulty events of the network is $\Sigma_f = \{$fault$_0$, fault$_1$, fault$_2\}$.

Event ele_0 triggers the transition from the initial mode of the root to mode *Cl*. Moreover, such an event electrifies the network. In fact, (observable) event ele_0 is synchronous with both (unobservable) events ele_1 and ele_2, which trigger the analogous transitions in child 1 and child 2, respectively. When all components are in mode *Cl*, the power flows from the root to both children. However, when a component is operating in mode *Cl*, a command can be sent to it in order to make it open. Command open_0 makes the root switch to mode *Op*. Such an event (open_0) is synchronous with both events de-ele_1 and de-ele_2; this means that, if command open_0 is dispatched when all components are closed, the root switches to mode *Op* while both children switch back to their initial mode, that is, they are de-electrified. While the root is open, it can repeatedly be de-electrified, reaching mode *OpNoE*, and electrified again, coming back to mode *Op*. When the root is in mode *Open* and the command close_0 is issued, the root switches to mode *Cl*, and the same holds also for the two children since event close_0 is synchronous with both events ele_1 and ele_2. If, instead, when all components are closed, either command open_1 or open_2 is sent in order to open one of the children, such a child switches to mode *Op*, without affecting the modes of the other components. If, while such a component is in mode *Op*, the network is repeatedly de-electrified and electrified again, the component at each iteration reaches mode *OpNoE* and comes back to mode *Op*. When a child is in mode *Op* and the relevant command to make it close is issued, the child switches to mode *Cl*. When a component is in mode *Cl*, it may get faulty.

When the root is electrified, its input current is sampled. Some observable discrete events (high-current_0, low-current_0, high-current_1, low-current_1, high-current_2, low-current_2) model the abnormal values of such a current. After a fault has occurred in a component, its new mode is *F*, and an abnormally high value of the current is observed (possibly several times), which is modeled by a loop transition relevant to mode *F*. Finally, some protection mechanism makes the component reach mode *T*, starting from which an abnormally low current can be indefinitely observed.

10.5.2 Continuous Model

The continuous behavior of a component c within each mode where it is electrified is modeled with some state variables:

- *internal resistance*: $R_c := \hat{R}_c$ if the mode is *Cl*; $R_c := +\infty$ if the mode is *Op* or *T*; $R_c := 0$ (close to zero) if the mode is *F*.
- *subtree resistance*: $R_{<c} := R_c$ if the component has no child; $R_{<c} := R_c + \frac{1}{\sum_{h \in C} \frac{1}{R_{c,h}+R_h}}$ where C is the list of children of the component (each child is represented by h) and $R_{c,h}$ is the resistance of the line between c and h. This holds for all electrified modes (*Cl*, *Op*, *F*, and *T*).
- *current*: $I := V/R_{<root}$, where V is the (constant) voltage. Variable I is the input current of the root, which is affected by the subtree resistance of the root itself.

Table 10.1 Constants in the example

Name	Value
V	100
$\hat{R}_{\text{root}}$	10
$\hat{R}_{\text{child1}}$	20
$\hat{R}_{\text{child2}}$	25
$R_{\text{root, child1}}$	5
$R_{\text{root, child2}}$	5

> Since the value of such a subtree resistance depends on the modes of all the components, the values of current I change whenever a component changes its mode.

The constants of the network in Fig. 10.8 are given in Table 10.1 (no unit of measurements is provided out of simplicity).

When a fault occurs in a component, its internal resistance drops, which makes current I peak before the component trips. After it has tripped, the internal resistance of the component increases and current I decreases.

If the fault takes place in the root, first the internal resistance of the root drops close to zero (mode F) and then it goes to $+\infty$ (mode T). Hence, current I will peak very high and then decrease down to zero.

Assuming all components are closed, if the fault occurs in child i, when child i switches to mode F, the subtree resistance relevant to the root will drop to $R_{<root} := \hat{R}_{\text{root}} + \frac{1}{\frac{1}{R_{\text{root, child}i}} + \frac{1}{R_{\text{root, child}j} + \hat{R}_{\text{child}j}}}$, where child j is the other child. When child i reaches mode T, the subtree resistance relevant to the root will rise to $R'_{<root} := \hat{R}_{\text{root}} + \frac{1}{\frac{1}{R_{\text{root, child}j} + \hat{R}_{\text{child}j}}}$.

If the occurred fault is fault$_1$, the values are $R_{<root} = 100/7 \simeq 14.3$, when child$_1$ is in mode F, and $R'_{<root} = 40$, when child$_1$ is in mode T. The values of current I corresponding to these subtree resistances are 7 and 2.5, respectively. Hence, observable event high-current$_1$ takes place whenever the sampled value of I is 7, while observable event low-current$_1$ takes place whenever the sampled value of I is 2.5.

If, instead, fault$_2$ has occurred, the values are $R_{<root} = 85/6 \simeq 14.17$, when child$_2$ is in mode F, and $R'_{<root} = 35$, when child$_1$ is in mode T. The values of current I corresponding to these subtree resistances are $120/17 \simeq 7.05$ and $20/7 \simeq 2.86$, respectively. Hence, observable event high-current$_2$ takes place whenever the sampled value of I is 7.05, while observable event low-current$_2$ takes place whenever the sampled value of I is 2.86.

10.5.3 Diagnosability Analysis

The methods to perform the diagnosability analysis of a DES with uncertain observations described in the previous sections are centralized, that is, they rely on a single DES representing the behavior of the whole system. Hence, in order to analyze the diagnosability of the network considered as an example in the current section by exploiting the methods described in Sects. 10.4.1–10.4.3, we should, first of all, carry out the parallel composition of the discrete models of the three components (see Sect. 10.5.1). The resulting DES corresponds to DES D mentioned in the various definitions given in those sections. We leave the construction of D and the subsequent application of the methods to the reader as a useful exercise. Instead, here we will deal with an intuitive diagnosability analysis in case of logically uncertain observations.

Observable events high-current$_i$ and low-current$_i$ specifically identify the occurrence of fault$_i$, which makes each fault in the network diagnosable (and the DES diagnosable too) in case there is no uncertainty in the observation.

Let us assume that, according to the distance matrix specific to the network, the distance between high-current$_1$ and high-current$_2$ is 1, since their corresponding continuous values of I are very close, while the distance between low-current$_1$ and low-current$_2$ is 2, since these discrete events correspond to continuous values of I that are slightly further apart. The interesting scenario is the case where all the network components are closed and one of the two children experiences a fault. In such a case, the sequence of observable events high-current$_1$.low-current$_1$, relevant to fault$_1$, needs to be distinguished from the sequence high-current$_2$.low-current$_2$, relevant to fault$_2$. If the logical uncertainty degree of the observation is $d = 2$, then the second observable event of one sequence can always be distinguished from the second observable event of the other, and the diagnosis will precisely identify the fault (and hence the faulty component). Therefore the system is $||^{\#2}$-diagnosable. However, it is not $||^{\#3}$-diagnosable as the observation of the two traces cannot be distinguished if the logical uncertainty degree increases to 3.

10.6 Conclusions

This chapter investigates how uncertainty in observations can affect the diagnosability of a DES, i.e. the ability of detecting a fault without any ambiguity within a finite number of observable events after the fault has occurred. The analysis is carried out in a scenario where the considered DES is diagnosable according to the original definition of DES diagnosability in the literature (that is, it is diagnosable given a certain observation) and the diagnostic algorithm is exact. In particular, the chapter deals with the above topic in the context of event-based approaches to fault modeling [12] and a relaxation of the temporal constraints and/or the logical constraints between observed events is considered.

In the case of temporal uncertainty only, the observation becomes a sequence of compound temporal events. If the maximum cardinality of the considered temporally compound events is ℓ (which is called the *temporal uncertainty level* of the observation), this means that, after ℓ (single) observable events have been recorded, we may be unable to find out in which temporal order they were produced by the DES. This is the case, for instance, when the observable events are conveyed to the observer through distinct channels, having distinct clocks or delays, and the synchronization error of these clocks (or the—possibly varying—length of these delays) is such that at most ℓ events can be received without the observer being able to disclose their reciprocal temporal order.

In the case of logical uncertainty only, the observation becomes a sequence of compound logical events. If the *logical uncertainty degree* is d, each observable event occurred in the DES may be confused with some other observable event(s) or even with pure noise, provided that its distance from them is less than d. This is the case, for instance, when the observable events are conveyed to the observer through channels affected by noise. Hence, both temporal and logical relaxations are quite meaningful and representative of real-world situations.

In the case both temporal and logical uncertainty can affect the observation, this becomes a sequence of compound combined events, having both a temporal uncertainty level and a logical uncertainty degree.

This chapter provides a definition of DES diagnosability that extends the original definition [21] in the literature, as well as a method to check whether the newly defined property holds for a given DES, that extends the original twin plant method [13] considering temporal and logical uncertainty and their combination. If a DES is diagnosable even if the observation has a temporal uncertainty level of value $\ell > 1$ and/or a logical uncertainty degree $d > 1$, the diagnosis task can be performed without any loss in the ability to disambiguate candidates although the available measuring equipment cannot get a certain observation. The higher the uncertainty level/degree that still guarantees diagnosability, the less expensive the needed measuring equipment and its design.

Future research can follow two orthogonal directions, one focused on the distribution and the other on the extension of the proposed conceptual framework. Such a framework is currently based on a global model of the DES at hand and on its monolithic processing. Instead, the model can be compositional and a distributed processing method can be adopted. As to the second research direction, the new definition of DES diagnosability and the proposed method to check it could be adapted to state-based approaches of fault modeling. In addition, all kinds of temporal uncertainty should be addressed including the relaxations of temporal constraints that are not sequences of temporally compound events but bring to uncertain observations. A further challenge is to define diagnosability in the frame of (temporal and/or logical) uncertainty of observations and approximate diagnostic algorithms altogether.

Finally, it is worthwhile underlining that investigating diagnosability with uncertain observations is especially meaningful for HSs. In fact, in an HS some observable events triggering discrete transitions, instead of being events perceivable by an external observer, as it is usually the case with DESs, can indeed be events whose values are computed based on processing relevant to the dynamics within states. For instance, let us assume that, given (the model of) a state having two exiting transitions, the diagnostic engine sometimes is not able to find out with certainty the transition the system is following when leaving the state. This situation can be represented as a logically uncertain observation relevant to the underlying DES, according to which the value of the observed event ranges over the pair of values marking such transitions. Hence, observable discrete events can be exploited as an artifice to model the interaction between continuous and discrete dynamics within HSs, and the diagnosability analysis of the DES accounts also for the reasoning on continuous processes. Moreover, HSs justify the investigation on diagnosability with several types of uncertainty in the observations since new types of uncertainty can be envisaged (besides temporal and logical) if some observable events of the underlying DES are actually meant to model the outcomes of reasoning on continuous processes within states.

References

1. P. Baroni, G. Lamperti, P. Pogliano, M. Zanella, Diagnosis of large active systems. Artif. Intell. **110**(1), 135–183 (1999)
2. A. Bauer, A. Botea, A. Grastien, P. Haslum, J. Rintanen, Alarm processing with model-based diagnosis of discrete event systems, in *AI for an Intelligent Planet (AIIP-11)* (2011), pp. 2:1–2:8
3. M. Bayoudh, L. Travé-Massuyès, Diagnosability analysis of hybrid systems cast in a discrete-event framework. J. Discrete Event Dyn. Syst. **24**(3), 309–338 (2014)
4. C. Cassandras, S. Lafortune, *Introduction to Discrete Event Systems*, 2nd edn. (Springer, New York, 2008)
5. A. Cimatti, C. Pecheur, R. Cavada, Formal verification of diagnosability via symbolic model checking, in *18th International Joint Conference on Artificial Intelligence (IJCAI-03)* (2003), pp. 363–369
6. O. Contant, S. Lafortune, D. Teneketzis, Diagnosability of discrete event systems with modular structure. J. Discrete Event Dyn. Syst. **16**, 9–37 (2006)
7. A. Grastien, M.O. Cordier, C. Largouët, Extending decentralized discrete-event modelling to diagnose reconfigurable systems, in *15th International Workshop on Principles of Diagnosis (DX-04)* (2004)
8. A. Grastien, A. Anbulagan, J. Rintanen, E. Kelareva, Diagnosis of discrete-event systems using satisfiability algorithms, in *22nd Conference on Artificial Intelligence (AAAI-07)* (2007), pp. 305–310
9. A. Grastien, P. Haslum, S. Thiébaux, Conflict-based diagnosis of discrete event systems: theory and practice, in *13th International Conference on the Principles of Knowledge Representation and Reasoning (KR-12)* (2012)
10. A. Grastien, L. Travé-Massuyès, V. Puig, Solving diagnosability of hybrid systems via abstraction and discrete event techniques, in *20th World Congress of the International Federation of Automatic Control (WC-17)* (2017), pp. 5174–5179
11. T.A. Henzinger, The theory of hybrid automata, in *Verification of Digital and Hybrid Systems*, ed. by M.K. Inan, R.P. Kurshan. NATO ASI Series (Series F: Computer and Systems Sciences), vol. 170 (Springer, Berlin, 2000), pp. 265–292

12. T. Jéron, H. Marchand, S. Pinchinat, M.O. Cordier, Supervision patterns in discrete-event systems diagnosis, in *17th International Workshop on Principles of Diagnosis (DX-06)* (2006), pp. 117–124
13. S. Jiang, Z. Huang, V. Chandra, R. Kumar, A polynomial algorithm for testing diagnosability of discrete event systems. IEEE Trans. Autom. Control **46**(8), 1318–1321 (2001)
14. G. Lamperti, M. Zanella, Diagnosis of discrete-event systems from uncertain temporal observations. Artif. Intell. **137**(1–2), 91–163 (2002)
15. G. Lamperti, M. Zanella, *Diagnosis of Active Systems – Principles and Techniques*. The Kluwer International Series in Engineering and Computer Science, vol. 741 (Kluwer Academic Publishers, Dordrecht, 2003)
16. G. Lamperti, M. Zanella, On processing temporal observations in monitoring of discrete-event systems, in *Enterprise Information Systems VIII*, ed. by Y. Manolopoulos, J. Filipe, P. Constantopoulos, J. Cordeiro. Lecture Notes in Business Information Processing, vol. 3 (Springer, Berlin, 2008), pp. 135–146
17. G. Lamperti, M. Zanella, Diagnosis of active systems by lazy techniques, in *12th International Conference on Enterprise Information Systems (ICEIS-10)*, Funchal, Madeira (2010), pp. 171–180
18. G. Lamperti, M. Zanella, Monitoring of active systems with stratified uncertain observations. IEEE Trans. Syst. Man Cybern. Part A Syst. Humans **41**(2), 356–369 (2011)
19. Y. Pencolé, M.O. Cordier, A formal framework for the decentralised diagnosis of large scale discrete event systems and its application to telecommunication networks. Artif. Intell. **164**(1–2), 121–170 (2005)
20. B. Pulido, C. Alonso González, Possible conflicts: a compilation technique for consistency-based diagnosis. IEEE Trans. Syst. Man Cybern. Part B Cybern. **34**(5), 2192–2206 (2004)
21. M. Sampath, R. Sengupta, S. Lafortune, K. Sinnamohideen, D. Teneketzis, Diagnosability of discrete-event systems. IEEE Trans. Autom. Control **40**(9), 1555–1575 (1995)
22. A. Schumann, Y. Pencolé, S. Thiébaux, A spectrum of symbolic on-line diagnosis approaches, in *22nd Conference on Artificial Intelligence (AAAI-07)* (2007), pp. 335–340
23. R. Sengupta, S. Tripakis, Decentralized diagnosability of regular languages is undecidable, in *43rd IEEE Conference on Decision and Control* (2004), pp. 423–428
24. M. Staroswiecki, G. Comtet-Varga, Analytical redundancy relations for fault detection and isolation in algebraic dynamic systems. Automatica **37**(5), 687–699 (2001)
25. X. Su, A. Grastien, Verifying the precision of diagnostic algorithms, in *21st European Conference on Artificial Intelligence* (2014), pp. 861–866
26. R. Su, W. Wonham, Global and local consistencies in distributed fault diagnosis for discrete-event systems. IEEE Trans. Autom. Control **50**(12), 1923–1935 (2005)
27. X. Su, M. Zanella, A. Grastien, Diagnosability of discrete-event systems with uncertain observations, in *25th International Joint Conference on Artificial Intelligence (IJCAI-16)* (2016), pp. 1265–1271
28. T. Yoo, S. Lafortune, Polynomial-time verification of diagnosability of partially observed discrete-event systems. IEEE Trans. Autom. Control **47**(9), 1491–1495 (2002)
29. H. Zaatiti, L. Ye, P. Dague, J.P. Gallois, Counter example guided abstraction refinement for hybrid systems diagnosability analysis, in *28th International Workshop on Principles of Diagnosis (DX-17)* (2017)

Chapter 11
Abstractions Refinement for Hybrid Systems Diagnosability Analysis

Hadi Zaatiti, Lina Ye, Philippe Dague, Jean-Pierre Gallois, and Louise Travé-Massuyès

11.1 Introduction

The increasing complexity of systems makes it challenging to detect and isolate faults. Hybrid systems are no exception, combining both discrete and continuous behaviors. Verifying behavioral or safety properties of such systems, either at design stage such as state reachability, diagnosability, and predictability or on-line such as fault detection and isolation is a challenging task. Actually, computing the reachable set of states of a hybrid system is an undecidable matter due to the infinite state space

H. Zaatiti
CEA, LIST, Laboratory of Model Driven Engineering for Embedded Systems, Gif-sur-Yvette, France

LRI, Univ. Paris-Sud & CNRS, Univ. Paris-Saclay, Gif-sur-Yvette, France
e-mail: hadi.zaatiti@cea.fr; hadi.zaatiti@lri.fr

L. Ye
CentraleSupélec, Univ. Paris-Saclay, Gif-sur-Yvette, France

LRI, Univ. Paris-Sud & CNRS, Univ. Paris-Saclay, Gif-sur-Yvette, France
e-mail: lina.ye@lri.fr

P. Dague (✉)
LRI, Univ. Paris-Sud & CNRS, Univ. Paris-Saclay, Gif-sur-Yvette, France
e-mail: philippe.dague@lri.fr

J. -P. Gallois
CEA, LIST, Laboratory of Model Driven Engineering for Embedded Systems, Gif-sur-Yvette, France
e-mail: jean-pierre.gallois@cea.fr

L. Travé-Massuyès
LAAS-CNRS, Univ. de Toulouse, Toulouse, France
e-mail: louise@laas.fr

M. Sayed-Mouchaweh (ed.), *Diagnosability, Security and Safety of Hybrid Dynamic and Cyber-Physical Systems*, https://doi.org/10.1007/978-3-319-74962-4_11

of continuous systems. One way to verify those properties over such systems is by computing discrete abstractions and inferring them from the abstract system back to the original system. Methods have been proposed for diagnosability verification for continuous and discrete systems separately, few of them handle hybrid automata [21, 35, 62]. Diagnosability is a property describing the system ability to determine whether a fault has effectively occurred based on the observations, which has received considerable attention in the literature [16, 32, 48, 52, 53, 60]. However, most of the existing works are applied on discrete event systems.

In this chapter, we are concerned with abstractions oriented towards hybrid systems diagnosability checking. Our goal is to create discrete abstractions in order to verify, at design stage, if a fault that would occur at runtime could be unambiguously detected in finite time (or within a given finite time bound for bounded diagnosability) by the diagnoser using only the allowed observations. This verification can be done on the abstraction by classical methods developed for discrete event systems, which provides a counterexample in case of non-diagnosability. The absence of such a counterexample proves the diagnosability of the original hybrid system. In presence of a counterexample, the first step is to check if it is not a spurious effect of the abstraction and actually exists for the hybrid system, witnessing thus non-diagnosability. Otherwise, we show how to refine the abstraction, guided by the elimination of the counterexample, and continue the process of looking for another counterexample until either a final result is obtained or we reach an inconclusive verdict. We make use of qualitative modeling and reasoning to compute discrete abstractions and we define several refinement strategies. Abstractions as timed automata are particularly studied as they allow one to capture qualitative temporal constraints [10, 13]. The chapter is organized as follows. We first present the hybrid automata formalism and define diagnosability for hybrid systems. We then introduce a formal framework for constructing hybrid automata abstractions while defining the refinement relation. Lastly, we detail the counterexample guided abstraction refinement (CEGAR) scheme adapted for diagnosability verification and a case study example illustrating this scheme.

11.2 Hybrid Dynamical Systems

In this section, we start with a brief general description of hybrid systems and then move on to propose a formal representation framework for hybrid automata that is adopted throughout the chapter. Later on, we provide an example of a practical system modeled as a hybrid automaton. Lastly, we introduce various classes of hybrid automata, among which timed automata are our primary interest.

11.2.1 Hybrid Automata Definition

Hybrid systems are dynamical systems that include discrete and continuous behaviors [38]. **Hybrid automaton** (HA) is a mean to model such systems; it is an infinite state machine frequently used for this purpose among the scientific community. Each state of the hybrid automaton is twofold with a discrete and a continuous part. The discrete part ranges over a finite domain while the continuous part ranges over the Euclidean space $\mathbb{R}^n$.

Definition 1 (Hybrid Automaton (HA)) An n-dimensional hybrid automaton (HA) is a tuple $H = (Q, X, S_0, \Sigma, F, Inv, T)$ where:

- Q is a finite set of modes (or locations), that can be possibly defined as the valuations set of a finite number of finite valued variables, and represents the *discrete* part of H. X is a set of n real-valued variables (which are continuously differentiable functions of time), whose valuations set $\mathbf{X} \subseteq \mathbb{R}^n$ represents the *continuous* part of H. $S = Q \times \mathbf{X}$ is the state space of H, whose elements, called states, are noted $(q, \mathbf{x})$ with q and $\mathbf{x}$ the respective discrete and continuous parts of the state.
- $S_0 \subseteq S$ is the set of initial states. If unique, the initial state is noted $(q_0, \mathbf{x}_0)$.
- Σ is a finite set of events.
- $F : S \to 2^{\mathbb{R}^n}$ is a mapping assigning to each state $(q, \mathbf{x}) \in S$ a set $F(q, \mathbf{x}) \subseteq \mathbb{R}^n$ constraining the time derivative $\dot{\mathbf{x}}$ of the continuous part of the mode q by $\dot{\mathbf{x}} \in F(q, \mathbf{x})$. If there is no uncertainty on the derivative, then F is a function $S \to \mathbb{R}^n$ specifying the flow condition $\dot{\mathbf{x}} = F(q, \mathbf{x})$ in each mode q (the dynamics in each mode is thus given by a set of n first-order ordinary differential equations (ODEs)).
- $Inv : Q \to 2^{\mathbf{X}}$ assigns to each mode q an invariant set $Inv(q) \subseteq \mathbf{X}$, which constrains the value of the continuous part of the state while the discrete part is q. We require, for all $q \in Q$, that $\{\mathbf{x} \mid (q, \mathbf{x}) \in S_0\} \subseteq Inv(q)$.
- $T \subseteq S \times \Sigma \times S$ is a relation capturing discontinuous state changes, i.e., instantaneous discrete transitions from one mode to another one. Precisely, $t = (q, \mathbf{x}, \sigma, q', \mathbf{x}') \in T$ represents a transition whose source and destination states are $(q, \mathbf{x})$ with $\mathbf{x} \in Inv(q)$ and $(q', \mathbf{x}')$ with $\mathbf{x}' \in Inv(q')$, respectively, and labeled by the event σ. It represents a jump from $\mathbf{x}$ in mode q to $\mathbf{x}'$ in mode q'.

We will call *(concrete) behavior* of H any sequence of continuous solution flows and discrete jumps, rooted in an initial state, satisfying all the constraints above defining H. Hybrid systems are typically represented as finite automata with (discrete, i.e., modes) states Q, initial states $Q_0 = \{q \in Q \mid \exists \mathbf{x} \in Inv(q)(q, \mathbf{x}) \in S_0\}$ and transitions δ defined by $\delta = \{(q, \sigma, q') \in Q \times \Sigma \times Q \mid \exists \mathbf{x}, \mathbf{x}'(q, \mathbf{x}, \sigma, q', \mathbf{x}') \in T\}$. To each state $q \in Q_0$ is associated an initial (continuous) nonempty set $Init(q) = \{\mathbf{x} \in Inv(q) \mid (q, \mathbf{x}) \in S_0\}$. To each transition $\tau = (q, \sigma, q') \in \delta$ are associated a nonempty guard set $G(\tau) = \{\mathbf{x} \mid \exists \mathbf{x}'(q, \mathbf{x}, \sigma, q', \mathbf{x}') \in T\} \subseteq Inv(q)$ and a set-valued reset map $R(\tau) : G(\tau) \rightarrow 2^{Inv(q')}$ given by $R(\tau)(\mathbf{x}) = \{\mathbf{x}' \mid (q, \mathbf{x}, \sigma, q', \mathbf{x}') \in T\}$. It is actually equivalent in the definition to provide either T or δ, G and R. In the last case, H is denoted by $(Q, X, S_0, \Sigma, F, \delta, Inv, G, R)$ and we have: $\forall (q, \mathbf{x}), (q', \mathbf{x}') \in S$, $\forall \sigma \in \Sigma$, $((q, \mathbf{x}, \sigma, q', \mathbf{x}') \in T \Leftrightarrow \tau = (q, \sigma, q') \in \delta \wedge \mathbf{x} \in G(\tau) \wedge \mathbf{x}' \in R(\tau)(\mathbf{x}))$.

It can be in some cases more convenient to adopt a relational-based representation than a set-based representation and to use predicates instead of subsets. By a slight abuse of notation, for each mode q, $Init(q)$ (for $q \in Q_0$), $F(q)$ and $Inv(q)$ indicate then predicates whose free variables are respectively from X, $X \times \dot{X}$ and X and $Init(q)(\mathbf{x})$, $F(q)(\mathbf{x}, \dot{\mathbf{x}})$ and $Inv(q)(\mathbf{x})$ being true means respectively $\mathbf{x} \in Init(q)$, $\dot{\mathbf{x}} \in F(q, \mathbf{x})$ and $\mathbf{x} \in Inv(q)$. In the same way, for each mode transition τ, $G(\tau)$ and $R(\tau)$ indicate predicates whose free variables are respectively from X and $X \times X$ and $G(\tau)(\mathbf{x})$ and $R(\tau)(\mathbf{x}, \mathbf{x}')$ being true means respectively $\mathbf{x} \in G(\tau)$ and $\mathbf{x}' \in R(\tau)(\mathbf{x})$. We will make use equally of both representations.

Guards in any mode q will be assumed non-intersecting: $\forall q \in Q, \forall \tau_1 = (q, \sigma_1, q_1) \in \delta, \forall \tau_2 = (q, \sigma_2, q_2) \in \delta, (\tau_1 \neq \tau_2 \Rightarrow G(\tau_1) \cap G(\tau_2) = \emptyset)$. Thus, at any moment of its continuous evolution in a mode q, the system may jump to at most one another mode and by a unique event. Nevertheless, a HA is generally non-deterministic: the continuous dynamics in each mode may be non-deterministic, the moment where a jump occurs is non-deterministic (as long as $Inv(q)(\mathbf{x})$ and $G(\tau)(\mathbf{x})$ are true, where q is the source mode of the mode transition τ, the system may continue to continuously evolve in q or make the transition τ) and the reset after a jump may be non-deterministic.

11.2.2 Modeling with Hybrid Automata

Hybrid automata represent an intuitive modeling framework. They are used in various domains to model complex hybrid systems. Here is a practical case where a hybrid automaton is used for modeling a system.

Example 1 A simple thermostat system maintains the temperature of an object quasi-constant by turning on and off a heater device. In practice such system contains at least, a temperature sensor, a heater device and logic control electronic circuits. The circuitry decides, given the actual measured temperature of the object, to activate or not the heater. A hybrid automaton $H = (Q, X, S_0, \Sigma, F, \delta, Inv, G, R)$ models the behavior of such system (see graphical representation on Fig. 11.1):

- $Q = \{on, off\}, X = \{x\}, S_0 = (off, [80, 90])$
- $\Sigma = \{B_{on}, B_{off}\}, F(on) = \{\dot{x} = -x + 100\}, F(off) = \{\dot{x} = -x\}$
- $\delta = \{\tau_1 = (off, B_{on}, on), \tau_2 = (on, B_{off}, off)\}, Inv(off) = x \geq 68, Inv(on) = x \leq 82$
- $G(\tau_1) = x \leq 70, G(\tau_2) = x \geq 80, R(\tau_1) = R(\tau_2) = (\mathbf{x} = \mathbf{x}')$

The assigned hybrid automaton H is one dimensional, where x represents the sensed temperature.

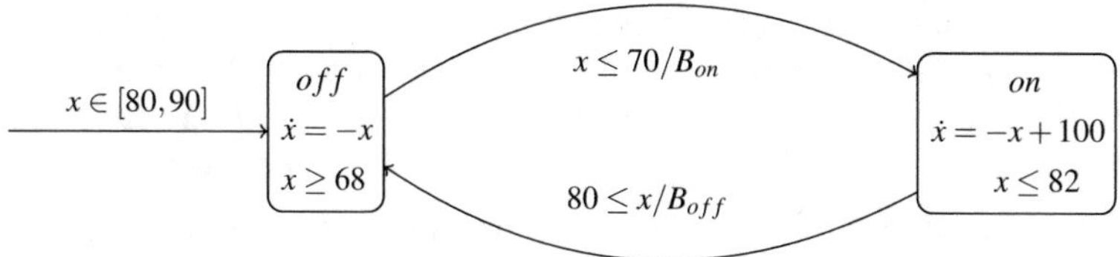

Fig. 11.1 One-dimensional hybrid automaton modeling a thermostat

11.2.3 Hybrid Automata Semantics

We denote by $q_0, q_1, \ldots$ the modes of Q, and by $x_1, x_2, \ldots, x_n$ the variables of X.

Definition 2 (Hybrid Automaton Semantics) The semantics of a hybrid automaton H, denoted by $[[H]]$, is the set of all executions, which are labeled sequences of states from S with labels in $L = \Sigma \cup \mathbb{R}_+$: $(q_0, \mathbf{x}_0) \xrightarrow{l_0} (q_1, \mathbf{x}_1) \ldots (q_i, \mathbf{x}_i) \xrightarrow{l_i} \ldots$ such that $(q_0, \mathbf{x}_0) \in S_0$ and, for any two successive states $(q_i, \mathbf{x}_i) \xrightarrow{l_i} (q_{i+1}, \mathbf{x}_{i+1})$ in the sequence, one of the following is true:

- $l_i = \sigma_i \in \Sigma$ and $(q_i, \mathbf{x}_i, \sigma_i, q_{i+1}, \mathbf{x}_{i+1}) \in T$;
- $l_i = d_i \in \mathbb{R}_+$, $q_i = q_{i+1}$, $\mathbf{x}_i, \mathbf{x}_{i+1} \in Inv(q_i)$ and $\exists x : [0, d_i] \rightarrow \mathbf{X}$ continuously differentiable function, with $x(0) = \mathbf{x}_i$, $x(d_i) = \mathbf{x}_{i+1}$ and $\forall t \in (0, d_i) \dot{x}(t) \in F(q_i, x(t))$ and $x(t) \in Inv(q_i)$.

In the first case, the system executes a discrete transition (also called discrete jump) $\tau_i = (q_i, \sigma_i, q_{i+1})$ from the mode q_i to the destination mode q_{i+1}. Such a transition is possible (enabled) as soon and as long as $\mathbf{x}_i \in G(\tau_i)$. After the jump, the system may follow the new dynamics given by $F(q_{i+1})$, starting from the continuous state $\mathbf{x}_{i+1} \in R(\tau_i)(\mathbf{x}_i)$. Notice that no time elapses during a discrete jump, which is instantaneous. In the second case, the system performs a continuous transition (also called continuous flow) of duration d_i inside the mode q_i, constrained by the dynamics $F(q_i)$ and the invariant set $Inv(q_i)$. The sequence $h = (\mathit{off}, 80) \xrightarrow{0.15} (\mathit{off}, 69) \xrightarrow{B_{on}} (on, 69) \xrightarrow{0.5} (on, 81) \xrightarrow{B_{\mathit{off}}} (\mathit{off}, 81) \ldots$ is valid for the thermostat example (Fig. 11.1), thus $h \in [[H]]$. The trace of an execution h, i.e., the sequence of its labels, is a word from $L^\star$ (or L^ω for infinite h), denoted as $trace(h)$. We denote the total time duration of h by $time(h) \in \mathbb{R}_+ \cup \{+\infty\}$, which is calculated as the sum of all time periods in the trace of h: $time(h) = \sum d_i$.

Let $\overline{S} = \bigcup_{q \in Q}(\{q\} \times Inv(q)) \subseteq S$ the (infinite) set of invariant satisfying states of H, $\overline{S}_0 = \bigcup_{q \in Q_0}(\{q\} \times Inv(q)) \subseteq S_0$ the subset of invariant satisfying initial states and $\rightarrow\ \subseteq \overline{S} \times L \times \overline{S}$ the transition relation defined by one or the other condition in Definition 2. The semantics of H is actually given by the labeled transition system $S_H^t = (\overline{S}, \overline{S}_0, L, \rightarrow)$, i.e., $[[H]]$ is the set of all paths of S_H^t issued from an initial state. S_H^t, called the **timed transition system of** H, is thus a discretization of H with infinite sets of states and of transition labels. It just abstracts continuous flows by timed transitions retaining only information about the source, the target and the duration of each flow and constitutes the finest abstraction of H we will consider.

The timeless abstraction of S^t_H, called the **timeless transition system of** H, is obtained by ignoring also the duration of flows and thus defined as $S_H = (\overline{S}, \overline{S}_0, \Sigma \cup \{\varepsilon\}, \rightarrow)$, obtained from S^t_H by replacing any timed transition $(q_i, \mathbf{x}_i) \xrightarrow{d_i} (q_{i+1}, \mathbf{x}_{i+1})$ with $d_i \in \mathbb{R}_+$ by the ε transition $(q_i, \mathbf{x}_i) \xrightarrow{\varepsilon} (q_{i+1}, \mathbf{x}_{i+1})$, that can be considered as a silent transition. It has infinite set of states but finite set of transition labels. It constitutes the finest timeless abstraction of H we will consider.

Theorem 1 (Correction and Completeness of the Semantics) *Any concrete behavior of H is timed (resp. timeless) abstracted into an $\overline{S}_0$ rooted path in S^t_H (resp. S_H). Conversely, any path in S^t_H (resp. S_H) that alternates continuous and discrete transitions (in particular any single transition) abstracts a part of a concrete behavior of H and, if F is a singleton function (i.e., deterministic derivative), any $\overline{S}_0$ rooted path in S^t_H (resp. S_H) abstracts a concrete behavior of H. In this latter case, there is thus no spurious abstract behavior in S^t_H (resp. S_H), which expresses faithfully the behavior of H.*

11.2.4 Hybrid Automata Classes and Particular Cases

Definition 3 (Discrete Automaton (DA)) It is the case where there is no continuous part. Thus, a (finite) discrete automaton (DA) is a tuple $D = (Q, Q_0, \Sigma, \delta)$ where Q is a finite set of discrete states (modes), $Q_0 \subseteq Q$ is the set of initial states, Σ is a finite set of events, and $\delta \subseteq Q \times \Sigma \times Q$ is a set of transitions of the form $\tau = (q, \sigma, q')$.

The semantics $[[D]]$ of D is given by the set of sequences (called paths) made up of successive states transitions labeled by events and rooted in an initial state. The trace of such a path is the word in $\Sigma^\star$ whose letters are the successive labels of the path.

Definition 4 (Continuous System (CS)) It is the case where there is no discrete part. Thus, an n-dimensional continuous system (CS) is a particular hybrid automaton C with only one mode ($|Q| = 1$) and $\Sigma, T = \emptyset$ (and thus δ, G, R too). It can thus be denoted as $C = (X, S_0, F, Inv)$ with $S_0 \subseteq Inv$.

The semantics $[[C]]$ of C is the set of all time labeled sequences of continuous states, rooted in an initial state, corresponding to the continuous transitions of a hybrid automaton, constrained by the dynamics F and the invariant set Inv.

The form of the dynamics F determines primarily the class of the hybrid automaton. The rectangular class, for which the dynamics valuations are a cartesian product of intervals, lies on the boundary of the decidability over reachability problem with some restrictions [39]. We will present some classes of hybrid automata starting from the particular to the more general classes.

Timed Automata We are particularly interested in timed automata, a class of hybrid automata where the continuous variables x_i, $1 \leq i \leq n$, with values in $\mathbb{R}_+$, called clocks, have all first order derivatives equal to one. So time elapses identically for all clocks. The set $C(X)$ of constraints over a set of clocks X is defined as follows: a constraint is either a primitive constraint of the form $x_i \ op \ c_i$, where $c_i \in \mathbb{R}_+$ (at the theoretical level because, in practice, $\mathbb{Q}$ is used instead of $\mathbb{R}$ for computer implementation reasons) and op is one of the operators $<, \leq, =, \geq, >$, or a finite conjunction of primitive constraints. The satisfiability set of a constraint is thus a rectangle in $\mathbb{R}^n_+$, i.e., the product of n intervals of the half real line $\mathbb{R}_+$, and we will identify $C(X)$ to the set of rectangles.

Definition 5 (Timed Automaton (TA)) A timed automaton (TA) is a hybrid automaton $T = (Q, X, S_0, \Sigma, F, \delta, Inv, G, R)$ such that:

- $\mathbf{X} = \mathbb{R}^n_+$.
- $S_0 = Q_0 \times \{\mathbf{0}\}$.
- $\forall q \in Q \ F(q, .) = \mathbf{1}$, which means that the dynamics of clocks evolution in each mode q is given by $\dot{x}_i = 1$.
- $Inv : Q \rightarrow C(X)$ associates to each mode q a rectangle invariant in $\mathbf{X}$. We require $\mathbf{0} \in Inv(q_0)$.
- $G : \delta \rightarrow C(X)$ associates to each discrete transition (q, σ, q') a rectangle guard in $Inv(q)$.
- $\forall \tau \in \delta \exists Y(\tau) \subseteq X \forall \mathbf{x} \in G(\tau) R(\tau)(\mathbf{x}) = \{\mathbf{x}'\}$ with $\mathbf{x}'_i = 0$ if $x_i \in Y(\tau)$ and $\mathbf{x}'_i = \mathbf{x}_i$ otherwise, i.e., clocks in $Y(\tau)$ are reset to zero with transition τ, the others keeping their values.

The notation of a timed automaton T is generally simplified as $T = (Q, X, Q_0, \Sigma, Inv, (\delta, G, Y))$. The semantics of T as a hybrid automaton, given by Definition 2, can be simplified by merging together in an execution successive timed transitions between two discrete transitions and summing up their time period labels. An execution in $[[T]]$ is thus a sequence h of alternating time steps (possibly with 0 time period) and discrete steps of the form $(q_0, \mathbf{x}_0) \xrightarrow{d_1} (q_0, \mathbf{x}_0 + d_1) \xrightarrow{\sigma_1} (q_1, \mathbf{x}_1) \xrightarrow{d_2} \ldots$ whose trace $trace(h)$ is the timed word $d_1 \sigma_1 d_2 \ldots \in \mathbb{R}_+(\Sigma \mathbb{R}_+)^*$ and duration is $time(h) = \sum d_i$.

The class of timed automata is particularly interesting as the reachability and language emptiness problems are decidable for that class [2]. Actually, decidability still holds for the larger class of **rectangular automata (RA)**, where the unique flow condition $F(.,.)$, the same for all modes, is given by a rectangle in $\mathbb{R}^n$ (instead of the singleton $\mathbf{1}$), $Init(q)$ is a bounded rectangle and $R(\tau)(\mathbf{x}) = \{\mathbf{x}' \mid \mathbf{x}'_i \in I_i$ if $x_i \in Y$ and $\mathbf{x}'_i = \mathbf{x}_i$ else$\}$, where the I_i's are bounded intervals depending only on τ [38]. And thus holds for the subclass of **singular automata**, where the flow rectangle is reduced to a singleton, whose timed automata are a particular case.

But decidability does not hold any more if the flow condition is allowed to change from one mode to another one. The simplest example of this is the generalization of timed automata where we allow the presence of stopwatches. A **stopwatch** is a clock which can switch from active (turned on) to inactive (turned off), or vice versa, when transiting between two modes. The generalized flow condition is thus given by: $\forall q \in Q\, \exists \mathbf{c} \in \{0, 1\}^n\; F(q, .) = \mathbf{c}$, which means that the dynamics of clocks evolution in each mode q is given by $\dot{x}_i = 1$ for those clocks active in q and $\dot{x}_i = 0$ for those clocks inactive in q. Thus, during inactivity, a stopwatch holds its last valuation when it was active (or 0 in case of reset). It happens that reachability decidability does not hold for a generalized timed automaton if only one clock is allowed to be a stopwatch, i.e., when the flow conditions are not independent of the mode (and thus also for the more general classes of **multisingular automata**, where the flow singletons depend on the mode, and of the **multirectangular automata**, where the flow rectangles depend on the mode). Notice nevertheless that, allowing changes of flow conditions with changes of modes may remain manageable if, e.g., we require a reset of the variables concerned when it occurs. That is how **initialized multi-rectangular automata**, i.e., where for each discrete jump, each variable whose flow interval is changed in this jump has to be reset (reinitialized), can be translated to rectangular automata. Another case where decidability is lost in general for a hybrid system (but not for a timed automaton, for which it does not change the expressivity) is when the set $C(X)$ of constraints is extended to contain primitive constraints of the form $(x_i - x_j)$ op c_{ij}, i.e., if variables are not pairwise independent (and thus also for the class of **triangular automata** which generalize rectangular automata by adding such constraints). **Linear automata** generalize both multirectangular and triangular automata by allowing sets $F(q), Init(q), Inv(q), G(\tau)$ to be any convex polyhedra in $\mathbb{R}^n$ (instead of just rectangles or triangles) and different flows conditions for different modes. And **polynomial automata** generalize linear automata by allowing those sets to be defined no longer by just linear constraints but by polynomial constraints.

11.3 Diagnosability of Hybrid Dynamical Systems

We will now introduce the model of hybrid systems that is used for diagnosability analysis and remind some methods from the literature aimed at verifying this property.

11.3.1 Hybrid System Model for Diagnosability Analysis

Fault diagnosis is a crucial and challenging task in the automatic control of complex systems, whose efficiency depends on the system property called diagnosability. This is a property describing the system ability to determine without ambiguity whether a fault of a given type has effectively occurred based on the observations. Diagnosability analysis has already received considerable attention in the literature over latest decades. However, most of the existing works refer to discrete event systems [16, 32, 34, 48, 52, 53, 60] with stochastic and fuzzy variants [41, 44, 54] or continuous systems [5, 15, 47, 57]. Diagnosability was also studied in the framework of decentralized and distributed architectures [46, 49, 50, 59].

But many modern technological processes exhibit both continuous evolution and discrete transitions whose combination is at the origin of complex behavior and important phenomena of such systems. To the best of our knowledge, very few works handle diagnosability of hybrid systems with satisfactory results.

As a first step, Travé-Massuyès et al. [56] proved that the existing definitions of diagnosability for discrete event systems and for continuous systems can be stated as a property of the system fault signatures, and a unified definition of diagnosability was established. However hybrid system diagnosability was not considered.

Among the contributions concerned with hybrid diagnosability, we can mention [11] that slightly modified the classical necessary and sufficient condition for diagnosability of a discrete event system of [52] and expressed it in terms of reachability. Reference [29] generalized this condition requiring more restrictive hypotheses. Despite the claim that the two above methods deal with hybrid systems, these works do not really account for the hybrid nature of the system as they use only a very high level discrete abstraction and ignore the continuous dynamics. On the other hand, in [19], diagnosability is expressed in terms of mode discernability (also called distinguishability by other authors) and is only based on the continuous dynamics.

Reference [8] was among the early works that coped with actual hybrid systems, introducing the idea to consider a hybrid model as a twofold mathematical object. A hybrid system is modeled as a hybrid automaton whose discrete states represent its operation modes for which the continuous dynamics are specified. The discrete event part (automaton) constrains the possible transitions among modes and is referred to as the *underlying DES*. The restriction of the hybrid system to the continuously-valued part of the model is defined as the *multimode system*.

Considering the analytical redundancy approach to define a set of residuals [33] for every mode, Bayoudh et al. [8] introduced the concept of *mode signature* which refines the classical concept of fault signature. Mode signatures determine mode distinguishability. The key idea of [8] is to abstract the continuous dynamics by defining a set of "diagnosis-aware" events, called *signature-events*, associated to mode signature changes across modes. Signature-events are used to enrich appropriately the underlying DES. The behavior of the abstract system is then modeled by a prefix-closed language over the alphabet enriched by these additional events. The finite state machine generating this language is called the *behavior automaton*. Based on the *abstract language*, the diagnosability analysis of the hybrid system is cast into a discrete event framework and standard methods of this field can be used.

The approach of [8] later consolidated in [6] can be compared to the approach proposed in [22, 23] which uses fault signatures to capture the continuous dynamics. The fault signatures of [22, 23] are based on fault transients and they directly express the expected dynamic behavior of measured variables after the fault abstracted in qualitative terms. The approach of [6, 8] differs in that it uses mode signatures that are specifically built for diagnosis, based on standard analytical redundancy residual methods of the FDI control field [27]. Its originality relies in that it proposes a way to integrate these methods with equally standard methods of the DES diagnosis field [63]. Bayoudh et al. [6, 8] adopt the diagnoser approach [52] because it has the advantage to also support straightforwardly online diagnosis. Diene et al. [25] repeats these ideas differing by the fact that the diagnoser is directly built from the underlying DES and mode distinguishability is used to cluster its state labels. This method leads to a so-called *clustered diagnoser*. Let us note that this method only applies to a restricted class of hybrid systems for which transitions triggered by continuous dynamics are not allowed.

Checking DES diagnosability with methods based on the construction of diagnosers has exponential complexity with the size of the underlying DES automaton. Hence, approaches based on *verifiers*, also known as *twin plant* approaches, are generally preferred. This is because, although a twin plant cannot be used for online diagnosis, it can be constructed in polynomial time. Methods integrating a twin plant approach with mode distinguishability checking for assessing hybrid system diagnosability are recent. The reader can refer to [26] as a first piece of work in this direction. Later, Grastien et al. [35] indicated that mode distinguishability could be complemented by another property of the continuous dynamics named *ephemerality*. Ephemerality states when the system cannot stay forever in a given set of modes. The continuous dynamics are hence abstracted remembering only these two pieces of information. In addition to this, Grastien et al. [35] checks diagnosability in an incremental way. It starts by generating the most abstract DES model of the hybrid system and checking diagnosability of this DES model. A "counterexample" that negates diagnosability is possibly provided based on the twin plant. The model is then refined to try to invalidate the counterexample and the procedure repeats as far as diagnosability is not proved. This approach hence uses just the necessary information about continuous dynamics, in an "on request" manner, hence making the best out of computation.

In the most recent literature concerned with hybrid system diagnosability like [35] and also [24], which characterizes the maximum delay for diagnosing faults given measurement uncertainty, abstraction is key. Abstraction is also at the core of other methods to check other properties of hybrid systems.

This is why this chapter reviews different ways of abstracting hybrid automata in the next section, then elaborates from the diagnosability procedure proposed in [35] that uses the counterexample guided abstraction refinement (CEGAR) as initially introduced in [3, 28]. The algorithmic basis refers to CEGAR considering that the verification problem for even very simple hybrid systems is undecidable [39]. The method abstracts the hybrid automaton and then refines the abstraction while being guided by a diagnosability counterexample found at this abstract level. A first discrete abstraction of the hybrid system is computed, diagnosability is then verified using classical discrete methods. The verification could either yield the abstraction as diagnosable, which infers the diagnosability property back to the hybrid system, or non-diagnosable with a generated counterexample that validates this decision. The produced counterexample is either present in the original system in which case the hybrid system is not diagnosable, or it is a spurious effect of the abstraction. In the latter case, the counterexample is analyzed and the current abstraction is refined according to this analysis and the diagnosability verification task is iterated.

11.3.2 Observations and Faults

Remind that diagnosability is a system property allowing one to determine with certainty, at the design stage, a fault occurrence, based on available observations. Precisely, in a given system model, the existence of two infinite behaviors, with the same observations but exactly one containing the considered fault, violates diagnosability. Hence, to be able to analyze such property, it is necessary to define what can be observed for given systems as well as what are considered as faults. In practice, the observations are partial, only parts of the system are known and are usually obtained from sensors. In whole generality we will consider that both some discrete jumps between modes and some continuous variables inside a mode may be observable. The sets of observable events and variables are assumed to be time invariant, the second one being also assumed to be independent of the mode for the sake of simplicity. Events are observed together with their instantaneous occurrence time and variables values are assumed to be observed at any moment. E.g., for our thermostat system in Fig. 11.1, transitions B_{on}, B_{off} and temperature x are assumed to be observable. For what concerns faults, we will suppose that they are modeled, as for discrete event systems, by some unobservable discrete jumps, between precisely a normal mode and a faulty mode, translating often in a change of dynamics. This is well adapted for abrupt faults but progressive faults or degraded modes (as a shift of parameter) can be also represented in this way, the designer abstracting a slow evolution in a sudden change when he estimates that the behavior variation induced (that he will model by means of the invariant and the guard) cannot any more let consider the given mode as normal. To sum up, we obtain the following definition.

Definition 6 (Partially Observable Hybrid Automaton (POHA)) A partially observable hybrid automaton (POHA) is a hybrid automaton H according to Definition 1 where:

- $\Sigma = \Sigma_o \uplus \Sigma_u \uplus \Sigma_f$, i.e., the set of events is the disjoint union of the set Σ_o of observable (normal) events, the set Σ_u of unobservable normal events and the set Σ_f of unobservable fault events.
- $X = X_o \uplus X_u$, i.e., the set of continuous real-valued variables is the disjoint union of the set X_o of observable variables and the set X_u of unobservable variables.

Definition 7 (Execution (Timed) Observation) Given an execution $h \in [[H]]$ of a POHA H, $h = (q_0, \mathbf{x}_0) \xrightarrow{l_0} (q_1, \mathbf{x}_1) \ldots (q_i, \mathbf{x}_i) \xrightarrow{l_i} \ldots$, with $l_i \in \Sigma \cup \mathbb{R}_+$, the (timed) observation of h is defined as $Obs(h) = \mathbf{x}_0^o, l_0^o, \mathbf{x}_1^o \ldots \mathbf{x}_i^o, l_i^o, \ldots$, where:

- $\mathbf{x}_i^o$ is obtained by projecting $\mathbf{x}_i$ on variables in X_o.
- $l_i^o = l_i$ if $l_i \in \Sigma_o \cup \mathbb{R}_+$. Otherwise, $l_i^o = \varepsilon$, which is then removed from $Obs(h)$.

Note that all durations labels $l_i = d_i$ in h are present in $Obs(h)$. Thus, any observable event $l_i = \sigma_i$ in h is present in $Obs(h)$ together with its occurrence time, obtained by adding up all durations d_j in $Obs(h)$ from the origin up to the event σ_i. In the same way, any observable variable x has its value known in $Obs(h)$ at all those instants t obtained as the sums of consecutive durations in $Obs(h)$ from the origin. If t is the occurrence time of an (observable or unobservable) event σ and if x is reset by the discrete transition σ, then the value of x changes instantaneously after this transition and the new value will be noted $x^+(t)$ to distinguish it from the value $x(t)$ before the transition (a reset observable variable may thus identify the presence of an unobservable event). Similarly, one can define observation for timed automata. The difference is that we do not assume any information about continuous clocks, so there is no $\mathbf{x}_i^o$. Then, the observation is obtained from the trace (a timed word) by erasing all unobservable events and by adding up the periods between any two successive observable events in the resulting sequence. We have thus defined what is the observation of a POHA H at the level of its timed transition system S_H^t. Defining its observation at the level of its timeless transition system S_H is similar, with $l_i \in \Sigma \cup \{\varepsilon\}$ and $l_i^o = l_i$ if $l_i \in \Sigma_o$ and removed otherwise. This means that the timeless observation is obtained from the timed observation $Obs(h)$ above by removing all durations d_i, keeping thus only observable events in Σ_o and values $\mathbf{x}_i^o$ of observable variables at each transition step as an ordered sequence without any occurrence time attached.

11.3.3 System Diagnosability Definition

As we just explained, a fault is modeled as a fault event that alters the system from a normal mode to an abnormal mode. There may exist different fault events in a given system. For the sake of reducing complexity (from exponential to linear in the number of different fault events) and of simplicity, in the following only one fault type, i.e. fault event, at a time is considered but multiple occurrences of this event are allowed, and the other types of fault events are thus processed as unobservable normal events. However, this framework can be extended in a straightforward way such that a number of different faults can be considered simultaneously. Now we adapt to hybrid systems the diagnosability definition [52] introduced for discrete event systems (the bounded one and the unbounded one in terms of executions lengths). h^F denotes a finite execution whose last label is a first occurrence of the fault event F considered. Given a finite execution $h \in [[H]]$ such that $h = (q_0, \mathbf{x}_0) \xrightarrow{l_0} (q_1, \mathbf{x}_1) \dots (q_i, \mathbf{x}_i)$, the set of post-executions of h in $[[H]]$ is defined as $[[H]]/h = \{h' = (q_i, \mathbf{x}_i) \xrightarrow{l_i} \dots \mid h.h' \in [[H]]\}$, where $h.h'$ is obtained by merging the final state of h and the first state of h', both should be the same.

Definition 8 ((Δ-)Faulty Executions) Given a hybrid automaton H and F a fault event, a faulty execution is an execution $h \in [[H]]$ such that $F \in trace(h)$. Thus

$h = h^F h'$ where h^F is the prefix of h whose last label is the first occurrence of F. We denote the *period from* (the first occurrence of) *fault F in h* by $time(h, F) = time(h')$. Given a positive real number $\Delta \in \mathbb{R}_+^*$, we say that at least Δ time units pass after the first occurrence of F in h, or, in short, that h is *Δ-faulty*, if $time(h, F) \geq \Delta$.

Definition 9 (Hybrid Automaton (Time Bounded and Unbounded) Diagnosability) Given $\Delta \in \mathbb{R}_+^*$, a fault F is said Δ-diagnosable in a POHA H iff (we abbreviate $F \in trace(h)$ by $F \in h$)

$$\forall h \in [[H]](h\Delta\text{–faulty}$$
$$\Rightarrow \forall h' \in [[H]](Obs(h') = Obs(h) \Rightarrow F \in h')).$$

i.e.,

$$\forall h^F \in [[H]] \forall h \in [[H]]/h^F(time(h) \geq \Delta$$
$$\Rightarrow \forall h' \in [[H]](Obs(h') = Obs(h^F.h) \Rightarrow F \in h')).$$

A fault F is said diagnosable in H iff

$$\exists \Delta \in \mathbb{R}_+^* (F\Delta\text{-diagnosable in } H).$$

This definition states that F is Δ-diagnosable (resp., diagnosable) iff, for each execution h^F in $[[H]]$, for each post-execution h' of h^F with time at least Δ (resp., with enough long time, depending only on F), then every execution in $[[H]]$ that is observably equivalent to $h^F.h'$ should contain F. Precisely, the existence of two indistinguishable behaviors, i.e., executions holding the same observations, with exactly one containing F and time long enough after F, i.e., whose time after F is at least Δ (resp., is arbitrarily long), violates the Δ-diagnosability (resp., diagnosability) property for hybrid automata. Inspired from the framework of discrete event systems, we define critical pairs for partially observable hybrid automata taking into account both continuous and discrete dynamics.

Definition 10 (Δ-Critical Pair) A pair of executions $h, h' \in [[H]]$ is called a Δ-critical pair with respect to F iff: $F \in h$ and $F \notin h'$ and $Obs(h) = Obs(h')$ and $time(h, F) \geq \Delta$.

Theorem 2 *A fault F is Δ-diagnosable in a POHA H iff there is no Δ-critical pair in $[[H]]$ with respect to F. F is diagnosable in H iff, for some Δ, there is no Δ-critical pair in $[[H]]$ with respect to F (i.e., there is no arbitrarily long time after F critical pair).*

Note that all above definitions (e.g., observable projection, post-executions, diagnosability, critical pairs, etc.) are applicable in a similar way to timed automata, which can be considered as a special type of hybrid automata. The only difference is that the set of continuous variables is the set of clock variables whose derivative is always 1 [7, 9, 20, 21, 35]. And, as for automata the existence of arbitrarily long (in terms of transitions number) after F faulty executions implies the existence of an infinite faulty execution, in the same way it has been proved [58] that for timed automata the existence of arbitrarily long time after F faulty executions implies the existence of a $+\infty$-faulty execution (extending the definition above to $\Delta = +\infty$) and thus that non-diagnosability is witnessed by the existence of a $+\infty$-critical pair and its checking is PSPACE-complete.

Theorem 3 *A fault F is diagnosable in a partially observable timed automaton T iff there is no $+\infty$-critical pair in $[[T]]$ with respect to F. Checking diagnosability of T is PSPACE-complete.*

We will rest on this result as diagnosability checking of a POHA H will be done on a time automaton T abstracting H.

11.4 Abstracting Hybrid Automata

For continuous systems, verifying the most basic properties such as "Is this state reachable?" is not decidable, due in particular to the uncountability of continuous domains [37]. A fortiori, for a hybrid system, a simple computation of the reachable set of states starting from an initial state is not a decidable matter except for few unpractical classes [39]. This is why a common practice is to partition infinite domains into a finite number of subsets, abstracting the system behavior in each of those subsets. The abstraction of the domain into representative sets is usable in computations to possibly reason about the infinite domain. In this section, we thus focus on abstractions that discretize the infinite state space defined by continuous variables into finite sets. We show the targeted class of properties we wish to verify. As our study considers abstractions of complex hybrid dynamical systems for diagnosability analysis, it is therefore crucial to first introduce abstractions for continuous dynamical systems which are a particular case of hybrid dynamical systems. Utilizing these abstractions, we aim at verifying temporal properties and bounding the time for fault detection and isolation. Abstractions retain less but important information regarding a property that we wish to verify about a complex system. Due to the uncountability of the continuous state space domain, verifying simple properties of continuous and more generally hybrid systems via abstraction becomes challenging. The challenging part about abstractions is the choice made to select the representative sets and the criterion for choosing them. This choice relies entirely on the class of the properties one wishes to verify and on the structure of the hybrid system itself. Here, we are interested in hybrid systems abstraction aimed towards diagnosability checking. Abstractions that can be refined if necessary are of our concern, as refinement allows adding more information into the abstracted system while being always guided by the property to check. In this section, we will discuss abstractions in general and focus on those that capture time constraints in particular.

11.4.1 Different Abstraction Strategies

Using Qualitative Principles We first take a look at abstractions using qualitative concepts. Given a set of ODEs, these concepts classically discretize the infinite state space of the continuous variables into finite sets. The discretization of $\mathbb{R}^n$ is often achieved by rectangles, i.e., is built by product from a discretization of $\mathbb{R}$. And this one is obtained by fixing a finite number of (rational) landmarks l_i, resulting in a finite partition in terms of open intervals (l_i, l_{i+1}) (with possibly infinite endpoints) and singleton intervals $[l_i]$, allowing what is called absolute order of magnitude reasoning. The coarsest partition (except $\mathbb{R}$ itself) is obtained from the single landmark 0 and corresponds to the sign partition: $(-\infty, 0)$, $[0]$, $(0, +\infty)$, giving rise to a partition of size 3^n of $\mathbb{R}^n$. It is particularly interesting when applied to the valuations of the variables derivatives, as it corresponds to discretize according to the sign of the derivative, which is constant within each set of the partition, and thus to the change direction of the variable itself (decreasing, constant, increasing) . The variables being continuously differentiable, it is not possible for the sign of the derivative to pass from negative to positive without crossing zero. Exploiting this feature, it is possible to draw a rough scheme of the behavior of the variables called "qualitative simulation" in the literature and to obtain a loose overview about how the system will behave [30, 43, 55].

Example 2 Consider this simple linear continuous system:

$$\begin{aligned} \dot{x} &= 3x \\ \dot{y} &= y - 1 \end{aligned} \tag{11.1}$$

Adopting the partition of the state space given by the signs of the derivatives, the abstract state space of size 9 is thus: $(\dot{x} > 0 \vee \dot{x} < 0 \vee \dot{x} = 0) \wedge (\dot{y} > 0 \vee \dot{y} < 0 \vee \dot{y} = 0)$. The transitions between the abstract states are computed according to the laws of evolution given the signs of the derivatives. The abstract state $(\dot{x} > 0 \wedge \dot{y} > 0)$ corresponds to the region $\{x, y \mid x > 0 \wedge y > 1\}$ in the state space. From this state, no transition is possible to another abstract state. Suppose we wish to verify a basic reachability property: starting from the state $(1, 3)$ is it possible to reach the state $(-5, -4)$? The answer would be no, the proof is given using the previous abstraction method and inferring the property back to the original system. Such abstraction is **sound**: from any initial state (x_0, y_0) the solutions of the differential equation system (11.1) will always satisfy the constraints imposed by the abstract system rules, i.e., the possible transitions.

Abstractions for the Verification of Temporal Properties The above abstraction partitions the state space into sets with a constant sign of the derivative. This abstraction is useful to trace the future evolution of the state given the initial one to prove a safety property of avoiding an unwanted state. Nonetheless, for proving a more complex property that involves the notion of time, classically expressed using temporal logic, the above abstraction is not sufficient. One needs to add time as a separate state variable and correlate the variables changes to changes in time.

Example 3 We consider the same continuous system and suppose the initial set of states I such that $I = \{(x, y) \mid 1 < x < 2 \wedge 1 < y < 2 \wedge x < y\}$ and the property $F(x > y)$ where F is the "eventually" linear temporal logic (LTL) operator. Fp, where p is a Boolean proposition, is equivalent to $\exists t_0 \in \mathbb{R}^+, \forall t > t_0, p = true$. It is obvious that the rate at which x is increasing with respect to time is much larger than that of y. Hence, for all the initial states within I the property is true. The previous abstraction method however does not capture the rate at which the derivative of x is changing and is thus useless for establishing the proof. Actually, changing the first equation in (11.1) by $\dot{x} = 0.5x$ would keep the abstract system unchanged and nevertheless change the truth value of the property. In our case, the system can be written as $\dot{\mathbf{x}} = A\mathbf{x} + \mathbf{b}$ where $\mathbf{x} = (x, y)^T$ and A is the corresponding matrix. We then deduce by computing the eigenvalues of A which are 3 and 1 in our example that the rate at which x increases is larger than the rate at which y increases, which provides a sufficient proof that the above property holds when the system is initiated from I.

The previous example illustrates the simple case of a linear dynamics where the eigenvectors are not rotated by the linear transformation and are thus invariant for the continuous system. Therefore, taking these two vectors into account during the abstraction process is an obvious choice. However, in the more general case of nonlinear dynamics, the invariant takes a more complex form. Some technique encodes the hybrid system, the property to verify and a specific parametric form of the invariant into an SMT (Satisfiability Modulo Theories) based solver and evaluates the unknown parameters of the invariant automatically. Once computed, the invariant is incorporated to make a finer and more representative abstraction [36].

11.4.2 Geometric Decomposition of the State Space

We now introduce finite state space decomposition of a hybrid automaton. We will then present an abstraction based on different decompositions that incorporates reachability and time constraints. Later on, in the next section, we will discuss the refinement of the abstraction yielding constraints with better precision than before refinement.

Definition 11 (Continuous Space Partition) A (finite) partition P of the Euclidean space $\mathbb{R}^n$ is a finite set of nonempty connected subsets of $\mathbb{R}^n$ such that every point $x \in \mathbb{R}^n$ is in one and only one of those subsets. We can write $\mathbb{R}^n = \biguplus_{p \in P} p$. An element p of P will be referred to as a partition element and we will call it a region. For a subset E of $\mathbb{R}^n$, we will denote by $P(E)$ the subset of regions of P that have a nonempty intersection with E.

The only smoothness hypothesis we will impose for the moment over a partition is that any (finite) continuous path crosses only a finite number of times each region, more precisely, $\forall x : [0, 1] \rightarrow \mathbb{R}^n$ a continuous function, $\forall p \in P$ a region, $x^{-1}(p)$ is a finite union of intervals. In practice, partitions are chosen enough regular and smooth, with regions in any dimension from 1 to n such as (from simpler to more complex) rectangles, zonotopes, polytopes or defined by a set of polynomial inequalities. The choice among the different partitions is guided by the property we wish to verify. For example, consider a continuous system with dynamics F. A coarse but helpful way to obtain a high level reachability mapping would be to identify regions of the state space that conserve the sign of F. E.g., in one dimension, for all elements of the same region the derivative signs would be all either negative or positive or null. Thus, the regions would be the connected components of the three subsets p_1, p_2, p_3 defined by:

$$p_1 = \{\mathbf{x} \in \mathbf{X} | \dot{x} > 0\}, p_2 = \{\mathbf{x} \in \mathbf{X} | \dot{x} < 0\}, p_3 = \{\mathbf{x} \in \mathbf{X} | \dot{x} = 0\} \tag{11.2}$$

For n dimensional systems, the regions would be the connected components of the 3^n subsets E_s of $\mathbb{R}^n$ parametrized by sign vectors $s \in \{-1, 0, +1\}^n$: $E_s = \{\mathbf{x} \in \mathbf{X} \mid \forall i, 1 \leq i \leq n, \dot{\mathbf{x}}^i < 0 \text{ if} s^i = -1, \dot{\mathbf{x}}^i = 0 \text{ if} s^i = 0, \dot{\mathbf{x}}^i > 0 \text{ if} s^i = +1\}$.

Example 4 (The Brusselator) We now illustrate some of the introduced concepts on a mathematical model used for representing chemical reactions: the brusselator whose dynamics is nonlinear. Consider a two-dimensional continuous system $C = (X, S_0, F, Inv)$ such that $X = \{x, y\}$ are two continuously differentiable variables and F given by [4]:

$$\begin{aligned} \dot{x} &= 1 - (b+1)x + ax^2y \\ \dot{y} &= bx - ax^2y \end{aligned} \tag{11.3}$$

where $a, b \in \mathbb{R}$ are two real constants. The stationary point for which $\dot{x} = \dot{y} = 0$ is $M_0(1, \frac{b}{a})$. If $b < a + 1$, then M_0 is an attractor and all trajectories converge towards M_0; if $b > a+1$ then it is a repeller and all trajectories close to M_0 converge towards an orbit. We consider two cases where $b = 1, a = 2$ and $b = 3, a = 1$ illustrated respectively in Fig. 11.2a, b. To characterize the dynamic behavior qualitatively as in Eq. (11.2), consider a partition P yielding nine regions $p_1, \ldots, p_9$ illustrated in the repeller case in Fig. 11.3 [30].

If the considered system is a hybrid automaton, it is practical to allow different partitions in different modes. In the following, we will assume that the sets $Init(q)$, $Inv(q)$, $G(\tau)$ and $R(\tau)(p)$ (for p connected subset of $G(\tau)$) can be expressed as finite unions of connected subsets (if this is not the case, we will over-approximate parts of them). We define thus a decomposition of the hybrid state space as follows.

Definition 12 (Hybrid State Space Decomposition) Given a hybrid automaton $H = (Q, X, S_0, \Sigma, F, \delta, Inv, G, R)$ and a set $\mathfrak{P}$ of partitions of the valuations set of X, $\mathbf{X} \subseteq \mathbb{R}^n$, we say that $\mathfrak{P}$ decomposes H if there is an onto function $d : Q \rightarrow \mathfrak{P}$ which associates to each $q \in Q$ a partition $d(q) \in \mathfrak{P}$.

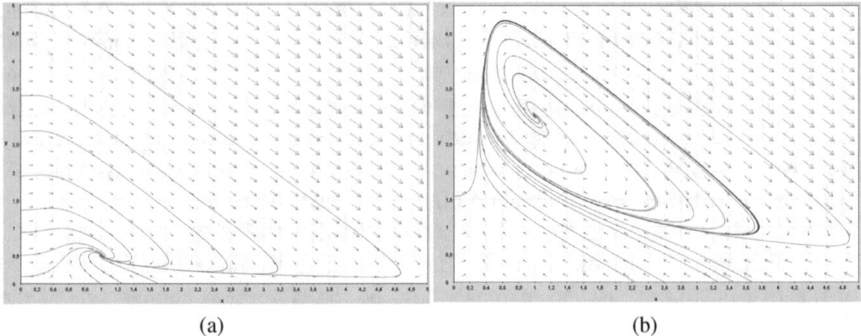

(a) (b)

Fig. 11.2 Brusselator phase plane: (**a**) Attractor, (**b**) Repeller

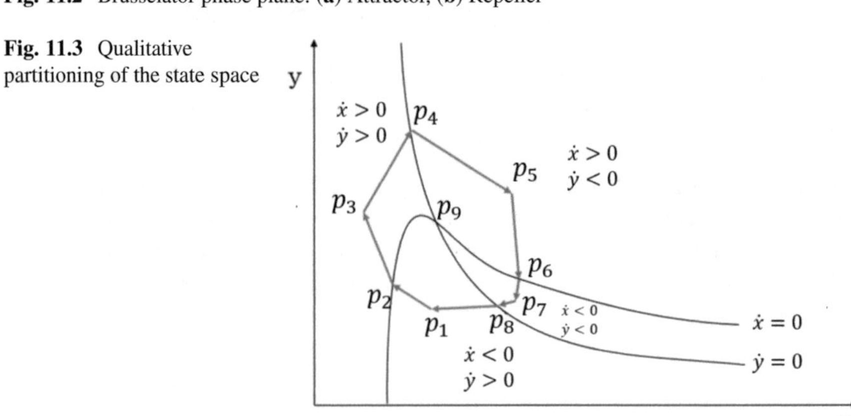

Fig. 11.3 Qualitative partitioning of the state space

The initial and invariant sets and the guards satisfiability domains and variables reset domains are primary elements to take into consideration while abstracting. For $q \in Q$ and $\tau = (q, \sigma, q') \in \delta$, we denote the regions families $d(q)(Init(q))$, $d(q)(Inv(q))$, $d(q)(G(\tau))$ by $d_{Init}(q)$, $d_{Inv}(q)$, $d_G(q, \tau) \subseteq d(q)$ and, for a region $p \in d_G(q, \tau)$, we denote $d(q')(R(\tau)(p \cap G(\tau)))$ by $d_R(q', \tau, p) \subseteq d(q')$. When possible, we will try to define d such that $Init(q)$, $Inv(q)$, $G(\tau)$, and $R(\tau)(p)$ are exactly the unions of the regions in those families (if not, those regions families over-approximate them).

11.4.3 Encoding Hybrid Automata Reachability Constraints

We have defined in Sect. 11.2.3, the timeless transition system S_H of a hybrid automaton H as the finest timeless abstraction that can be obtained. However, in practice S_H can only be computed for very restricted classes of hybrid automata. We will define a less granular time-abstract transition system based on a set of partitions and define the relations between adjacent regions.

Definition 13 (Adjacent Regions) Two distinct regions p_1, p_2 of a partition P of $\mathbb{R}^n$ are adjacent if one intersects the boundary of the other: $p_1 \cap \overline{p_2} \neq \emptyset$ or $\overline{p_1} \cap p_2 \neq \emptyset$, where $\overline{p}$ refers to the closure of p.

Definition 14 (Decomposition-Based Timeless Abstract Automaton of a Hybrid Automaton) Given a hybrid automaton $H = (Q, X, S_0, \Sigma, F, Inv, T)$ and a decomposition $(\mathfrak{P}, d)$ of H, we define the timeless abstract (finite) automaton of H with respect to $\mathfrak{P}$ as $DH_{\mathfrak{P}} = (Q_{DH}, Q_{0_{DH}}, \Sigma_{DH}, \delta_{DH})$ with:

- $Q_{DH} = \{(q, p) | q \in Q, p \in d(q)\}$.
- $Q_{0_{DH}} = \{(q, p_{Init}) | q \in Q_0, p_{Init} \in d_{Init}(q)\}$.
- $\Sigma_{DH} = \Sigma \cup \{\varepsilon\}$.
- $((q_i, p_k), \sigma, (q_j, p_l)) \in \delta_{DH}$ iff one of both is true:
  - $q_i \neq q_j$ and $\sigma \in \Sigma$ and $p_k \in d_G(q_i, \tau)$ and $p_l \in d_R(q_j, \tau, p_k)$ where $\tau = (q_i, \sigma, q_j)$ and $\exists t = (q_i, \mathbf{x}, \sigma, \mathbf{x}', q_j) \in T$ such that $\mathbf{x} \in p_k$ and $\mathbf{x}' \in p_l$.
  - $q_i = q_j$ and $\sigma = \varepsilon$ and $p_k, p_l \in d_{Inv}(q_i)$ are adjacent regions and $\exists d \in \mathbb{R}^*_+$ and $\exists x : [0, d] \to \mathbf{X}$ continuously differentiable function such that $\forall t \in (0, d)$ $\dot{x}(t) \in F(q_i, x(t))$, $\forall t \in [0, d]$ $x(t) \in Inv(q_i)$, $x(0) \in p_k$, $x(d) \in p_l$, $\exists c$ $0 \leq c \leq d$ $\forall t \in (0, c)$ $x(t) \in p_k$ $\forall t \in (c, d)$ $x(t) \in p_l$ and $x(c) \in p_k \cup p_l$.

The defined timeless abstract automaton encodes reachability with adjacent regions of the state space, the events in Σ witnessing mode changes and ε transitions representing a continuous evolution between adjacent regions in the same mode. Notice that $((q_i, p_k), \sigma, (q_j, p_l)) \in \delta_{DH} \Rightarrow \exists \mathbf{x}_k \in p_k\ \exists \mathbf{x}_l \in p_l\ (q_i, \mathbf{x}_k) \xrightarrow{\sigma} (q_j, \mathbf{x}_l)$ in S_H, the converse being true for $\sigma \in \Sigma$. The mapping $\alpha_{\mathfrak{P}}$ defined by $\alpha_{\mathfrak{P}}((q, \mathbf{x})) = (q, p)$ with $p \in d(q)$ and $\mathbf{x} \in p$ defines an onto timeless abstraction function $\alpha_{\mathfrak{P}} : \overline{S} \to Q_{DH}$. If the flow condition F is a singleton, $\alpha_{\mathfrak{P}}$ maps any transition of S_H to a unique path in $DH_{\mathfrak{P}}$. The coarsest timeless abstract automaton is obtained when partitions of $\mathfrak{P}$ have all a unique region $p = \mathbf{X}$ and is thus (Q, Q_0, Σ, δ), i.e., the discrete part of H without its continuous part. It corresponds to the coarsest timeless abstraction function $\alpha_{\{\{\mathbf{X}\}\}}((q, \mathbf{x})) = q$. For our previous thermostat example, this gives $(\{off, on\}, \{off\}, \{B_{on}, B_{off}\}, \{(off, B_{on}, on), (on, B_{off}, off)\})$ and the abstraction of the execution h given previously (11.2.3) is just $off \xrightarrow{B_{on}} on \xrightarrow{B_{off}} off \ldots$

Theorem 4 (Timeless Abstraction Completeness) *Given a decomposition $\mathfrak{P}$ of H, any concrete behavior of H is timeless abstracted into a $Q_{0_{DH}}$ rooted path in $DH_{\mathfrak{P}}$ and any transition of $DH_{\mathfrak{P}}$ abstracts a part of a concrete behavior of H. If the flow condition F is a singleton function then the timeless abstraction function $\alpha_{\mathfrak{P}}$ defines a trace preserving mapping (still denoted by $\alpha_{\mathfrak{P}}$) from $\overline{S}_0$ rooted paths in S_H (i.e., timeless executions of H) to $Q_{0_{DH}}$ rooted paths in $DH_{\mathfrak{P}}$ and thus the language defined by S_H is included in the language defined by $DH_{\mathfrak{P}}$.*

Obviously, a path in $DH_{\mathfrak{P}}$ does not abstract in general a concrete behavior of H (as the behaviors parts abstracted by the individual transitions do not connect in general) which expresses that abstraction creates spurious behaviors.

If now H is a POHA, in the same way we defined the observation of a concrete execution in Definition 7 we define the observation of its timeless abstraction.

Definition 15 (Timeless Abstraction Observation) Given a POHA H and $h = (q_0, p_0) \xrightarrow{\sigma_0} (q_1, p_1) \ldots (q_i, p_i) \xrightarrow{\sigma_i} \ldots$, with $\sigma_i \in \Sigma \cup \{\varepsilon\}$, a timeless abstract path in $DH_{\mathfrak{P}}$, the observation of h is defined as $Obs(h) = p_0^o, \sigma_0^o, p_1^o \ldots p_i^o, \sigma_i^o, \ldots$, where

- p_i^o is obtained by projecting p_i on variables in X_o.
- $\sigma_i^o = \sigma_i$ if $\sigma_i \in \Sigma_o$. Otherwise, $\sigma_i^o = \varepsilon$, which is then removed from $Obs(h)$.

Consider an execution $h \in [[H]]$ of the POHA H, $h = (q_0, \mathbf{x}_0) \xrightarrow{l_0} (q_1, \mathbf{x}_1) \ldots (q_i, \mathbf{x}_i) \xrightarrow{l_i} \ldots$, with $l_i \in \Sigma \cup \mathbb{R}_+$, its (timed) observation $Obs(h) = \mathbf{x}_0^o, l_0^o, \mathbf{x}_1^o \ldots \mathbf{x}_i^o, l_i^o, \ldots$ as in Definition 7, its timeless abstraction $\alpha_{\mathfrak{P}}(h) = (q_0, p_0) \xrightarrow{\sigma_0} (q_1, p_1) \ldots (q_i, p_i) \xrightarrow{\sigma_i} \ldots$, with $\sigma_i \in \Sigma \cup \{\varepsilon\}$, as in Theorem 4 (assuming F a singleton) and the observation of this last one $Obs(\alpha_{\mathfrak{P}}(h)) = p_0^o, \sigma_0^o, p_1^o \ldots p_i^o, \sigma_i^o, \ldots$ as in Definition 15. We could try to define the timeless abstraction of the observation $Obs(h)$. A natural definition would be $\alpha_{\mathfrak{P}}(Obs(h)) = p_0', \sigma_0^o, p_1' \ldots p_i', \sigma_i^o, \ldots$, with $\sigma_i^o = l_i^o$ if $l_i^o \in \Sigma_o$ (and $= \varepsilon$, which is removed, otherwise), i.e., the same σ_i^o's as in $Obs(\alpha_{\mathfrak{P}}(h))$, and $p_i' = \biguplus_{\{p | \mathbf{x}_i^o \in p^o\}} p^o$ the union of the projections on X_o of all regions containing a value whose projection on X_o is equal to $\mathbf{x}_i^o$ (assuming to simplify the same partition for each mode, as the mode may be unknown from observation). So, we notice that $Obs(\alpha_{\mathfrak{P}}(h))$ is more precise than $\alpha_{\mathfrak{P}}(Obs(h))$, as $p_i^o \subseteq p_i'$, which we denote by $Obs(\alpha_{\mathfrak{P}}(h)) \sqsubseteq \alpha_{\mathfrak{P}}(Obs(h))$ to mean that both sequences have common events and there is inclusion of the qualitative space values as subsets of $\mathbf{X}_o$, the valuations set corresponding to the observable variables.

11.4.4 Encoding Hybrid Automata Time Constraints

We are concerned with verifying temporal properties of hybrid systems and checking the diagnosability property using time constraints. For this reason, we define in this subsection, always related to a decomposition of the state space into partitions, an abstraction of the hybrid automaton as a timed automaton that partly captures the time constraints at the level of the regions. We will first introduce some intuitive ideas. Consider a partition P of the $\mathbb{R}^n$ state space of a continuous system with arbitrary dynamics F, the set of trajectories (i.e., the continuous solution flows) entering a region $p \in P$ is in one of these two cases: either at least one of the trajectories ends up trapped inside p for all future times or all of them exit p to an adjacent region within a bounded time under the continuity assumption. In the first case, no time constraint can be associated with the region p unless a reshaping of p is applied; in the latter, it is possible to compute time constraints satisfied by all trajectories entering and leaving the region p. We will give a formal definition of the timed automaton constructed from given hybrid automaton and partitions set and then discuss some cases where a time bound can be practically computed.

Definition 16 (Region Time Interval and Time Bounds) Given a continuous system CS, a partition P of $\mathbb{R}^n$ and $p \in P$ one of its regions, we say that $I_p = [t_{min}, t_{max}]$, with $t_{min}, t_{max} \in \mathbb{R}_+ \cup \{+\infty\}$, is a region time interval of p for CS if all trajectories of the CS entering p at time t leave p at time $t + t_{min}$ at least and $t + t_{max}$ at most. t_{min} and t_{max} are lower and upper time bounds of p.

For a hybrid automaton, we denote the time interval relative to the region p in mode q as $I_{(q,p)}$.

Definition 17 (Decomposition-Based Timed Abstract Automaton of a Hybrid Automaton) Given a hybrid automaton $H = (Q, X, S_0, \Sigma, F, \delta, Inv, G, R)$, a decomposition $(\mathfrak{P}, d)$ and the timeless abstract automaton $DH_{\mathfrak{P}} = (Q_{DH}, Q_{0_{DH}}, \Sigma_{DH}, \delta_{DH})$ of H with respect to $\mathfrak{P}$, we define the timed abstract automaton of H with respect to $\mathfrak{P}$ as $TH_{\mathfrak{P}} = (Q_{DH}, \{c\}, Q_{0_{DH}}, \Sigma_{DH}, Inv_{TH}, (\delta_{DH}, G_{TH}, Y_{TH}))$ such that, $\forall (q, p) \in Q_{DH}$ with a region time interval $I_{(q,p)} = [t_{min}, t_{max}]$:

- $Inv_{TH}((q, p)) = [0, t_{max}]$.
- $\forall \tau = ((q, p), \sigma, (q_1, p_1)) \in \delta_{DH}$, $G_{TH}(\tau) = [t_{min}, +\infty)$ if $\sigma = \varepsilon$ and p does not intersect any reset set (i.e., $\forall \tau' = (q', \sigma', q) \in \delta_{DH}\ p \notin d(q)(R(\tau')(G(\tau')))$) or $[0, +\infty)$ else.

and, $\forall \tau \in \delta_{TH}$, $Y_{TH}(\tau) = \{c\}$.

The timed abstract automaton adds time constraints to those states (q, p) of the timeless abstract automaton for which an interval $I_{(q,p)}$ is computable as non-trivial (i.e., $I_{(q,p)} \neq [0, +\infty)$), by using one local clock c (reset at 0 in each state) that measures the sojourn duration t in each state (q, p), i.e., in each region p, and coding these constraints by means of invariant and guard of c in each state. The invariant codes the maximum sojourn duration as the upper time bound of the region p and the guard codes the minimum sojourn duration as the lower time bound of the region p when both entering and leaving the region are not the result of discrete jumps (controlled here directly for the out-transition and by requiring that p does not intersect any reset set for all possible in-transitions). In the thermostat example, consider the partition into two regions associated to the mode *off* given by the initial states set $(\mathit{off}, [80, 90])$ and by $(\mathit{off}, [68, 80))$. Then we take as time bounds for $(\mathit{off}, [80, 90])$ $t_{min} = 0$ and $t_{max} = 0.12$ (the exact upper bound, i.e., the time for the temperature to decrease from 90 to 80 is $Log(\frac{9}{8})$). It means that we define in the timed abstract automaton $Inv_{TH}((\mathit{off}, [80, 90])) = [0, 0.12]$. A beginning of execution of the timed abstract automaton is, for example, $(\mathit{off}, [80, 90]) \xrightarrow{0.08} (\mathit{off}, [68, 80])$.

Theorem 5 (Timed Abstraction Completeness) *Given a decomposition $\mathfrak{P}$ of H, any concrete behavior of H is timed abstracted into an execution in $TH_{\mathfrak{P}}$. If the flow condition F is a singleton function then the abstraction function $\alpha_{\mathfrak{P}}$ defines a mapping, denoted by $\alpha^t_{\mathfrak{P}}$, from $\overline{S}_0$ rooted paths in S^t_H (i.e., executions of H) to executions in $TH_{\mathfrak{P}}$. This mapping is trace preserving once ϵ labels are erased from executions traces in $TH_{\mathfrak{P}}$ and time period labels are added up between two consecutive events labels in both executions traces in S^t_H and in $TH_{\mathfrak{P}}$. This means* *that, for any execution* $(q_0, \mathbf{x}_0) \xrightarrow{w}_* (q_i, \mathbf{x}_i) \in [[H]]$, *with* $w \in L^*$ *(where* $L = \Sigma \cup \mathbb{R}_+$*), it exists a unique execution* $(q_0, p_0) \xrightarrow{w'}_* (q_j, p_j) \in [[TH_{\mathfrak{P}}]]$, *with* $w' \in L'*$ *(where* $L' = L \cup \{\epsilon\}$*),* $\mathbf{x}_0 \in p_0$, $q_j = q_i$, $\mathbf{x}_i \in p_j$, $w'_{|\Sigma} = w_{|\Sigma}$ *(where* $_{|\Sigma}$ *is the projection of timed words on words on* Σ^**) and, for any two successive events* $w_l = w'_{l'}$ *and* $w_m = w'_{m'}$ *of* $w_{|\Sigma}$, $\sum_{l'<k'<m', w'_{k'} \neq \epsilon} w'_{k'} = \sum_{l<k<m} w_k$.

Forgetting time, i.e., removing the clock, provides a natural abstraction function α from $TH_{\mathfrak{P}}$ to $DH_{\mathfrak{P}}$ which maps an execution $(q_0, p_0) \xrightarrow{l_0} (q_1, p_1) \ldots (q_i, p_i) \xrightarrow{l_i} \ldots$, with $l_i \in \Sigma \cup \{\varepsilon\} \cup \mathbb{R}_+$, in $TH_{\mathfrak{P}}$ into the execution $(q_0, p_0) \xrightarrow{\sigma_0} (q_1, p_1) \ldots (q_i, p_i) \xrightarrow{\sigma_i} \ldots$, with $\sigma_i \in \Sigma \cup \{\varepsilon\}$, in $DH_{\mathfrak{P}}$, with $\sigma_i = l_i$ if $l_i \in \Sigma \cup \{\varepsilon\}$ and continuous transitions labeled by $l_i = d_i \in \mathbb{R}_+$ are suppressed. We have: $\alpha_{\mathfrak{P}} = \alpha \circ \alpha^t_{\mathfrak{P}}$.

Definition 18 (Timed Abstraction Observation) Given a POHA H and $h = (q_0, p_0) \xrightarrow{l_0} (q_1, p_1) \ldots (q_i, p_i) \xrightarrow{l_i} \ldots$, with $l_i \in \Sigma \cup \{\varepsilon\} \cup \mathbb{R}_+$, an execution in $TH_{\mathfrak{P}}$, i.e., a timed abstract path, the observation of h is defined as $Obs(h) = p^o_0, l^o_0, p^o_1 \ldots p^o_i, l^o_i, \ldots$, where

- p^o_i is obtained by projecting p_i on variables in X_o.
- $l^o_i = l_i$ if $l_i \in \Sigma_o \cup \mathbb{R}_+$. Otherwise, $l^o_i = \varepsilon$, which is then removed from $Obs(h)$.

As for the timeless case, we can define the timed abstraction of an observation (of an execution $h \in [[H]]$) and we obtain: $Obs(\alpha^t_{\mathfrak{P}}(h)) \sqsubseteq \alpha^t_{\mathfrak{P}}(Obs(h))$.

From another side, the abstraction function α that forgets time maps a timed abstract observation $p^o_0, l^o_0, p^o_1 \ldots p^o_i, l^o_i, \ldots$, with $l^o_i \in \Sigma_o \cup \mathbb{R}_+$, into the timeless abstract observation $p^o_0, \sigma^o_0, p^o_1 \ldots p^o_i, \sigma^o_i, \ldots$, with $\sigma^o_i \in \Sigma_o$, suppressing duration labels $l_i = d_i \in \mathbb{R}_+$. For any concrete execution $h \in [[H]]$, we have: $Obs(\alpha_{\mathfrak{P}}(h)) = \alpha(Obs(\alpha^t_{\mathfrak{P}}(h)))$, i.e., $Obs \circ \alpha_{\mathfrak{P}} = \alpha \circ Obs \circ \alpha^t_{\mathfrak{P}}$.

11.4.5 Computing Time Bounds

In this subsection, we present and discuss some situations for which time bounds can be computed for the continuous evolution of the hybrid system. In the following, CS is a continuous system, and P is a partition of $\mathbb{R}^n$, $p \in P$ and $I_p = [t_{min}, t_{max}]$ the associated region time interval of p.

Proposition 1 (Sojourn Bounds) *A sufficient but not necessary condition for the region* p *to have finite time bounds (*t_{max} *finite, thus real nonnegative constant) is that* $\exists i\ 1 \leq i \leq n \quad \forall \mathbf{x} \in \overline{p} \quad \dot{\mathbf{x}}_i \neq 0$.

If the condition in Proposition 1 is verified over p for a dimension i, then all trajectories should respect finite time bounds for staying in the region p. On the other hand, a trajectory making a finite number of orbital spins once inside p then exiting p does not satisfy this condition while having finite time bounds. One way to look for a finite time bound is to refine the partition with the objective that the regions become small enough for the condition to hold for each of them. This idea will be developed in Sect. 11.5. If the derivatives along each axis take each a finite number of null values inside the partition, but never all at the same time, there is no problem with refining the partition to have each null derivative point alone in one region. In the other cases, we present now some examples for which a time bound can be nevertheless obtained.

Finite Number of Equilibrium Points The derivatives along each coordinate are null in a finite number of points X_{eq} at the same time, we repartition p such that each new region p_i has exactly one point of X_{eq} on its boundary. In one dimension, suppose the null derivative point is at x_{eq} and the initial value of $x(t)$ is x_0. The time, which we denote by $\mathscr{T}$, taken by the continuous variable $x(t)$ to cross from x_0 to x_{eq} can in some cases be finite. If $\dot{x}$ is of the form x^k where $k \in \mathbb{R}$, then by studying $\mathscr{T}$ in a neighborhood around $k = 1$ we obtain:

$$\mathscr{T} = \int_0^{t_{eq}} \mathrm{d}t = \int_{x_0}^{x_{eq}} \frac{1}{\dot{x}}\,\mathrm{d}x = \int_{x_0}^{0} \frac{1}{x^k}\,\mathrm{d}x = \left[\frac{x^{-k+1}}{(-k+1)}\right]_{x_0}^{0} \tag{11.4}$$

$\mathscr{T}$ is convergent if $k < 1$. Thus a time bound can be computed for all dynamics of the form $\dot{x} = x^k$ with $k < 1$, for example for the square root $\dot{x} = \sqrt{x}$. In two dimensions, let r be the distance from the equilibrium point $M_{eq}(x_{eq}, y_{eq})$ to a point $M(x, y)$ which is initially in region p. Let $X = x_{eq} - x$ and $Y = y_{eq} - y$. Since $r^2 = X^2 + Y^2$ then $2r\dot{r} = 2X\dot{X} + 2Y\dot{Y}$ and if $\dot{r} \neq 0$ the time to reach M_{eq} is :

$$\mathscr{T} = \int_0^{t_{eq}} \mathrm{d}t = \int_{r_0}^{0} \frac{1}{\dot{r}}\,\mathrm{d}r = \int_{r_0}^{0} \frac{r}{X\dot{X} + Y\dot{Y}}\,\mathrm{d}r \tag{11.5}$$

Example 5 Consider the two dimensional continuous system $\dot{x} = -x^2$ and $\dot{y} = -y$ where $(x, y) \in \mathbb{R}^+ \times \mathbb{R}^+$. The equilibrium point is $M_{eq}(0, 0)$. In polar coordinates $x = r\cos(\theta)$ and $y = r\sin(\theta)$, then $r\dot{r} = x\dot{x} + y\dot{y} = -x^3 - y^2 = -x^2 - y^2 + x^2 - x^3 = -r^2 + x^2(1 - x) = r^2(-1 + \cos^2(\theta)(1 - r\cos(\theta)))$. In a neighborhood around $(0, 0)$:

$$\mathscr{T} = \int_{r_0}^{0} \frac{1}{r(-1 + \cos^2(\theta)(1 - r\cos(\theta)))}\,\mathrm{d}r \geq \int_{r_0}^{0} \frac{-1}{r}\,\mathrm{d}r \tag{11.6}$$

Thus $\mathscr{T}$ is infinite, the equilibrium point is never reached. This reasoning can be extended to dimension n by evaluating $r\dot{r}$ and using spherical (or hyperspherical) coordinates and to polynomial with real exponents. We can take an example of square root, for instance $\dot{x} = \sqrt{x}$ and demonstrate the time $\mathscr{T}$ is finite, then the equilibrium is reached.

Infinite Number of Null Derivatives Studying a case where at least one derivative along an axis takes an infinite number of null values in a connected set can be done by extending the previous method. For the particular class of continuous systems where the dynamics are only allowed multi-affine function form, Maler and Batt [45] showed how it is possible to capture time constraints by decomposing the infinite state space $\mathbb{R}^n$ into hypercubes and evaluate the time elapsed between entering and exiting each cube by bounding the dynamics.

Brusselator Time Bounds For our Example 4 of the brusselator dynamics, consider, for the repeller case with $b = 3, a = 1$, a ring set R that excludes M_0 such that $R = \{(x, y) \mid (-1 + x)^2 + (-3 + y)^2 > 0.09 \wedge (-1 + x)^2 + (-3 + y)^2 < 0.5625\}$. Let $v = \sqrt{\dot{x}^2 + \dot{y}^2}$ then $v^2 = (1 - 4x + x^2y)^2 + (3x - x^2y)^2$ and, using a solver, we compute a lower bound $v_{low}^2 = 0.0051$ of v^2. It has been proven that all trajectories initially in $S_0 = \mathbb{R}^2 \setminus \{M_0\}$ converge towards a fixed orbit of the phase plane contained within R. Suppose we split the region R into two connected sets R_1 and R_2 such that $R = R_1 \uplus R_2$, for each of which the maximal sojourn duration t_{max} is a positive real constant. This is possible since v admits a lower bound v_{low}. This states that all trajectories initiated from S_0 will cross R_1 and R_2 sequentially infinitely often. In practice, the presence of the system in either R_1 or R_2 can correspond to two different visible colors of the chemical reaction.

11.5 Hybrid Automata Abstraction Refinement

We will now explain and formalize the refinement process of the previously defined abstraction. For this purpose, we construct a finer couple of discrete and timed automata by defining a more granular decomposition for regions and give the necessary assumptions to compute such refinement. By making the partition more granular in regions of interest, tighter time bounds are also obtained. The refinement is a necessary step for the CEGAR scheme, it is a required step when a proof for the verification of a property could not be made at a given abstraction level.

Definition 19 (Partition Refinement) Given two partitions P and P' of $\mathbb{R}^n$, we say that P' is a refining partition of P iff $\forall p' \in P' \; \exists p \in P \; p' \subseteq p$. This implies: $\forall p \in P \; \exists P'_p \subseteq P' \; p = \biguplus_{p' \in P'_p} p'$.

Definition 20 (Hybrid State Space Decomposition Refinement) Given two decompositions $(\mathfrak{P}, d)$ and $(\mathfrak{P}', d')$ of a hybrid automaton $H = (Q, X, S_0, \Sigma, F, Inv, T)$, we say that $\mathfrak{P}'$ refines $\mathfrak{P}$, denoted by $\mathfrak{P}' \preceq \mathfrak{P}$, if $\forall q \in Q \; d'(q)$ is a refining partition of $d(q)$.

11.5.1 Refined Timeless Model

Definition 21 (Refined Timeless Abstract Automaton) Given a hybrid automaton H and two abstract timeless automata $DH_{\mathfrak{P}}$ and $DH_{\mathfrak{P}'}$ of H with respect to two decompositions $(\mathfrak{P}, d)$ and $(\mathfrak{P}', d')$ respectively, we say that $DH_{\mathfrak{P}'}$ is a timeless refinement of $DH_{\mathfrak{P}}$ abstracting H if $\mathfrak{P}' \prec \mathfrak{P}$, which we denote by $DH_{\mathfrak{P}'} \prec DH_{\mathfrak{P}}$.

Definition 22 (State Split Operation) Given an abstract timeless automaton $DH_{\mathfrak{P}}$ of a hybrid automaton H, a split operation of the state $(q,p) \in Q_{DH}$ is defined by a partition $\{p_1, p_2\}$ of p, $p = p_1 \uplus p_2$, and results in two states (q,p_1) and (q,p_2) and in the refined abstract timeless automaton $DH_{\mathfrak{P}'}$ with $\mathfrak{P}'$ obtained from $\mathfrak{P}$ by replacing $d(q)$ by $d'(q) = d(q) \backslash \{p\} \cup \{p_1, p_2\}$.

The construction of $DH_{\mathfrak{P}'}$ from $DH_{\mathfrak{P}}$ after a (q,p) state split is a local operation as only the transitions of δ_{DH} having as source or as destination the state (q,p) have to be recomputed from H. In practice, the refined model is obtained by performing a finite number of state split operations. After having performed the split operations and in order for the obtained automaton to satisfy Definition 14, it is only required to recompute some of its transitions, while inheriting the rest from $DH_{\mathfrak{P}}$. Let $Q_{split} \subseteq Q_{DH}$ be the set of split states and $post(Q_{split}) = \{q \in Q_{DH} \mid \exists q_s \in Q_{split} \exists (q_s, \sigma, q) \in \delta_{DH}\}$ and $pre(Q_{split}) = \{q \in Q_{DH} \mid \exists q_s \in Q_{split} \exists (q, \sigma, q_s) \in \delta_{DH}\}$. Then to obtain $DH_{\mathfrak{P}'}$ it is sufficient to only recompute transitions (q, σ, q') such that $q, q' \in Q_{split} \cup post(Q_{split}) \cup pre(Q_{split})$.

The onto abstraction function $\alpha_{\mathfrak{P}',\mathfrak{P}} : Q_{DH}^{\mathfrak{P}'} \rightarrow Q_{DH}^{\mathfrak{P}}$ defined by $\alpha_{\mathfrak{P}',\mathfrak{P}}((q,p')) = (q,p)$ with $p' \subseteq p$ defines a trace preserving mapping $\alpha_{\mathfrak{P}',\mathfrak{P}}$ from $DH_{\mathfrak{P}'}$ to $DH_{\mathfrak{P}}$ (and thus the language defined by $DH_{\mathfrak{P}'}$ is included in the language defined by $DH_{\mathfrak{P}}$) and we have: $\alpha_{\mathfrak{P}} = \alpha_{\mathfrak{P}',\mathfrak{P}} \circ \alpha_{\mathfrak{P}'}$. Defining in a natural way as previously the $\mathfrak{P}$-abstraction $\alpha_{\mathfrak{P}',\mathfrak{P}}$ of the observation of a timeless $\mathfrak{P}'$-abstract execution h, we obtain: $Obs(\alpha_{\mathfrak{P}',\mathfrak{P}}(h)) \sqsubseteq \alpha_{\mathfrak{P}',\mathfrak{P}}(Obs(h))$.

11.5.2 Refined Timed Model

Definition 23 (Refined Timed Abstract Automaton) Given a hybrid automaton H and two abstract timed automata $TH_{\mathfrak{P}}$ and $TH_{\mathfrak{P}'}$ of H with respect to two decompositions $(\mathfrak{P}, d)$ and $(\mathfrak{P}', d')$ respectively, we say that $TH_{\mathfrak{P}'}$ is a timed refinement of $TH_{\mathfrak{P}}$ abstracting H if $\mathfrak{P}' \prec \mathfrak{P}$. We denote it similarly by $TH_{\mathfrak{P}'} \prec TH_{\mathfrak{P}}$.

Concerning the refined abstract timed automaton $TH_{\mathfrak{P}'}$ resulting from a split of (q,p) into (q,p_1) and (q,p_2), if $I_{(q,p)} = [t_{min}, t_{max}]$ the region time intervals $I_{(q,p_1)} = I_{(q,p_2)} = [0, t_{max}]$ can be adopted in first approximation as they are safe, but in general new tighter time bounds are recomputed from H for the sojourn duration in the regions p_1 and p_2. Thus, the refined timed model is obtained by a finite sequence of the two operations:

- **State split**: similar as before, the state split of (q,p) whose time interval is $I_{(q,p)} = [t_{min}, t_{max}]$ yields (q,p_1) and (q,p_2). The time intervals for the new split regions are set as $I_{(q,p_1)} = I_{(q,p_2)} = [0, t_{max}]$ (t_{max} stays a safe upper bound of the sojourn duration but t_{min} is reset to 0 since the split induces a distance shrink).
- **Time bounds refinement:** in this case, more precise time bounds are obtained for a given region of a discrete state, i.e., if $I_{(q,p)} = [t_{min}, t_{max}]$ then $I'_{(q,p)} = [t'_{min}, t'_{max}]$ with $t'_{min} \geq t_{min}$ and $t'_{max} \leq t_{max}$, at least one of both being a strict inequality.

The onto abstraction function $\alpha_{\mathfrak{P}',\mathfrak{P}}$ defines a trace preserving mapping $\alpha^t_{\mathfrak{P}',\mathfrak{P}}$ from $TH_{\mathfrak{P}'}$ to $TH_{\mathfrak{P}}$, after trace simplification as in Theorem 5 and provided the time bounds used in $TH_{\mathfrak{P}'}$, once added for all regions p' included in a given region p, are at least as tight as the time bounds used in $TH_{\mathfrak{P}}$. And we have: $\alpha^t_{\mathfrak{P}} = \alpha^t_{\mathfrak{P}',\mathfrak{P}} \circ \alpha^t_{\mathfrak{P}'}$. Finally, for any timed $\mathfrak{P}'$-abstract execution h, we obtain: $Obs(\alpha^t_{\mathfrak{P}',\mathfrak{P}}(h)) \sqsubseteq \alpha^t_{\mathfrak{P}',\mathfrak{P}}(Obs(h))$.

11.5.3 Refinement Guided by Reachability Analysis

We give here some general mathematical properties, in particular about conservation of the connectivity property of the regions when following the solution flow, that are useful for the refinement process when guided by the dynamics of the hybrid system and reachability conditions.

Reachability from a Connected Set Let $CS = (X, S_0, F, Inv)$ be a continuous system with deterministic flow condition $F : \mathbf{X} \to \mathbb{R}^n$ and $\mathbf{K} \subset \mathbb{R}^n$ a set, such that the following hypotheses are verified:

- $\mathbf{K}$ is a connected and bounded closed (i.e., compact) set and $\forall \mathbf{x} \in \mathbf{K}, F(\mathbf{x}) \neq \mathbf{0}$.
- The flow solution function $x(t, \mathbf{x}_0)$ initially starting at $t = 0$ from $\mathbf{x}_0 \in \mathbf{K}$ is continuous with respect to $\mathbf{x}_0 \in \mathbf{K}$ and of class C^1 with respect to the time t (this property is true in the case of polynomial dynamics).

With these hypotheses, trajectories issued from $\mathbf{x}_0$ are continuous with respect to $\mathbf{x}_0$ (proof can be made using the uniform continuity deduced from the continuity as $\mathbf{K}$ is a compact set).

Let $\mathbf{y}$ be a reachable element from $\mathbf{K}$ and $x(t)$ the trajectory reaching $\mathbf{y}$ at time t_0. We define the successor trajectory $post(\mathbf{y})$ and the predecessor trajectory $pre(\mathbf{y})$ by:

$$post(\mathbf{y}) = \{\mathbf{x} \in \mathbf{X} \mid \exists t > t_0 \mathbf{x} = x(t)\} pre(\mathbf{y}) = \{\mathbf{x} \in \mathbf{X} \mid \exists 0 < t < t_0 \mathbf{x} = x(t)\} \tag{11.7}$$

We extend this definition to the set $\mathbf{K}$:

$$post(\mathbf{K}) = \bigcup_{k \in \mathbf{K}} post(k) \quad pre(\mathbf{K}) = \bigcup_{k \in \mathbf{K}} pre(k) \tag{11.8}$$

With continuity argument from the hypotheses, we have the following result.

Theorem 6 *$post(\mathbf{K})$ and $pre(\mathbf{K})$ are connected sets.*

This result shows that our connectivity property assumed for all regions is conservative along trajectories (backward and forward) issued from set $\mathbf{K}$.

Reachability from Initial State to Guard Set Let H be a hybrid automaton with polynomial dynamics and $\mathbf{X}_0 = \{\mathbf{x} \mid (q_0, \mathbf{x}) \in S_0\}$ the set of its initial states in a mode q_0. We suppose we have built an abstract automaton of H with respect to a given partition. By construction of this partition, we consider that for each region p, all incoming trajectories must exit and become outgoing trajectories after having passed a bounded time in p. Then we can define $in(p)$ as the set of all trajectories restricted to p, which is equal to $post(\mathbf{X}_0) \cap p$. We define also the subset $pre(p)$ of the incoming points of $in(p)$ and the subset $post(p)$ of the outgoing points of $in(p)$. It can be demonstrated that $pre(p)$, $in(p)$, and $post(p)$ are unions of a finite number of connected sets. The proof can be made by induction starting from the initial set assumed to be defined by convex linear predicates and using the fact the dynamics is polynomial.

If we consider a guard set $G(\tau)$, assumed to be a convex linear predicate set, the set of outgoing points of the trajectories crossing p and verifying this guard is $post(p) \cap G(\tau)$ and it can be proved that this set is again a union of a finite number of connected sets. So the time passed by a trajectory inside p with outgoing points verifying $G(\tau)$ can be represented by a union of finite number of intervals, as internal trajectories form a union of finite number of connected sets (the set of instants passed in trajectories belonging to a connected set is a time interval if the trajectories are continuous). Outside this union of time intervals for internal trajectories in p, the guard $G(\tau)$ cannot be verified by outgoing points. This analysis will provide important results for the diagnosability verification.

Brusselator Example This result applies to the polynomial dynamics of the brusselator, more specifically the two defined regions R_1 and R_2 corresponding to the color change of the system. We consider the repeller case of the brusselator as a discrete hybrid mode. With the previous result, we demonstrate that any inconsistency in observing the color change within the computed maximal time bound could be diagnosed with a change of the current repellor mode. If we model the fault as a discrete jump to the attractor case, the observations in terms of color change are sufficient to diagnose this fault, since in the attractor case all trajectories converge asymptotically to M_0, thus there is no color change.

11.6 CEGAR Adaptation for Diagnosability Verification

Since hybrid systems have an infinite state space due to continuous dynamics, verifying formally their properties often rests on using ordinary model checking together with a finite-state abstraction. Model-checking can be inconclusive, in which case the abstraction must be refined. In this section, we adapt counterexample guided abstraction refinement (CEGAR) that was originally proposed to verify safety properties [3, 28].

Note that to verify safety properties, it is sufficient to check one execution at a time and verify whether the execution can reach an unsafe state. Verifying diagnosability reveals a more complex task as one is required to simultaneously analyze two executions at a time, i.e., to verify whether or not the two executions have the same observations while only one of them contains the considered fault.

In our abstraction method, time constraints are used explicitly. When an abstraction refinement is required, tighter time bounds are obtained over the new regions of the refined decomposition. The proposed abstraction method hence differs from the one proposed in [35]. In [35], the abstraction consists in retaining properties of the continuous dynamics, namely mode distinguishability and ephemerality, which are directly checked on the concrete hybrid system when necessary. On the contrary, in our approach the abstractions refer directly to the continuous state space and the continuous dynamics are interpreted with increasing levels of granularity, which results in finer and finer state space decompositions to which time constraints are associated. These abstractions take the form of timed automata.

The adaptation of CEGAR to verify diagnosability of a hybrid automaton H consists in three steps described as follows and to be detailed in the next subsections:

- **Diagnosability checking** of a timed abstract automaton of H using the twin plant method, which generates a counterexample $C.E$ when diagnosability is not verified.
- **Validation of the** $\boldsymbol{C.E}$ by checking whether the $C.E$ is valid or spurious.
- **Refinement** of the timed abstract automaton by using a finer hybrid state space decomposition.

11.6.1 CEGAR Scheme for Hybrid Automata Diagnosability Verification

Verifying diagnosability of a hybrid automaton by checking it on abstractions of this automaton is justified because if the diagnosability property is verified for an abstraction, then it is verified also for the concrete hybrid system. This can be established by showing that a concrete counterexample of diagnosability lifts up into an abstract counterexample of diagnosability. Actually, given a hybrid automaton $H = (Q, X, S_0, \Sigma, F, Inv, T)$, two executions h, $h' \in [[H]]$, $h = (q_0, \mathbf{x}_0) \xrightarrow{l_0} (q_1, \mathbf{x}_1) \ldots (q_i, \mathbf{x}_i) \xrightarrow{l_i} \ldots$, $h' = (q'_0, \mathbf{x}'_0) \xrightarrow{l'_0} (q'_1, \mathbf{x}'_1) \ldots (q'_i, \mathbf{x}'_i) \xrightarrow{l'_i} \ldots$ are called a counterexample of diagnosability in H with respect to the fault F if they satisfy the three conditions defined in Definition 10, i.e., if h and h' constitute a critical pair of H.

We will denote each state $(q_i, \mathbf{x}_i)$ by s_i and $(q'_i, \mathbf{x}'_i)$ by s'_i. We assume that the flow condition F is a singleton function (deterministic). Then, from Theorem 5, given a timed abstract automaton $TH_{\mathfrak{P}}$ of H with abstraction function $\alpha^t_{\mathfrak{P}}$, h and h' are mapped by $\alpha^t_{\mathfrak{P}}$ into executions $\hat{h}, \hat{h}' \in [[TH_{\mathfrak{P}}]]$, $\hat{h} = \hat{s}_0 \xrightarrow{\hat{l}_0} \hat{s}_1 \ldots \hat{s}_i \xrightarrow{\hat{l}_i} \ldots$, $\hat{h}' = \hat{s}'_0 \xrightarrow{\hat{l}'_0} \hat{s}'_1 \ldots \hat{s}'_i \xrightarrow{\hat{l}'_i} \ldots$ and, as (h, h') is a critical pair in H with respect to F, so is

Algorithm 1 CEGAR scheme for hybrid automata diagnosability verification

INPUT: hybrid automaton H; considered fault F; constant positive integer *precision*
OUTPUT: *decision* := H is diagnosable (*true*) | H is not diagnosable (*false*) | precision is reached (*max_reached*)

```
TH ← Initial Timed Abstract Automaton of H
C.E ← Diagnosability Check (TH, F)
abstraction_level ← 0
while C.E ≠ ∅ ∧ abstraction_level < precision do
   if Validate(C.E, H) then
      decision ← false EXIT
   else
      TH ← Refine(TH, C.E, H)
      C.E ← Diagnosability Check(TH, F)
      abstraction_level ← abstraction_level + 1
   end if
end while
if C.E = ∅ then
   decision ← true EXIT
else
   decision ← max_reached EXIT
end if
```

$(\hat{h}, \hat{h}')$ in $TH_{\mathfrak{P}}$, which establishes thus a counterexample of diagnosability in $TH_{\mathfrak{P}}$: $C.E = (\hat{h}, \hat{h}')$. This proves the following result.

Theorem 7 *Given a hybrid automaton H with singleton flow condition, a timed abstract automaton $TH_{\mathfrak{P}}$ of H with abstraction function $\alpha^t_{\mathfrak{P}}$ and a modeled fault F in H, if F is diagnosable in $TH_{\mathfrak{P}}$ then F is diagnosable in H.*

Now, with this result, Algorithm 1 illustrates the CEGAR scheme adaptation for hybrid automata diagnosability verification.

11.6.2 Twin Plant Based Diagnosability Checking

Diagnosability checking of a discrete event system, modeled as an automaton, based on the twin plant method [40, 61] is polynomial in the number of states (actually it has been proved it is NLOGSPACE-complete [51]). The idea is to construct a non-deterministic automaton, called pre-diagnoser or verifier, that preserves only all observable information and appends to every state the knowledge about past fault occurrence. The twin plant is then obtained by synchronizing the pre-diagnoser with itself based on observable events to get as paths in the twin plant all pairs of executions with the same observations in the original system. Each state of the twin plant is a pair of pre-diagnoser states that provide two possible diagnoses. A twin plant state is called an ambiguous one if the corresponding two pre-diagnoser states give two different diagnoses (presence for one and absence for the other of a past fault occurrence). A *critical path* is a path in the twin plant with at least one ambiguous state cycle. It corresponds to a critical pair and it has thus been proved that the existence of a critical path is equivalent to non-diagnosability. The twin plant method has been adapted to be applied to timed automata [58], where a twin plant is constructed in a similar way except that the time constraints of two executions are explicitly taken into account using clock variables. The idea is to verify whether the time constraints can further distinguish two executions by comparing the occurrence time of observable events. The definition of a critical path in the twin plant is analog, except that ambiguous state cycle is replaced by infinite time ambiguous path.

Lemma 1 *A fault is diagnosable in a timed automaton iff its twin plant contains no critical path [58].*

For timed automata, checking diagnosability is PSPACE-complete.

11.6.3 Counterexample Validation or Refusal

After applying the twin plant method on a timed abstract automaton $TH_{\mathfrak{P}}$ of H as described in [58], suppose that a critical pair $C.E = (\hat{h}, \hat{h}')$ is returned (if not, it means that $TH_{\mathfrak{P}}$, and thus H, is diagnosable). Whether we find or not two concrete executions $h, h' \in [[H]]$ whose abstractions by $\alpha^t_{\mathfrak{P}}$ are $\hat{h}$ and $\hat{h}'$ and form a concrete critical pair decides if $C.E$ is validated or refuted. We detail below both procedures for validation or refusal and the reasons for which, in the latter case, a critical pair can be assumed spurious.

Validated Counterexample If it exists $h, h' \in [[H]]$, whose abstractions by $\alpha^t_{\mathfrak{P}}$ (according to Theorem 5) are $\hat{h}, \hat{h}'$ and such that $Obs(h) = Obs(h')$, then (h, h') constitutes a concrete counterexample realizing $C.E$ and proves thus the non-diagnosability of H. If not, the abstract counterexample $C.E$ is said spurious. In practice, this step involves computing reachable sets of states using safe over approximations such as ellipsoids and zonotopes for complex dynamics or hypercubes for simpler ones [1, 12]. Obviously, due to inherent undecidability of reachability problem at the concrete level of the hybrid automaton, it can happen that a concrete critical pair realizing $C.E$ does actually exist but this existence will not be proved and $C.E$ will be declared spurious, with new chance to discover a concrete critical pair at the next refinement loop.

Refuted Counterexample In case of spurious $C.E = (\hat{h}, \hat{h}')$, the idea is to construct longest finite executions $h, h' \in [[H]]$, that abstract by $\alpha^t_{\mathfrak{P}}$ into finite prefixes of $\hat{h}, \hat{h}'$ and such that $Obs(h) = Obs(h')$. The fact they cannot be extended means that $\forall \overline{h} \in [[H]]/h,\ \overline{h'} \in [[H]]/h'$ one step executions, either (i) $\hat{s}_{|h|+1} \neq \alpha^t_{\mathfrak{P}}(s_{|h|+1})$ (or $\hat{s}'_{|h'|+1} \neq \alpha^t_{\mathfrak{P}}(s'_{|h'|+1})$) or (ii) $\hat{l}_{|h|} \neq l_{|h|}$ (or $\hat{l}'_{|h|} \neq l'_{|h|}$) or (iii) $Obs(\overline{h}) \neq Obs(\overline{h'})$. In this case, $s^{reach}_{|h|+1}$ and $s'^{reach}_{|h'|+1}$ are returned, that represent the two sets of reachable concrete states that are the first ones to disagree with the abstract *C.E*. We summarize below the reasons resulting in the *C.E* being spurious.

Spurious State Reachability There is no concrete execution in H whose abstraction is one of $\hat{h}$ or $\hat{h}'$, as one of the set of states of H whose abstraction is an abstract state $\hat{s}_i$ or $\hat{s}'_i$ is not reachable in H starting from the initial states of H. Note that care will have to be taken when refining $TH_{\mathfrak{P}}$ (see next subsection). E.g., a possible case is that there exist two executions (h, h') reaching (s_1, s'_1) and then (s_2, s'_2) but not reaching (s_3, s'_3) (and none passing by (s_1, s'_1) and (s_2, s'_2) reaches (s_3, s'_3)), and two other executions (u, u') reaching (s_2, s'_2) from $(s, s') \neq (s_1, s'_1)$, with $\alpha^t_{\mathfrak{P}}(s) = \alpha^t_{\mathfrak{P}}(s_1)$ and $\alpha^t_{\mathfrak{P}}(s') = \alpha^t_{\mathfrak{P}}(s'_1)$, and then reaching (s_3, s'_3), all with time periods compatible to those of the abstract executions. If the refined model simply eliminated the transition from $\hat{s}_2$ to $\hat{s}_3$ or from $\hat{s}'_2$ to $\hat{s}'_3$ then it could no longer be considered an abstraction of H, since some concrete execution in H would have no abstract counterpart. Thus the refinement has to apply the split operation as previously described, so that preserving the abstraction while eliminating the spurious counterexample.

Spurious Time Constraints Satisfaction The abstract critical pair, when considered timeless, owns a concrete critical pair realization in H but none verifying the time bounds imposed by the abstract timed automaton. In this case it is not a spurious state reachability problem but a spurious timed state reachability problem. Actually the time constraints of the abstract critical pair cannot be satisfied by any concrete critical pair realizing it in H.

Spurious Observation Undistinguishability The two executions of the abstract critical pair share the same observations (observable events with their occurrence times and snapshots of the values of observable continuous variables at arrival times in each abstract state) but actually any two concrete executions realizing this critical pair in H do distinguish themselves by the observation of some observable continuous variable.

11.6.4 Refinement of the Abstraction

If it reveals that abstract counterexample $C.E = (\hat{h}, \hat{h}')$ is spurious, then one refines the timed abstract automaton $TH_{\mathfrak{P}}$ to get $TH_{\mathfrak{P}'}$, guided by the information from *C.E*. The first step is analyzing *C.E* to identify the reasons why it is spurious (as classified previously). The idea is to avoid getting relatively close spurious abstract counterexample when applying twin plant method on the refined timed abstract automaton $TH_{\mathfrak{P}'}$. The refinement procedure is described as follows and will be illustrated on our example in the next section.

1. Suppose that $C.E$ is refuted due to an illegal stay, i.e., the corresponding invariant is not respected. The consequence could be $s^{reach}_{|h|+1} = \emptyset$, i.e., an illegal transition. To eliminate such spurious counterexample next time, one can partition the region containing $\hat{s}_{|h|}$ to get a new region representing the legal stay such that the refinement can be done based on this partition. The idea is to eliminate illegal (unobservable) transitions between the new region and others by tightening time constraints. In a similar way, one can handle spurious counterexamples with illegal transitions due to the unsatisfiability of the corresponding guards by the evolution of continuous variables, but with a legal stay this time.
2. Suppose that the refutation of $C.E$ is due to different observations from $s_{|h|+1}$ and $s'_{|h'|+1}$ without reachability problem. The idea is to calculate the exact moment, denoted $t_{spurious}$, before which it is still possible to get the same observations while after it the observations will diverge. With $t_{spurious}$, one can partition $\hat{s}_{|h|+1}$ and $\hat{s}'_{|h'|+1}$ to get a new region whose legal stay is limited by $t_{spurious}$ and transition to another region gives birth to a new refined observation by means of an observable continuous variable if any.

11.7 Case Study Example

The CEGAR scheme for diagnosability checking of hybrid automata will be illustrated by the following case study example.

Example 6 (Fault Tolerant Thermostat Model) The two observable events B_{on} and B_{off} allow one to witness mode changes and the continuous variable x is assumed to be observable. The system starts from $x \in [80, 90]$. Two faults are modeled as unobservable events F_1 and $F_2 \in \Sigma_f$ shown in Fig. 11.4. In practice, the fault F_1 models a bad calibration of the temperature sensor. As for fault F_2, it represents a defect in the heater impacting its efficiency and is modeled by a parametric change of a constant in the expression of the first order derivative of x.

11.7.1 CEGAR Scheme for Fault F_1

Initial Abstraction We consider an initial decomposition $\mathfrak{P} = \{P_{off}, P_{on}, P^{F_1}_{off}, P^{F_1}_{on}$ $P^{F_2}_{off}, P^{F_2}_{on}\}$ of the hybrid state space. Each partition $P \in \mathfrak{P}$ is made up of only one region representing the reals $\mathbb{R}$. Hence computing t_{min} and t_{max} for each region p yields $I_p = [0, +\infty)$, in other words the initial abstraction contains no time constraints.

Diagnosability Check The diagnosability check using the twin plant method generates a counterexample $C.E = (\hat{h}, \hat{h}')$ such that:

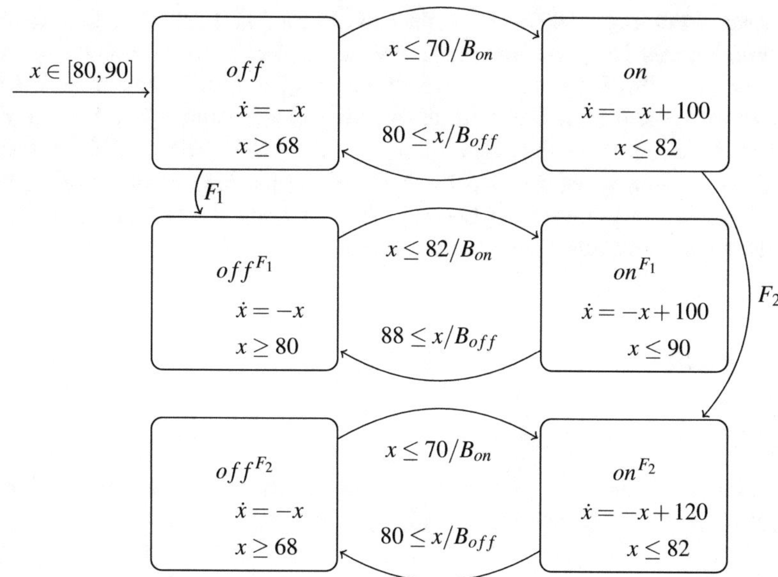

Fig. 11.4 One-dimensional hybrid automaton modeling a faulty thermostat

$$\hat{h} = (off, p_{off}) \xrightarrow{0.15} (off, p_{off}) \xrightarrow{Bon} (on, p_{on}) \xrightarrow{0.3} (on, p_{on}) \xrightarrow{B_{off}} (off, p_{off}) \dots$$

$$\hat{h}' = (off, p_{off}) \xrightarrow{0.1} (off, p_{off}) \xrightarrow{F_1} (off^{F_1}, p_{off}^{F_1}) \xrightarrow{0.05} (off^{F_1}, p_{off}^{F_1}) \xrightarrow{B_{on}}$$
$$(on^{F_1}, p_{on}^{F_1}) \xrightarrow{0.3} (on^{F_1}, p_{on}^{F_1}) \xrightarrow{B_{off}} (off^{F_1}, p_{off}^{F_1}) \dots$$

Validation or Refusal The computation of the set of concrete executions $\{h\}$ (resp. $\{h'\}$) whose abstraction is $\hat{h}$ (resp. $\hat{h}'$) yields an approximation as follows:

$$\{h\} = (off, [80, 90]) \xrightarrow{0.15} (off, [69, 77]) \xrightarrow{Bon} (on, [69, 70]) \dots$$

$$\{h'\} = (off, [80, 90]) \xrightarrow{0.1} (off, [72, 81]) \xrightarrow{F_1} (off^{F_1}, [80, 81]) \xrightarrow{0.05}$$
$$(off^{F_1}, [76, 77] : invalid)$$

The concrete state computations show that it is not possible to stay 0.05 time units in mode off^{F_1} as the temperature reached would be [76,77] violating the invariant $x \geq 80$. The *C.E* is thus refuted.

Refinement of the State Space The refinement aims at eliminating the previous spurious *C.E*. From this *C.E*, it is possible to compute the exact time constraint for staying in mode off^{F_1} and then triggering the transition B_{on} and refine the hybrid state space accordingly. Once refined, the new abstraction should not contain similar counterexamples. The validation process reveals that all trajectories entering mode off^{F_1} with $x \in [80, 81]$ cannot stay more than $t_{max} = 0.0124$ time units but it is possible for some trajectories to instantaneously change from *off* to off^{F_1} to on^{F_1} in which case $t_{min} = 0$, thus $I_{(off^{F_1},[80,81])} = [0, 0.0124]$. The refined abstraction would carry this new information by updating the partition of mode off^{F_1}, from $\mathbb{R}$ to $(-\infty, 80) \uplus [80, 81] \uplus (81, +\infty)$, thus ensuring that all future generated counterexamples would satisfy this constraint.

11.7.2 CEGAR Scheme for Fault F_2

Initial Abstraction The same as for F_1.

Diagnosability Check The diagnosability check of the initial abstraction using the twin plant method generates a $C.E = (\hat{h}, \hat{h}')$:

$$\hat{h} = (off, p_{off}) \xrightarrow{0.5} (off, p_{off}) \xrightarrow{B_{on}} (on, p_{on}) \xrightarrow{0.5} (on, p_{on}) \xrightarrow{B_{off}} (off, p_{off}) \dots$$

$$\hat{h}' = (off, p_{off}) \xrightarrow{0.5} (off, p_{off}) \xrightarrow{B_{on}} (on, p_{on}) \xrightarrow{0.4} (on, p_{on}) \xrightarrow{F_2} (on^{F_2}, p_{on}^{F_2}) \xrightarrow{0.1}$$

$$(on^{F_2}, p_{on}^{F_2}) \xrightarrow{B_{off}} (off^{F_2}, p_{off}^{F_2}) \dots$$

Validation or Refusal The computation of the set of concrete executions $\{h\}$ (resp. $\{h'\}$) whose abstraction is $\hat{h}$ (resp. $\hat{h}'$) yields an approximation as follows:

$$\{h\} = (off, [80, 90]) \xrightarrow{0.5} (off, [48.5, 54.58] : invalid)$$

$$\{h'\} = (off, [80, 90]) \xrightarrow{0.5} (off, [48.5, 54.58] : invalid)$$

This $C.E$ is refuted due to illegal stay in the mode *off* violating the corresponding invariant. In other words the trajectories are not feasible: if the system stays in mode *off* for 0.5 time units, then the state invariant is no longer true. Thus, if B_{on} is observed, then the duration of stay in *off* should be smaller.

Refinement of the State Space To prevent future similar spurious counterexamples, a refinement is applied to the initial abstraction. The refined model considers new regions in mode *off*: $p_{off_1} = [80, 90]$ (initial region) and $p_{off_2} = [68, 80)$ (legal region). The computation of the time intervals relative to each region are: $I_{(off,[68,80))} = [0, 0.16]$ and $I_{(off,[80,90])} = [0, 0.12]$. The refined abstraction will encode these time constraints and ensure that a set of similar counterexamples (including this one) are eliminated. Regions that are not reachable will be eliminated, such as $[0, 68)$.

Second Abstraction

Diagnosability Check The second $C.E$ generated from the refined twin plant is:

$$\hat{h} = (\mathit{off}, p_{\mathit{off}_1}) \xrightarrow{0.08} (\mathit{off}, p_{\mathit{off}_1}) \xrightarrow{\varepsilon} (\mathit{off}, p_{\mathit{off}_2}) \xrightarrow{0.07} (\mathit{off}, p_{\mathit{off}_2}) \xrightarrow{B_{on}} (on, p_{on})$$
$$\xrightarrow{0.5} (on, p_{on}) \xrightarrow{B_{\mathit{off}}} (\mathit{off}, p_{\mathit{off}}) \ldots$$
$$\hat{h}' = (\mathit{off}, p_{\mathit{off}_1}) \xrightarrow{0.08} (\mathit{off}, p_{\mathit{off}_1}) \xrightarrow{\varepsilon} (\mathit{off}, p_{\mathit{off}_2}) \xrightarrow{0.07} (\mathit{off}, p_{\mathit{off}_2}) \xrightarrow{B_{on}} (on, p_{on})$$
$$\xrightarrow{0.4} (on, p_{on}) \xrightarrow{F_2} (on^{F_2}, p_{on}^{F_2}) \xrightarrow{0.1} (on^{F_2}, p_{on}^{F_2}) \xrightarrow{B_{\mathit{off}}} (\mathit{off}^{F_2}, p_{\mathit{off}}^{F_2}) \ldots$$

Validation or Refusal Note that the continuous transitions in the second $C.E$ respect the temporal constraints added during the refinement based on the first $C.E$. The corresponding concrete approximate executions of this $C.E$ are:

$$\{h\} = (\mathit{off}, [80, 90]) \xrightarrow{0.15} (\mathit{off}, [69, 77]) \xrightarrow{B_{on}} (on, [69, 70]) \xrightarrow{0.5} (on, [80.5, 81.8]) \ldots$$
$$\{h'\} = (\mathit{off}, [80, 90]) \xrightarrow{0.15} (\mathit{off}, [69, 77]) \xrightarrow{B_{on}} (on, [69, 70]) \xrightarrow{0.4} (on, [79, 79.9])$$
$$\xrightarrow{F_2} (on^{F_2}, [79, 79.9]) \xrightarrow{0.1} (on^{F_2}, [82.9, 83.7]) \ldots$$

In this case, this second $C.E$ is also considered as spurious because, given the time constraints, the two trajectories are different in the observations of x in the hybrid system since the last regions are disjoint, i.e., $[80.5, 81.8] \cap [82.9, 83.7] = \emptyset$.

Refinement of the State Space The counterexample analysis could identify the time boundary $t_{spurious}$, up to which the observations could be the same for at least two concrete trajectories, and after which the critical pair becomes spurious. In our example, suppose the fault occurred at t_f where $x \in [a, b]$, then $t_{spurious}$ is the time instant from which faulty and nominal sets of trajectories are disjoint:

$$t_{spurious} = \ln\left(\frac{b - a + 20}{20}\right) + t_f \tag{11.9}$$

For the second spurious $C.E$, $t_f = 0.4$ and $t_{spurious} - t_f = 0.044$. The two concrete nominal and faulty executions originating from $(on, [79, 79.9])$ will be in the following temperature range after 0.044 time units: $x \in [79.90, 80.7]$ in the mode on and $x \in [80.7, 81.6]$ in the mode on^{F_2}. Hence, at any future time, the observations are different. By incorporating the time constraint in the refined abstraction, we ensure that counterexamples that are spurious because of disjoint observations including the previous one cannot be generated again. For the sake of simplicity, we analyzed the two faults separately. One more sophisticated strategy is to analyze the next fault based on the refined abstraction obtained from the analysis of the precedent fault.

11.8 Conclusion

In this chapter, we were interested in verifying a given formal safety property on a hybrid system, based on discrete abstractions of this system, for which checking this property is decidable and which guarantee that the property is satisfied at the concrete hybrid level if it is satisfied at the abstract level. We focused on the diagnosability property, for its importance in safety analysis at design stage and the challenge it gives rise to. We presented elements from the literature regarding hybrid automata abstractions, however few works handle diagnosability verification, as this property deals with a pair of trajectories and partial observations of the system and is thus more complex to check than reachability. In order to handle time constraints at the abstract level, we chose abstractions of the hybrid automaton as timed automata, related to a decomposition of the state space into geometric regions, the abstract time constraints coming from the estimation of the sojourn time of trajectories in each region. Thus the abstractions over-approximate the regions of interest to which are added time constraints obtained from the dynamics of the concrete system. We adapted a CEGAR scheme for hybrid systems diagnosability verification, based on the counterexample provided at the abstract level by the twin plant based diagnosability checking when diagnosability is proved to be unsatisfied. We presented situations for which the produced counterexample is spurious and a refinement in finer regions and tighter time constraints is then required.

This preliminary work draws many perspectives. First of all, we have to develop refinement strategies by analysis of the counterexample and progress in the (partial) automation of the whole process and the integration of the algorithms for abstract diagnosability checking, for validation of the counterexample and for refinement. We plan in particular to use and extend existing tools for timed automata model checking and for over-approximation reachability at the continuous level. We want to investigate also the usage of SMT solvers [17, 18], in particular with theories including ODEs [31], to deal simultaneously with discrete and continuous variables. Moreover, we plan to apply this approach to other formal safety properties such as predictability, which guarantees to predict in advance the future unavoidable occurrence of a fault given some sequence of observations. With all this, we will be able to tackle real applications and get deeper experimental results.

Another promising aspect is the potential of this approach to deduce minimal concrete sets of observations for which the system is diagnosable. These observations specify a minimal (for set inclusion) needed set of events and continuous variables for which the system is diagnosable. If one element of this set is not considered as observable, the system becomes non-diagnosable. Thus, such a minimal set draws the boundary between the diagnosable and non-diagnosable systems [14] from the point of view of their observability.

Up to now we assume a continuous domain for both the values and the time stamps of the observable variables without taking into account sensor capability, i.e., the minimal interval (of value and of time) that can be captured. This is the reason why our current algorithm may not terminate, due to an infinite refinement process. A fundamental and essential future work perspective is to provide a general algorithm for diagnosability verification with ε-precision [42], for ε arbitrary small (for a given metric to be defined). This will ensure theoretical termination of the algorithm, as the number of refinement steps to reach the precision will then be finite. And this is actually justified in practice because both model parameters and observations cannot be infinitely accurate, thus the value ε for the precision would come from the precision of the model parameters and of the measurements, in space and time. In the same spirit, one interesting future work would be to demonstrate a bi-simulation relation between the concrete model and the final refined abstract model when considering this minimal precision imposed by the model and the sensors, where the termination of the refinement can thus be guaranteed. In other words, theoretically, we could always deduce the right verdict, either the system is diagnosable or it is not diagnosable with respect to a given minimal precision.

Acknowledgements The authors would like to thank Alban Grastien for his valuable comments and suggestions.

References

1. M. Althoff, O. Stursberg, M. Buss, Computing reachable sets of hybrid systems using a combination of zonotopes and polytopes. Nonlinear Anal. Hybrid Syst. **4**(2), 233–249 (2010)
2. R. Alur, D.L. Dill, A theory of timed automata. Theor. Comput. Sci. **126**(2), 183–235 (1994)
3. R. Alur, T. Dang, F. Ivančić, Counterexample-guided predicate abstraction of hybrid systems. Theor. Comput. Sci. **354**(2), 250–271 (2006)
4. S. Ault, E. Holmgreen, Dynamics of the Brusselator. Academia (2009)
5. M. Basseville, M. Kinnaert, M. Nyberg, On fault detectability and isolability. Eur. J. Control. **7**(6), 625–641 (2001)
6. M. Bayoudh, L. Travé-Massuyès, Diagnosability analysis of hybrid systems cast in a discrete-event framework. Discrete Event Dyn. Syst. **24**(3), 309–338 (2014)
7. M. Bayoudh, L. Travé-Massuyès, X. Olive, Hybrid systems diagnosability by abstracting faulty continuous dynamics, in *Proceedings of the 17th International Workshop on Principles of Diagnosis (DX'06)* (2006), pp. 9–15
8. M. Bayoudh, L. Travé-Massuyès, X. Olive, Coupling continuous and discrete event system techniques for hybrid systems diagnosability analysis, in *Proceedings of the 18th European Conference on Artificial Intelligence (ECAI-08)*, Patras (2008), pp. 219–223
9. M. Bayoudh, L. Travé-Massuyès, X. Olive, Active diagnosis of hybrid systems guided by diagnosability properties. IFAC Proc. Vol. **42**(8), 1498–1503 (2009)
10. G. Behrmann, A. David, K.G. Larsen, A tutorial on Uppaal, in *Formal Methods for the Design of Real-Time Systems* (Springer, Berlin, 2004), pp. 200–236
11. S. Biswas, D. Sarkar, S. Mukhopadhyay, A. Patra, Diagnosability analysis of real time hybrid systems, in *Proceedings of the IEEE International Conference on Industrial Technology (ICIT'06)*, Mumbai (2006), pp. 104–109
12. O. Botchkarev, S. Tripakis, Verification of hybrid systems with linear differential inclusions using ellipsoidal approximations, in *Proceedings of the 3rd International Workshop on Hybrid Systems: Computation and Control (HSCC'00)*. LNCS, vol. 1790 (Springer, Berlin, 2000), pp. 73–88

13. M. Bozga, C. Daws, O. Maler, A. Olivero, S. Tripakis, S. Yovine, Kronos: a model-checking tool for real-time systems, in *International Symposium on Formal Techniques in Real-Time and Fault-Tolerant Systems* (Springer, Berlin, 1998), pp. 298–302
14. L. Brandán Briones, A. Lazovik, P. Dague, Optimizing the system observability level for diagnosability, in *Proceedings of the 3rd International Symposium on Leveraging Applications of Formal Methods, Verification and Validation (ISoLA08)*, Chalkidiki, Kassandra (2008)
15. J. Chen, R. Patton, A re-examination of the relationship between parity space and observer-based approaches in fault diagnosis, in *Proceedings of the IFAC Symposium on Fault Detection, Supervision and Safety of Technical Systems Safeprocess'94*, Helsinki (1994), pp. 590–596
16. A. Cimatti, C. Pecheur, R. Cavada, Formal verification of diagnosability via symbolic model checking, in *Proceedings of the 18th International Joint Conference on Artificial Intelligence (IJCAI-03)* (2003), pp. 363–369
17. A. Cimatti, S. Mover, S. Tonetta, SMT-based scenario verification for hybrid systems. Formal Methods Syst. Des. **42**(1), 46–66 (2013)
18. A. Cimatti, A. Griggio, S. Mover, S. Tonetta, HyComp: an SMT-based model checker for hybrid systems, in *Proceedings of the 21st International Conference on Tools and Algorithms for the Construction and Analysis of Systems (TACAS-2015)*, London (2015), pp. 52–67
19. V. Cocquempot, T.E. Mezyani, M. Staroswiecki, Fault detection and isolation for hybrid systems using structured parity residuals, in *Proceedings of the IEEE/IFAC-ASCC: Asian Control Conference*, vol. 2, Melbourne (2004), pp. 1204–1212
20. M.J. Daigle, A qualitative event-based approach to fault diagnosis of hybrid systems. PhD thesis, Vanderbilt University, 2008
21. M. Daigle, X. Koutsoukos, G. Biswas, An event-based approach to hybrid systems diagnosability, in *Proceedings of the 19th International Workshop on Principles of Diagnosis (DX'08)* (2008), pp. 47–54
22. M.J. Daigle, D. Koutsoukos, G. Biswas, An event-based approach to integrated parametric and discrete fault diagnosis in hybrid systems. Trans. Inst. Meas. Control. (Special Issue on Hybrid and Switched Systems) **32**(5), 487–510 (2010)
23. M.J. Daigle, I. Roychoudhury, G. Biswas, D. Koutsoukos, A. Patterson-Hine, S. Poll, A comprehensive diagnosis methodology for complex hybrid systems: a case study on spacecraft power distribution systems. IEEE Trans. Syst. Man Cybern. Part A (Special Issue on Model-based Diagnosis: Facing Challenges in Real-world Applications) **4**(5), 917–931 (2010)
24. Y. Deng, A. D'Innocenzo, M.D. Di Benedetto, S. Di Gennaro, A.A. Julius, Verification of hybrid automata diagnosability with measurement uncertainty. IEEE Trans. Autom. Control **61**(4), 982–993 (2016)
25. O. Diene, E.R. Silva, M.V. Moreira, Analysis and verification of the diagnosability of hybrid systems, in *Proceedings of the 53rd IEEE Conference on Decision and Control (CDC-14)* (IEEE, New York, 2014), pp. 1–6
26. O. Diene, M.V. Moreira, V.R. Alvarez, E.R. Silva, Computational methods for diagnosability verification of hybrid systems, in *Proceedings of the IEEE Conference on Control Applications (CCA-15)* (IEEE, New York, 2015), pp. 382–387
27. S. Ding, *Model-Based Fault Diagnosis Techniques: Design Schemes, Algorithms, and Tools* (Springer, London, 2008)
28. C. Edmund, F. Ansgar, H. Zhi, K. Bruce, S. Olaf, T. Michael, Verification of hybrid systems based on counterexample-guided abstraction refinement, in *Proceedings of International Conference on Tools and Algorithms for the Construction and Analysis of Systems (TACAS-2003)*, ed. by H. Garavel, J. Hatcliff. Lecture Notes in Computer Science, vol. 2619 (Springer, Cham, 2003), pp. 192–207
29. G. Fourlas, K. Kyriakopoulos, N. Krikelis, Diagnosability of hybrid systems, in *Proceedings of the 10th Mediterranean Conference on Control and Automation (MED-2002)*, Lisbon (2002), pp. 3994–3999
30. J.-P. Gallois, J.-Y. Pierron, Qualitative simulation and validation of complex hybrid systems, in *Proceedings of the 8th European Congress on Embedded Real Time Software and Systems (ERTS-2016)*, Toulouse (2016)

31. S. Gao, S. Kong, E. Clarke, Satisfiability modulo ODEs, in *Formal Methods in Computer-Aided Design (FMCAD)* (2013)
32. V. Germanos, S. Haar, V. Khomenko, S. Schwoon, Diagnosability under weak fairness, in *Proceedings of the 14th International Conference on Application of Concurrency to System Design (ACSD'14)*, Tunis (IEEE Computer Society Press, New York, 2014)
33. J. Gertler, *Fault Detection and Diagnosis in Engineering Systems* (Marcel Dekker, New York, 1998)
34. A. Grastien, Symbolic testing of diagnosability, in *Proceedings of the 20th International Workshop on Principles of Diagnosis (DX-09)* (2009), pp. 131–138
35. A. Grastien, L. Travé-Massuyès, V. Puig, Solving diagnosability of hybrid systems via abstraction and discrete event techniques, in *Proceedings of the 27th International Workshop on Principles of Diagnosis (DX-16)* (2016)
36. S. Gulwani, A. Tiwari, Constraint-based approach for analysis of hybrid systems, in *Proceedings of the 20th International Conference on Computer Aided Verification (CAV-2008)* (2008), pp. 190–203
37. E. Hainry, Decidability and undecidability in dynamical systems. Rapport de recherche (CiteSeer, 2009). http://hal.inria.fr/inria-00429965/en/
38. T.A. Henzinger, The theory of hybrid automata, in *Proceedings of the 11th Annual IEEE Symposium on Logic in Computer Science* (IEEE Computer Society Press, Los Alamitos, CA, 1996), pp. 278–292
39. T.A. Henzinger, P.W. Kopke, A. Puri, P. Varaiya, What's decidable about hybrid automata?, in *Proceedings of the 27th Annual Symposium on Theory of Computing* (ACM Press, New York, 1995), pp. 373–382
40. S. Jiang, Z. Huang, V. Chandra, R. Kumar, A polynomial algorithm for testing diagnosability of discrete-event systems. IEEE Trans. Autom. Control **46**(8), 1318–1321 (2001)
41. E. Kilic, Diagnosability of fuzzy discrete event systems. Inf. Sci. **178**(3), 858–870 (2008)
42. K.-D. Kim, S. Mitra, P.R. Kumar, Computing bounded epsilon-reach set with finite precision computations for a class of linear hybrid automata, in *Proceedings of the ACM International Conference on Hybrid Systems: Computation and Control* (2011)
43. B. Kuipers, *Qualitative Reasoning: Modeling and Simulation with Incomplete Knowledge* (MIT Press, Cambridge, MA, 1994)
44. F. Liu, D. Qiu, Safe diagnosability of stochastic discrete-event systems. IEEE Trans. Autom. Control **53**(5), 1291–1296 (2008)
45. O. Maler, G. Batt, Approximating continuous systems by timed automata, in *Formal Methods in Systems Biology* (Springer, Berlin, 2008), pp. 77–89
46. T. Melliti, P. Dague, Generalizing diagnosability definition and checking for open systems: a Game structure approach, in *Proceedings of the 21st International Workshop on Principles of Diagnosis (DX'10)*, Portland, OR (2010), pp. 103–110
47. M. Nyberg, Criterions for detectability and strong detectability of faults in linear systems. Int. J. Control. **75**(7), 490–501 (2002)
48. Y. Pencolé, Diagnosability analysis of distributed discrete event systems, in *Proceedings of the 16th European Conference on Artificial Intelligent (ECAI-04)* (2004), pp. 43–47
49. Y. Pencolé, A. Subias, A chronicle-based diagnosability approach for discrete timed-event systems: application to web-services. J. Universal Comput. Sci. **15**(17), 3246–3272 (2009)
50. P. Ribot, Y. Pencolé, Design requirements for the diagnosability of distributed discrete event systems, in *Proceedings of the 19th International Workshop on Principles of Diagnosis (DX'08)*, Blue Mountains (2008), pp. 347–354
51. J. Rintanen, Diagnosers and diagnosability of succinct transition systems, in *Proceedings of the 20th International Joint Conference on Artificial Intelligence (IJCAI-07)*, Hyderabad (2007), pp. 538–544
52. M. Sampath, R. Sengupta, S. Lafortune, K. Sinnamohideen, D. Teneketzis, Diagnosability of discrete event systems. Trans. Autom. Control **40**(9), 1555–1575 (1995)
53. A. Schumann, J. Huang, A scalable jointree algorithm for diagnosability, in *Proceedings of the 23rd American National Conference on Artificial Intelligence (AAAI-08)* (2008), pp. 535–540

54. D. Thorsley, D. Teneketzis, Diagnosability of stochastic discrete-event systems. IEEE Trans. Autom. Control **50**(4), 476–492 (2005)
55. L. Travé-Massuyès, P. Dague, *Modèles et raisonnements qualitatifs* (Hermès, Paris, 2003)
56. L. Travé-Massuyès, M. Cordier, X. Pucel, Comparing diagnosability criterions in continuous systems and discrete events systems, in *Proceedings of the 6th IFAC Symposium on Fault Detection, Supervision and Safety of Technical Processes (Safeprocess'06)*, Beijing (2006), pp. 55–60
57. L. Travé-Massuyès, T. Escobet, X. Olive, Diagnosability analysis based on component-supported analytical redundancy relations. IEEE Trans. Syst. Man Cybern. Part A **36**(6), 1146–1160 (2006)
58. S. Tripakis, Fault diagnosis for timed automata, in *Proceedings of International Symposium on Formal Techniques in Real-Time and Fault-Tolerant Systems (FTRTFT-2002)*. Lecture Notes in Computer Science, vol. 2469 (Springer, Berlin, 2002), pp. 205–221
59. Y. Yan, L. Ye, P. Dague, Diagnosability for patterns in distributed discrete event systems, in *Proceedings of the 21st International Workshop on Principles of Diagnosis (DX'10)*, Portland, OR (2010), pp. 345–352
60. L. Ye, P. Dague, Diagnosability analysis of discrete event systems with autonomous components, in *Proceedings of the 19th European Conference on Artificial Intelligence (ECAI-10)* (2010), pp. 105–110
61. T.-S. Yoo, S. Lafortune, Polynomial-time verification of diagnosability of partially observed discrete-event systems. IEEE Trans. Autom. Control **47**(9), 1491–1495 (2002)
62. M. Yu, D. Wang, M. Luo, D. Zhang, Q. Chen, Fault detection, isolation and identification for hybrid systems with unknown mode changes and fault patterns. Expert Syst. Appl. **39**(11), 9955–9965 (2012)
63. J. Zaytoon, S. Lafortune, Overview of fault diagnosis methods for discrete event systems. Annu. Rev. Control. **37**(2), 308–320 (2013)

Index

M. Sayed-Mouchaweh (ed.), *Diagnosability, Security and Safety of Hybrid Dynamic and Cyber-Physical Systems*, https://doi.org/10.1007/978-3-319-74962-4

E

F

G

H

Printed in the United States
By Bookmasters